W9-BIM-614

# PROTECTING INDIVIDUAL PRIVACY IN THE STRUGGLE AGAINST TERRORISTS

## A Framework for Program Assessment

Committee on Technical and Privacy Dimensions of Information
for Terrorism Prevention and Other National Goals

Committee on Law and Justice and Committee on National Statistics
Division on Behavioral and Social Sciences and Education

Computer Science and Telecommunications Board
Division on Engineering and Physical Sciences

## NATIONAL RESEARCH COUNCIL
*OF THE NATIONAL ACADEMIES*

THE NATIONAL ACADEMIES PRESS
Washington, D.C.
**www.nap.edu**

THE NATIONAL ACADEMIES PRESS    500 Fifth Street, N.W.    Washington, DC 20001

NOTICE: The project that is the subject of this report was approved by the Governing Board of the National Research Council, whose members are drawn from the councils of the National Academy of Sciences, the National Academy of Engineering, and the Institute of Medicine. The members of the committee responsible for the report were chosen for their special competences and with regard for appropriate balance.

Support for this project was provided by the Bureau of Transportation Statistics, with assistance from the National Science Foundation under sponsor award number SES-0112521; the Department of Homeland Security, with assistance from the National Science Foundation under sponsor award number SES-0411897; the National Center for Education Statistics, with assistance from the National Science Foundation under sponsor award number SBR-0453930; and the National Science Foundation under sponsor award numbers SRS-0632055 and IIS-0441216. Additional funding was provided by the Presidents' Circle Communications Initiative of the National Academies.

**Library of Congress Cataloging-in-Publication Data**

Protecting individual privacy in the struggle against terrorists : a framework for program assessment.
   p. cm.
   Includes bibliographical references.
   ISBN 978-0-309-12488-1 (pbk.) — ISBN 978-0-309-12489-8 (pdf)  1.  Terrorism—United States—Prevention. 2.  Surveillance detection—United States. 3.  Privacy, Right of—United States. 4.  Technological innovations—Law and legislation—United States.
   HV6432.P76 2008
   363.325'163--dc22
                          2008033554

This report is available from

Committee on Law and Justice *or*
Computer Science and Telecommunications Board
National Research Council
500 Fifth Street, N.W.
Washington, DC 20001

Additional copies of this report are available from the National Academies Press, 500 Fifth Street, N.W., Lockbox 285, Washington, DC 20055; (800) 624-6242 or (202) 334-3313 (in the Washington metropolitan area); Internet, http://www.nap.edu.

# THE NATIONAL ACADEMIES
*Advisers to the Nation on Science, Engineering, and Medicine*

The **National Academy of Sciences** is a private, nonprofit, self-perpetuating society of distinguished scholars engaged in scientific and engineering research, dedicated to the furtherance of science and technology and to their use for the general welfare. Upon the authority of the charter granted to it by the Congress in 1863, the Academy has a mandate that requires it to advise the federal government on scientific and technical matters. Dr. Ralph J. Cicerone is president of the National Academy of Sciences.

The **National Academy of Engineering** was established in 1964, under the charter of the National Academy of Sciences, as a parallel organization of outstanding engineers. It is autonomous in its administration and in the selection of its members, sharing with the National Academy of Sciences the responsibility for advising the federal government. The National Academy of Engineering also sponsors engineering programs aimed at meeting national needs, encourages education and research, and recognizes the superior achievements of engineers. Dr. Charles M. Vest is president of the National Academy of Engineering.

The **Institute of Medicine** was established in 1970 by the National Academy of Sciences to secure the services of eminent members of appropriate professions in the examination of policy matters pertaining to the health of the public. The Institute acts under the responsibility given to the National Academy of Sciences by its congressional charter to be an adviser to the federal government and, upon its own initiative, to identify issues of medical care, research, and education. Dr. Harvey V. Fineberg is president of the Institute of Medicine.

The **National Research Council** was organized by the National Academy of Sciences in 1916 to associate the broad community of science and technology with the Academy's purposes of furthering knowledge and advising the federal government. Functioning in accordance with general policies determined by the Academy, the Council has become the principal operating agency of both the National Academy of Sciences and the National Academy of Engineering in providing services to the government, the public, and the scientific and engineering communities. The Council is administered jointly by both Academies and the Institute of Medicine. Dr. Ralph J. Cicerone and Dr. Charles M. Vest are chair and vice chair, respectively, of the National Research Council.

**www.national-academies.org**

*viii*

# Preface

In late 2005, the National Research Council (NRC) convened the Committee on Technical and Privacy Dimensions of Information for Terrorism Prevention and Other National Goals. Supported by the U.S. Department of Homeland Security and the National Science Foundation, the committee was charged with addressing information needs of the government that arise in its deployment of various forms of technology for broad access to and analysis of data as it faces the challenges of terrorism prevention and threats to public health and safety. Specifically of interest was the nexus between terrorism prevention, technology, privacy, and other policy issues and the implications and issues involved in deploying data mining, information fusion, and behavioral surveillance technologies. The study sought to develop a conceptual framework that policy makers and the public can use to consider the utility, appropriateness, and empirical validity of data generated and analyzed by various forms of technology currently in use or planned in the near future. The committee notes that the development of this framework did not include the development of systems for preventing terrorism. By design and in response to the charge for the study, this report focuses on data mining and behavioral surveillance as the primary techniques of interest.

The committee interpreted its charge as helping government policy makers to evaluate and make decisions about information-based programs to fight terrorism or serve other important national goals, and it thus sought to provide a guide for government officials, policy makers, and technology developers as they continue to explore new surveillance tools in the service of important national security goals. Chapter 1 scopes the issues involved and introduces key concepts that are explored in much greater depth in the appendixes. Chapter 2 outlines a framework for a systematic assessment of information-based programs being considered or already in use for counterterrorist purposes (and other important national needs, such as law enforcement and public health) in terms of each program's effectiveness and its consistency with U.S. laws and values. Chapter 3 provides the committee's conclusions and recommendations. The appendixes elaborate extensively on the scientific and technical foundations that underpin the committee's work and the legal and organizational context in which information-based programs necessarily operate. The committee regards the appendixes as essential elements of the report.

Note that although the committee heard from representatives from many government agencies, this report does not evaluate or critique any specific U.S. government program. Rather, it is intended to provide policy makers with a systematic framework for thinking about existing and future operational information-based programs, especially in a counterterrorist context.

Nowhere is the need for this study and the framework it proposes more apparent than in the history of the Total Information Awareness (TIA) program. Indeed, the TIA program and the issues it raised loomed large in the background when this committee was appointed, and although the TIA program was terminated in September 2003, it is safe to say that the issues raised by this program have not been resolved in any fundamental sense. Moreover, many other data mining activities supported by the U.S. government continue to raise the same issues: the potential utility of large-scale databases containing personal information for counterterrorist and law enforcement purposes and the potential privacy impact of law enforcement and national security authorities using such databases. A brief history of the TIA program is contained in Appendix J.

The committee consisted of 21 people with a broad range of expertise, including national security and counterterrorism, intelligence and counterintelligence, privacy law and information protection, organizations and organizational structure, law enforcement, statistics, information technology, cognitive psychology, terrorism, database architecture, public health, artificial intelligence, databases, cryptography, machine learning and statistics, and information retrieval.

From 2005 to 2007, the committee held six meetings, most of which were intended to enable it to explore a wide range of points of view. For example, briefings and other inputs were obtained from government officials at all levels, authorities on international law and practice relating to policy, social scientists and philosophers concerned with collection of personal data, experts on privacy-enhancing technologies, business representatives concerned with the gathering and uses of personal data, and researchers who use personal data in their work. Several papers were commissioned and received, as well as a number of contributed white papers.

Preparation of the report was undertaken on an unclassified basis. Although a number of classified programs of the U.S. government make use of data mining, the fundamental principles of data mining themselves are not classified, and these principles apply to both classified and unclassified applications. Thus, at the level of analysis presented in this report, the fact that some of the U.S. government's counterterrorist programs are classified does not materially affect the analysis provided here. In addition, the U.S. government operates a variety of classified programs intended to collect data that may be used for counterterrorist purposes. However, as collection programs, they are out of the scope of this report, and all that need be noted is that they produce data relevant to the counterterrorist mission and that data mining and information fusion technologies must process.

This study could not have been undertaken without the support of the government project officers, Larry Willis, U.S. Department of Homeland Security, and Larry Brandt and Brian D. Humes, National Science Foundation, who recognize the complex issues involved in developing and using new technologies to respond to terrorism and other national efforts, such as law enforcement and public health, and the need to think through how this might best be done.

Given the scope and breath of the study, the committee benefited greatly from the willingness of many individuals to share their perspectives and expertise. We are very grateful to the following individuals for their helpful briefings on technologies for data mining and detection of deception: Paul Ekman, University of California, San Francisco; Mark Frank, University of Buffalo; John Hollywood, RAND Corporation; David Jensen, University of Massachusetts; Jeff Jonas, IBM; David Scott, Rice University; John Woodward, RAND Corporation; and Thomas Zeffiro, Georgetown University. Useful insights on the use of these technologies in the private sector were provided by Scott Loftnesness, Glenbrook Partners, and Dan Schutzer, Financial Services Technical Consortium. William Winkler, Census Bureau, helped the committee understand the technologies' potential impact on federal statistical agencies.

Background briefings on relevant privacy law and policy were provided by Henry Greely, Stanford University; Barry Steinhardt, American Civil Liberties Union; Kim Taipale, Center for Advanced Studies in Science and Technology Policy; and Lee Tien, Electronic Frontier Foundation. We also benefited from the expert testimony of Whitfield Diffie, Sun Microsystems; John Pike, Global Security; and Jody Westby, Global Cyber Risk, on the role of information technologies in counterterrorism. In addition to counterterrorism, the impact and implications of data mining for law enforcement and public health were important foci of the committee's work. In the public health area, the following persons contributed to the committee's understanding: James Lawler, Homeland Security Council, White House; Farzad Mostashari, New York City Public Health Department; Patricia Quinlisk, State of Iowa; and Barry Rhodes and Lynn Steele, Centers for Disease Control and Prevention. Useful insights on the role of law enforcement in counterterrorism were provided in presentations made by Roy Apseloff, National Media Exploitation Center; Michael Fedarcyk, Federal Bureau of Investigation (retired); and Philip Reitinger, Microsoft. We found extremely helpful the international perspectives of Joe Connell, New Scotland Yard (retired), and Ravi Ron, former head of Israel's Ben Gurion Airport.

This study also benefited considerably from briefings by government officials involved on a daily basis with the issues at the heart of the study. We particularly want to thank Randy Ferryman and Admiral Scott Redd from the National Counter Terrorism Center and Clint C. Brooks (retired) from the National Security Agency, who shared their vision of how the nation should conduct its counterterrorism activities while maintaining its democratic ideals. Numerous staff members from the Department of Homeland Security (DHS) also shed important light on government activities relating to terrorism prevention, including Mel Bernstein, Timothy Keefer, Hyon Kim, Sandy Landsberg, John V. Lawler, Tiffany Lightbourn, Grace Mastalli, Allison Smith, and Lisa J. Walby. Toby Levin was particularly helpful in sharing timely and relevant information on the work of the DHS Privacy Office, and the committee appreciated the interest of the DHS Data Privacy and Integrity Advisory Committee in its work and their willingness to keep members abreast of their activities and role in protecting privacy.

The committee also thanks Michael D. Larsen of Iowa State University and Peter Swire of Ohio State University, who responded to its request for white papers, and Amy Corning and Eleanor Singer, University of Michigan, who prepared an informative paper on public opinion.

This study involved NRC staff from three different NRC units. We would like to thank them for their valuable assistance to this project as well as for their collegiality, which contributed to a far richer experience for all involved. Betty Chemers of the NRC's Committee on Law and Justice served as study director and organized and facilitated the meetings, Michael Cohen of the Committee on National Statistics provided technical expertise on statistical and data mining issues, and Herbert

Lin of the Computer Science and Telecommunications Board undertook the difficult job of turning the committee's writing contributions into a coherent whole and working with the co-chairs to mediate and resolve intellectual disagreements within the committee. Carol Petrie provided guidance and support throughout the study process. We would also like to thank Julie Schuck and Ted Schmitt for their research assistance and Jennifer Bishop, Barbara Boyd, Linda DePugh, and Janice Sabuda for their administrative support. Finally, we greatly appreciate the efforts undertaken by Eugenia Grohman, Susan Maurizi, Kirsten Sampson Snyder, and Yvonne Wise to complete the review and editing processes and bring this report to fruition.

> Charles M. Vest and William J. Perry, *Co-chairs*
> Committee on Technical and Privacy
> Dimensions of Information for Terrorism
> Prevention and Other National Goals

# Acknowledgment of Reviewers

This report has been reviewed in draft form by individuals chosen for their diverse perspectives and technical expertise, in accordance with procedures approved by the National Research Council's Report Review Committee. The purpose of this independent review is to provide candid and critical comments that will assist the institution in making its published report as sound as possible and to ensure that the report meets institutional standards for objectivity, evidence, and responsiveness to the study charge. The review comments and draft manuscript remain confidential to protect the integrity of the deliberative process. We wish to thank the following individuals for their review of this report:

Steve M. Bellovin, Columbia University,
R. Stephen Berry, University of Chicago,
David L. Carter, Michigan State University,
Richard F. Celeste, Colorado College,
Hermann Habermann, Bureau of the U.S. Census (retired),
David Jensen, University of Massachusetts, Amherst,
Alan F. Karr, National Institute of Statistical Sciences,
Diane Lambert, Google, Inc.,
Butler Lampson, Microsoft Corporation,
Michael D. Larsen, Iowa State University,
Lance Liebman, Columbia Law School,
Patricia Quinlisk, State of Iowa,
Jerome Reiter, Duke University,

Andrew P. Sage, George Mason University,
Paul Schwartz, University of California, Berkeley,
Eugene Spafford, Purdue University,
Robert D. Sparks, California Medical Association Foundation,
William O. Studeman, Northrop Grumman Mission Systems, and
Peter Weinberger, Google, Inc.

Although the reviewers listed above have provided many constructive comments and suggestions, they were not asked to endorse the conclusions or recommendations, nor did they see the final draft of the report before its release. The review of this report was overseen by William H. Press, University of Texas at Austin, and James G. March, Stanford University. Appointed by the National Research Council, they were responsible for making certain that an independent examination of this report was carried out in accordance with institutional procedures and that all review comments were carefully considered. Responsibility for the final content of this report rests entirely with the authoring committee and the institution.

# Contents

# Executive Summary

In a democratic society it is vitally important that citizens and their representatives be able to make an informed judgment on how to appropriately balance privacy with security. This report seeks to contribute to that informed judgment.

September 11, 2001, provided vivid proof to Americans of the damage that a determined, fanatical terrorist group can inflict on our society. Based on the available information about groups like Al Qaeda, most importantly their own statements, it seems clear that they will continue to try to attack us. Further attacks by such groups, and indeed by domestic terrorists like Timothy McVeigh, could be as serious as, or even more serious than, September 11 and Oklahoma City. Because future terrorist attacks on the United States could cause major casualties as well as severe economic and social disruption, the danger they pose is real, and it is serious. Thus, high priority should be given to developing programs to detect intended attacks before they occur so that there is a chance of preventing them.

At the same time, the nation must ensure that its institutions, information systems, and laws together constitute a trustworthy and accountable system that protects U.S. citizens' rights to privacy.

In this report, the Committee on Technical and Privacy Dimensions of Information for Terrorism Prevention and Other National Goals examines the role of data mining and behavioral surveillance technologies in counterterrorism programs,[1] and it provides a framework for making decisions about deploying and evaluating those and other information-based programs on the basis of their effectiveness and associated risks to personal privacy.

The most serious threat today comes from terrorist groups that are international in scope. These groups make use of the Internet to recruit, train, and plan operations, and they use public channels to communicate. Therefore, intercepting and analyzing these information streams might provide important clues regarding the nature of the terrorist threat. Important clues might also be found in commercial and government databases that record a wide range of information about individuals, organizations, and their transactions, movements, and behavior. But success in such efforts will be extremely difficult to achieve because:

• The information sought by analysts must be filtered out of the huge quantity of data available (the needle in the haystack problem); and

• Terrorist groups will make calculated efforts to conceal their identity and mask their behaviors, and will use various strategies such as encryption, code words, and multiple identities to obfuscate the data they are generating and exchanging.

Modern data collection and analysis techniques have had remarkable success in solving information-related problems in the commercial sector; for example, they have been successfully applied to detect consumer fraud. But such highly automated tools and techniques cannot be easily applied to the much more difficult problem of detecting and preempting a terrorist attack, and success in doing so may not be possible at all. Success, if it is indeed achievable, will require a determined research and development effort focused on this particular problem.

Detecting indications of ongoing terrorist activity in vast amounts of communications, transactions, and behavioral records will require technology-based counterterrorism tools. But even in well-managed programs such tools are likely to return significant rates of false positives, especially if the tools are highly automated. Because the data being analyzed are primarily about ordinary, law-abiding citizens and businesses, false positives can result in invasion of their privacy. Such intrusions raise valid concerns about the misuse and abuse of data, about the accuracy

---

[1]In this report, the term "program" refers to the system of technical, human, and organizational resources and activities required to execute a specific function. Humans—not computers—are always fully responsible for the actions of a program.

of data and the manner in which the data are aggregated, and about the possibility that the government could, through its collection and analysis of data, inappropriately influence individuals' conduct. Intruding on privacy also risks ignoring constitutional concerns about general search, as reflected in the Fourth Amendment. The committee strongly believes that such intrusion must be minimized through good management and good design, even if it cannot be totally eliminated.

The difficulty of detecting the activity of terrorist groups through their communications, transactions, and behaviors is hugely complicated by the ubiquity and enormity of electronic databases maintained by both government agencies and private-sector corporations. Retained data and communication streams concern financial transactions, medical records, travel, communications, legal proceedings, consumer preferences, Web searches, and, increasingly, behavioral and biological information. This is the essence of the information age—it provides us with convenience, choice, efficiency, knowledge, and entertainment; it supports education, health care, safety, and scientific discovery. Everyone leaves personal digital tracks in these systems whenever he or she makes a purchase, takes a trip, uses a bank account, makes a phone call, walks past a security camera, obtains a prescription, sends or receives a package, files income tax forms, applies for a loan, e-mails a friend, sends a fax, rents a video, or engages in just about any other activity. The proliferation of security cameras and means of tagging and tracking people and objects increases the scope and nature of available data. Law-abiding citizens leave extensive digital tracks, and so do criminals and terrorists.

Gathering and analyzing electronic, behavioral, biological, and other information can play major roles in the prevention, detection, and mitigation of terrorist attacks, just as they do against other criminal threats. In fact the U.S. government has increased its investment in counterterrorism programs based on communications surveillance, data mining, and information fusion. Counterterrorism agencies are particularly interested in merging several different databases (information fusion) and then probing the combined data to understand transactions and interactions of specific persons or organizations of interest (data mining). They would also like to identify individuals (through data mining and behavioral surveillance) whose transactions and behavior might indicate possible terrorist links.

Such techniques often work well in commercial settings, for example for fraud detection, where they are applied to highly structured databases and are honed through constant use and learning. But the problems confronting counterterrorism analysts are vastly more difficult. Automated identification of terrorists through data mining (or any other known methodology) is neither feasible as an objective nor desirable as a goal of technology development efforts.

One reason is that collecting and examining information to inhibit terrorists inevitably conflicts with efforts to protect individual privacy. And when privacy is breached, the damage is real. The degree to which privacy is compromised is fundamentally related to the sciences of database technology and statistics as well as to policy and process. For example, there is no way to make personal information in databases fully anonymous. Technical, operational, legal, policy, and oversight processes to minimize privacy intrusion and the damage it causes must be established and uniformly applied. Even under the pressure of threats as serious as terrorism, the privacy rights and civil liberties that are the cherished core values of our nation must not be destroyed.

The quality of the data used in the difficult task of preempting terrorism is also a substantial issue. Data of high quality are correct, current, complete, and relevant, and so they can be used effectively, economically, and rapidly to inform and evaluate decisions. Data derived by linking high-quality data with data of lesser quality will tend to be low-quality data. Because data of questionable quality are likely to be the norm in counterterrorism, analysts must be cognizant of their effects, especially in fused or linked databases, and officials must carefully consider the consequent likelihood of false positives and privacy intrusions.

The preliminary nature of the scientific evidence, the risk of false positives, and operational vulnerability to countermeasures argue for behavioral observation and physiological monitoring being used at most as a preliminary screening method for identifying individuals who merit additional follow-up investigation. Although laboratory research and development of techniques for automated, remote detection and assessment of anomalous behavior, for example deceptive behavior, may be justified, there is not a consensus within the relevant scientific community nor on the committee regarding whether any behavioral surveillance or physiological monitoring techniques are ready for use at all in the counterterrorist context given the present state of the science.

The committee has developed and provides in Chapter 2 a specific framework for evaluation and operation of information-based counterterrorism programs to guide deployment decisions and facilitate continual improvement of the programs.

National security authorities of course should always adhere to the law, but the committee recognizes that laws will have to be reviewed and revised from time to time to ensure that they are appropriate, up to date, and responsive to real needs and contemporary technologies.

With these several concerns and issues in mind, the committee makes the following recommendations.

**Recommendation 1. U.S. government agencies should be required to follow a systematic process (such as the one described in the framework proposed in Chapter 2) to evaluate the effectiveness, lawfulness, and consistency with U.S. values of every information-based program, whether classified or unclassified, for detecting and countering terrorists before it can be deployed, and periodically thereafter.** Under most circumstances, this evaluation should be required as a condition for deployment of information-based counterterrorism programs, but periodic evaluation and continual improvement should *always* be required when such programs are in use. The committee believes that the framework presented in Chapter 2 defines an appropriate process for this purpose.

**Periodically after a program has been operationally deployed, and in particular before a program enters a new phase in its life cycle, policy makers should apply a framework such as the one proposed in Chapter 2 to the program before allowing it to continue operations or to proceed to the next phase.** Consistency with relevant laws and regulations, and impact on individual privacy and civil liberties—as well as validity, effectiveness, and technical performance—should be rigorously assessed. Such review is especially necessary given that the committee found little evidence of any effective evaluation performed for current programs intended to detect terrorist activity by automated analysis of databases. (If such evidence does exist, it should be presented in the appropriate oversight forums as part of such review.) Periodic review may result in significant modification of a program or even its cancellation.

**Any information-based counterterrorism program of the U.S. government should be subjected to robust, independent oversight.** All three branches of government have important roles to play to ensure that such programs adhere to relevant laws. **All such programs should provide meaningful redress to any individuals inappropriately harmed by their operation.**

To protect the privacy of innocent people, the research and development of any information-based counterterrorism program should be conducted with synthetic population data. If and when a program meets the criteria for deployment in the committee's illustrative framework described in Chapter 2, it should be deployed only in a carefully phased manner, e.g., being field tested and evaluated at a modest number of sites before being scaled up for general use. At all stages of a phased deployment, data about individuals should be rigorously subjected to the full safeguards of the framework.

**Recommendation 2. The U.S. government should periodically review the nation's laws, policies, and procedures that protect individuals' private**

information for relevance and effectiveness in light of changing tech-nologies and circumstances. **In particular, Congress should reexamine existing law to consider how privacy should be protected in the context of information-based programs (e.g., data mining) for counterterrorism.** Such reviews should consider establishment of restrictions on how personal information can be used. Currently, legal restrictions are focused primarily on how records are collected and assessed, rather than on their use.

# 1

# Scoping the Issue:
# Terrorism, Privacy, and Technology

## 1.1 THE NATURE OF THE TERRORIST
## THREAT TO THE UNITED STATES

Since September 11, 2001, the United States has faced a real and serious threat from terrorist action. Although the primary political objectives of terrorist groups vary depending on the group (e.g., the political objectives of Al Qaeda differ from those of Aum Shinrikyo), terrorist actions throughout history have nevertheless shared certain common characteristics and objectives. First, they have targeted civilians or noncombatants for political purposes. Second, they are usually violent, send a message, and have symbolic significance. The common objectives of terrorists include seeking revenge, renown, and reaction; that is, terrorists generally seek to "pay back" those they see as repressing them or their people; to gain notoriety or social or spiritual recognition and reward; and to cause those they attack to respond with fear, an escalating spiral of violence, irrational reaction and thus self-inflicted damage (e.g., reactions that strengthen the hand of the terrorists), or capitulation. Third, terrorists often blend with the targeted population—and in particular, they can exploit the fundamental values of open societies, such as the United States, to cover and conceal their planning and execution.

Despite these commonalities, today's terrorist threat is fundamentally different from those of the past. First, the scale of damage to which modern terrorists aspire is much larger than in the past. The terrorist acts of September 11, 2001, took thousands of lives and caused hundreds of billions of dollars in economic damage. Second, the potential terrorist use of weapons of mass destruction (e.g., nuclear weapons, biological or chemical agents) poses a threat that is qualitatively different from a threat based on firearms or chemical explosives. Third, terrorists operate in a modern environment plentiful in the amount of available information and increasingly ubiquitous in its use of information technology.

Even as terrorist ambitions and actions have increased in scale, smaller bombings and attacks are also on the rise in many corners of the world. To date, all seem to have been planned and executed by groups or networks and therefore have required some level of interaction and communication to plan and execute.

Left unaddressed, this terrorist threat will create an environment of fear and anxiety for the nation's citizens. If people come to believe that they are infiltrated by enemies that they cannot identify and that have the power to bring death, destruction, and havoc to their lives, and that preventing that from happening is beyond the capability of their governments, then the quality of national life will be greatly depreciated as citizens refrain from fully participating in their everyday lives. That scenario would constitute a failure to "establish Justice, insure domestic Tranquility, provide for the common defense, promote the general Welfare, and secure the Blessings of Liberty to ourselves and our Posterity," as pledged in the Preamble to the Constitution.

To address this threat, new technologies have been created and are creating dramatic new ways to observe and identify people, keep track of their location, and perhaps even deduce things about their thoughts and behaviors. The task for policy makers now is to determine who should have access to these new data and capabilities and for what purposes they should be used. These new technologies, coupled with the unprecedented nature of the threat, are likely to bring great pressure to apply these technologies and measures, some of which might intrude on the fundamental rights of U.S. citizens.

Appendix B ("Terrorism and Terrorists") addresses the terrorist threat in greater detail.

## 1.2  COUNTERTERRORISM AND PRIVACY AS AN AMERICAN VALUE

In response to the mounting terrorist threat, the United States has increased its counterterrorist efforts with the aim of enhancing the ability of the government to prevent terrorist actions before they occur. These efforts have raised concerns about the potential negative impacts of counterterrorism programs on the privacy and other civil liberties of U.S. citizens, as well as the adequacy of relevant civil liberties protections. Because terrorists blend into law-abiding society, activities aimed at detecting and countering their actions before they occur inherently raise concerns that such efforts may damage a free, democratic society through well-intentioned steps intended to protect it. One such concern is that law-abiding citizens who come to believe that their behavior is watched too closely by government agencies and powerful private institutions may be unduly inhibited from participating in the democratic process, may be inhibited from contributing fully to the social and cultural life of their communities, and may even alter their purely private and perfectly legal behavior for fear that discovery of intimate details of their lives will be revealed and used against them in some manner.

Privacy is, and should continue to be, a fundamental dimension of living in a free, democratic society. An array of laws protect "government, credit, communications, education, bank, cable, video, motor vehicle, health, telecommunications, children's, and financial information; generally carve out exceptions for disclosure of personal information; and authorize use of warrants, subpoenas, and court orders to obtain information."[1] These laws usually create boundaries between individuals and institutions (or sometimes other individuals) that may limit what information is collected (as in the case of wiretapping or other types of surveillance) and how that information is handled (such as the fair information practices that seek care and openness in the management of personal information). They may establish rules governing the ultimate use of information (such as prohibitions on the use of certain health information for making employment decisions), access to the data by specific individuals or organizations, or aggregation of these data with other data sets. The great strength of the American ideal of privacy has been its robustness in the face of new social arrangements, new business practices, and new technologies. As surveillance technologies have expanded the technical capability of the government to intrude into personal lives, the law has sought to maintain a principled balance between the needs of law enforcement and democratic freedoms.

Public attitudes, as identified in public opinion polls, mirror this delicate balance.[2] For example, public support for counterterrorism measures appears to be strongly influenced by perceptions of the terrorist threat,

---

[1]U.S. Congressional Research Service, *Privacy: Total Information Awareness Programs and Related Information Access, Collection, and Protection Laws* (RL31730), updated March 21, 2003, by Gina Marie Stevens.

[2]See Appendix M ("Public Opinion Data on U.S. Attitudes Toward Government Counterterrorism Efforts") for more details. Among them are two caveats about the identification of public attitudes through public opinion surveys. The first one has to do with the framing of survey questions, in terms of both wording and context, which have been shown to strongly influence the opinions elicited. The second has to do with declining response rates to national sample surveys and the inability to detect or estimate nonresponse bias.

an assessment of government effectiveness in dealing with terrorism, and perceptions as to how these measures are affecting civil liberties. Thus, one finds that since 9/11, public opinion surveys reflect a diminishing acceptance of government surveillance measures, with people less willing to cede privacy and other civil liberties in the course of increased terrorism investigation and personally less willing to give up their freedoms and more pessimistic about protection of the right to privacy. Yet recent events, such as the London Underground bombings of July 2005 and reports in August 2006 that a major terrorist attack on transatlantic airliners had been averted, appeared to influence public attitudes; support increased for such surveillance measures as expanded camera surveillance, monitoring of chat rooms and other Internet forums, and expanded monitoring of cellular phones and e-mails. However, public attitudes toward recently revealed monitoring programs are mixed, with no clear consensus.

Public opinion polls also indicate that the public tends to defend civil liberties more vigorously in the abstract than in specific situations. At the same time, people seem to be less concerned about privacy in general (i.e., for others) but rather with protecting the privacy of information about themselves. In addition, most people are more tolerant of surveillance when it is aimed at specific racial or ethnic groups, when it concerns activities they do not engage in, or when they are not focusing on its potential personal impact. Thus the perception of threat might explain why passenger screening and searches both immediately after September 11, 2001, and continuing through 2006 consistently receive high levels of support while, at the same time, the possibility of personal impact reduces public support for government collection of personal information about travelers. The public is also ambivalent regarding biometric identification technologies and public health uses, such as prevention of bioterrorism and the sharing of medical information. For these, support increases with assurances of anonymity and personal benefits or when they demonstrate a high degree of reliability and are used with consent.

Legal analysts,[3] even courts,[4] if not the larger public, have long recognized that innovation in information and communications technologies often moves faster than the protections afforded by legislation, which is usually written without an understanding of new or emerging technologies, unanticipated terrorist tactics, or new analytical capabilities. Some of these developing technologies are described in Section 1.6 and in greater

---

[3]For example, see R.A. Pikowsky, "The need for revisions to the law of wiretapping and interception of email," *Michigan Telecommunications & Technology Law Review* 10(1), 2004.

[4]U.S. Court of Appeals. (No. 00-5212; June 28, 2001), p. 10. Available at http://www.esp.org/misc/legal/USCA-DC_00-5212.pdf.

detail in Appendixes C ("Information and Information Technology") and H ("Data Mining and Information Fusion"). The state of the law and its limitations are detailed in Appendix F ("Privacy-Related Law and Regulation: The State of the Law and Outstanding Issues"). As new technologies are brought to bear in national security and counterterrorism efforts, the challenge is no different from what has been faced in the past with respect to potential new surveillance powers: identify those new technologies that can be used effectively and establish specific rules that govern their use in accordance with basic constitutional privacy principles.[5]

## 1.3 THE ROLE OF INFORMATION

Information and information technology are ubiquitous in today's environment. Massive databases are maintained by both governments and private-sector businesses that include information about each person and about his or her activities. For example, public and private entities keep bank and credit card records; tax, health, and census records; and information about individuals' travel, purchases, viewing habits, Web search queries, and telephone calls. Merchants record what individuals look at, the books they buy and borrow, the movies they watch, the music they listen to, the games they play, and the places they visit. Other kinds of databases include imagery, such as surveillance video, or location information, such as tracking data obtained from bar code readers or RFID (radio frequency identification) tags. Through formal and informal relationships between government and private-sector entities, much of the data available to the private sector is also available to governments.

In addition, digital devices for paying tolls, computer diagnostic equipment in car engines, and global positioning services are increasingly common on passenger vehicles. Cellular telephones and personal digital assistants record not only call and appointment information, but also location, transmitting this information to service providers. Internet service providers record online activities, digital cable and satellite systems record what individuals watch and when, alarm systems record when people enter and leave their homes. People back up personal data files online and access online photo, e-mail, and music storage services. Global positioning technologies are appearing in more and more products, and RFID tags are beginning to be used to identify consumer goods, identification documents, pets, and even people.

Modern technology offers myriad options for communication

---

[5]"[T]he law must advance with the technology to ensure the continued vitality of the Fourth Amendment," Senate Judiciary Committee Report on the Electronic Communications Privacy Act of 1986 (S. 2575), Report 99-541, 99th Congress, 2nd Session, 1986, p. 5.

between individuals and among small groups, including cell phones, e-mail, chat rooms, text messaging, and various forms of mass media. With voice-over-IP telephone service, digital phone calls are becoming indistinguishable from digital documents: both can be stored and accessed remotely. New sensor technologies enable the tagging and tracking of information about individuals without their permission or awareness.

As noted earlier, the terrorists of today are embedded and operate in this environment. It is not unreasonable to believe that terrorists planning an attack might leave "tracks" or "signatures" in these digital databases and networks and might make use of the communications channels available to all. Extracting terrorist tracks from nonthreat tracks might be the goal, but this is nevertheless not easy. One could imagine that aspects of a terrorist signature may be information that is not easily available or easily linked to other information or that some signatures may garner suspicion but are really not threats. However, with appropriate investigative leads, the potential increases that examining these databases, monitoring the contents of terrorist communications, and using other techniques, such as tagging and tracking, may yield valuable clues to terrorist intentions.

These possibilities have not gone unnoticed by the U.S. government, which has increased the number of and investment in counterterrorism programs that collect and analyze information to protect America from terrorism and other threats to public health and safety.[6] The government collects information from many industry and government organizations, including telecommunications, electricity, transportation and shipping, law enforcement, customs agents, chemical and biological industries, finance, banking, and air transportation. The U.S. government also has the technical capability and, under some circumstances, the legal right to collect and hold information about U.S. citizens both at home and abroad. To improve the overall counterterrorism effort, the government has mandated interagency and interjurisdictional information sharing.[7] In short, the substantial power of the U.S. government's capability to collect information about individuals in the United States, as well as that of private-sector corporations and organizations, and the many ways that

---

[6]In this report, the term "program" refers to the resources required to execute a specific function—for example, a counterterrorism program, such as the Terrorist Information Awareness program. A program always involves people executing information-intensive processes. Frequently, a program involves an information system and other information systems with which it exchanges information. Humans are always fully responsible for the actions of a program.

[7]*Intelligence Reform and Terrorism Prevention Act of 2004*, Public Law 108-458, December 17, 2004.

advancing technology is improving that capability necessitate explicit steps to protect against its misuse.

If it were possible to automatically find the digital tracks of terrorists and automatically monitor only the communications of terrorists, public policy choices in this domain would be much simpler. But it is not possible to do so. All of the data contained in databases and on networks must be analyzed to attempt to distinguish between the data associated with terrorist activities and those associated with legitimate activities. Much of the analysis can be automated, a fact that provides some degree of protection for most personal information by having data manipulated within the system and restricted from human viewing. However, at some point, the outputs need to be considered and weighed, and some data associated with innocent individuals will necessarily and inevitably be examined by a human analyst—a fact that leads to some of the privacy concerns raised above. (Other privacy concerns, largely rooted in a technical definition of privacy described below, arise from the mere fact that certain individuals are singled out for further attention, regardless of whether a human being sees the data at all.)

In conceptualizing how information is used, it is helpful to consider what might be called the information life cycle. Addressed in greater detail in Appendix C, digital information typically goes through a seven-step information life cycle:

- *Collection.* Information, whether accurate or inaccurate, is collected by some means, whether in an automated manner (e.g., financial transactions at a point of sale terminal or on the Web, call data records in a telecommunications network) or a manual manner (e.g., a Federal Bureau of Investigation (FBI) agent conducting an interview with an informant). Information may often be collected or transmitted (or both) without the subject's awareness. In some instances, the party collecting the information may not be the end user of that information. This is especially relevant in government use of databases compiled by private parties, since laws that regulate government collection of information do not necessarily place comparable restrictions on government use of such information.

- *Correction.* Information determined to be erroneous, whether through automated or manual means, may be discarded or corrected. Information determined to be incomplete may be augmented with additional information. Under some circumstances, the person associated with the collected information can make corrections. Information correction is not trivial, especially when large volumes of data are involved. The most efficient and practical means of correcting information may reduce uncertainties but is not likely to eliminate them, and indeed error correction may itself sometimes introduce more error.

• *Storage.* Information is stored in data repositories—databases, data warehouses, or simple files.

• *Analysis and processing.* Information is used or analyzed, often using query languages, business intelligence tools, or analytical techniques, such as data mining. Analysis may require access to multiple data repositories, possibly distributed across the Internet.

• *Dissemination and sharing.* Results of information analysis and processing are published or shared with the intended customer or user community (which may consist of other analysts). Disseminated information may or may not be in a format compatible with users' applications.

• *Monitoring.* Information and analytical results are monitored and evaluated to ensure that technical and procedural requirements have been and are continuing to be met. Examples of important requirements include security (Are specified security levels being maintained?), authorization (Are all access authorized?), service level agreements (Is performance within promised levels?), and compliance with applicable government regulations.

• *Selective retention or deletion.* Information is retained or deleted on the basis of criteria (explicit or implicit) set for the information repository by the steward or by prevailing laws, regulations, or practices. The decreasing cost of storage and the increasing belief in the potential value to be mined from previously collected data are important factors enabling the increase in typical data retention periods. The benefits of retention and enhanced predictive power have to be balanced against the costs of reduced confidentiality. Data retention policies should therefore be regularly justified through an examination of this trade-off.

As described, these steps in the information life cycle can be regarded as a notional process for the handling of information. However, in practice, one or more of these steps may be omitted, or the sequencing may be altered or iterated. For example, in some instances, it may be that data are first stored and then corrected. Or the data may be stored with no correction at all or processed without being stored, which is what firewalls do.

Additional issues arise when information is assembled or collected from a variety of storage sources for presentation to an analysis application. Assembling such a collection generally entails linking records based on data fields, such as unique identifiers if present and available (identification numbers) or less perfect identifiers (combinations of name, address, and date of birth). The challenge of accurately linking large databases should not be underestimated. In practice, it is often the case that data may be linked with little or no control for accuracy or ability to correct errors in these fields, with the likely outcome that many records will be linked improperly and that many other records that should be linked are not linked. Without checks on the accuracy of such linkages, there is no way of understanding how errors resulting from linkage may affect the quality or provenance of the subsequent analysis.

Finally, different entities handle information differently because of the standards and regulations imposed on them. The types of information that can be collected, corrected, stored, disseminated, and retained and by whom, when, and for how long vary across private industries and government agencies. For example, three different kinds of agencies in the United States have some responsibility for combating terrorism: agencies in the intelligence community (IC), agencies of federal law enforcement (FLE), and agencies of state, local, and tribal law enforcement (SLTLE). The information-handling policies and practices of these different types of agency are governed by different laws and regulations. For example, the information collection policies and practices of SLTLE agencies require the existence of a "criminal predicate" to collect and retain information that identifies individuals and organizations; a criminal predicate refers to the possession of "reliable, fact-based information that reasonably infers that a particularly described . . . subject has committed, is committing or is about to commit a crime."[8] No such predicate is required for the collection of similar information by agencies in the intelligence community. Some FLE agencies (in particular, the FBI and the Drug Enforcement Agency) are also members of the intelligence community, and when (and only when) they are acting in this role, they are not required to have such predicates, either. The rules for information retention and storage are also more restricted for SLTLE agencies than for IC agencies (or FLE agencies acting in an IC role).

## 1.4 ORGANIZATIONAL MODELS FOR TERRORISM AND THE INTELLIGENCE PROCESS

A variety of models exists for how terrorist groups are organized, so it is helpful to consider two ends of a spectrum of organizational practices. At one end is a command-and-control model, which also characterizes traditional military organizations and multinational corporations. In this top-down structure, the leaders of the organization are responsible for planning, and they coordinate the activities of operational cells. At the other end of the spectrum is an entrepreneurial model, in which terrorist

---

[8]D.L. Carter, *Civil Rights and Privacy in the Law Enforcement Intelligence Process*, Intelligence Program, School of Criminal Justice, Michigan State University, March 2008.

cells form spontaneously and do their planning and execution without asking anybody's permission or obtaining external support, although they may be loosely coordinated with respect to some overall high-level objective (such as "kill Westerners in large numbers"). In practice, terrorist groups can be found at one end or the other of this spectrum, as well as somewhere in the middle. For example, a terrorist cell might form itself spontaneously but then make contact with a central organization in order to obtain some funding and technical support (such as a visit by a bomb-making expert).

The spectrum of organizational practice is important because the nature of the organization in question is closely related to the various information flows among elements of the organization. These flows are important, because they provide opportunities for disruption and exploitation in counterterrorist efforts. Exploitation in particular is important because that is what yields information that may be relevant to anticipating an attack.

Because it originates spontaneously and organically, the decentralized terrorist group, almost by definition, is usually composed of individuals who do blend very well and easily into the society in which they are embedded. Thus, their attack planning and preparation activities are likely to be largely invisible when undertaken against the background of normal, innocent activities of the population at large. Information on such activities is much more likely to come to the attention of the authorities through tips originating in the relevant neighborhoods or communities or through observations made by local law enforcement authorities. Although such tips and observations are also received in the context of many other tips and observations, some useful and others not, the amount of winnowing necessary in this case is very much smaller than the amount required when the full panoply of normal, innocent activities constitutes the background.

By contrast, the command-and-control terrorist group potentially leaves a more consistent and easily discernible information footprint in the aggregate (although the individual elements may be small, such as a single phone call or e-mail). By definition, a top-down command structure involves regular communication among various elements (e.g., between platoon leaders and company commanders). Against the background noise, such regularities are more easily detected and understood than if the communication had no such structure. In addition, such groups typically either "light up" with increased command traffic or "go dark" prior to conducting an attack. Under these circumstances, there is greater value in a centralized analysis function that assembles the elements together into a mosaic.

Although data mining techniques are defined and discussed below in Section 1.6.1, it is important to point out here that different kinds of analytical approaches are suitable in each situation. This report focuses on two general types of data mining techniques (described further in Appendix H): subject-based and pattern-based data mining. Subject-based data mining uses an initiating individual or other datum that is considered, based on other information, to be of high interest, and the goal is to determine what other persons or financial transactions or movements, etc., are related to that initiating datum. Pattern-based data mining looks for patterns (including anomalous data patterns) that might be associated with terrorist activity—these patterns might be regarded as small signals in a large ocean of noise.

In the case of the decentralized group, subject-based data mining is likely to augment and enhance traditional police investigations by making it possible to access larger volumes of data more quickly. Furthermore, communications networks can more easily be identified and mapped if one or a few individuals in the network are known with high confidence. By contrast, pattern-based data mining may be more useful in finding the larger information footprint that characterizes centrally organized terrorist groups.

Note that there is also a role for an analytical function after an attack occurs or a planned attack is uncovered and participants captured. Under these circumstances, plausible starting points are available to begin an investigation, and this kind of analytical activity follows quite closely the postincident activities in counterespionage: who were these people, who visited them, with whom were they communicating, where did the money come from, and so on. These efforts (often known as "rolling up the network") serve both a prosecutorial function in seeking to bring the perpetrators to justice and a prophylactic function in seeking to prevent others in the network from carrying out further terror attacks.

## 1.5  ACTIVITIES OF THE INTELLIGENCE COMMUNITY AND OF LAW ENFORCEMENT AGENCIES

The intelligence community is responsible for protecting U.S. national security from threats that have been defined by the executive branch. When threats are defined, further information is sought (i.e., "intelligence requirements") to understand the status and operations of the threat, from which intervention strategies are developed to prevent or mitigate the threat. The information collection and management process for the intelligence community is driven by presidential policy.

In contrast, law enforcement agencies identify threats based on behaviors that are specifically identified as criminal (i.e., with the Fourth Amendment requirement of particularity). The law enforcement approach to the threat is based on traditional criminal investigation and case building, a problem-solving intervention, or a hybrid of these two. The law enforcement agency information collection and management process is driven by crime. The parameters and policy of law enforcement activity to deal with the threat are stipulated constitutional law (notably the law of criminal evidence and procedure) and civil rights cases (42 USC 1983) particularly based on consent decrees related to the intelligence process in a number of cities. Two civil cases—*Handschu v. Special Services Division (NYPD)* and *American Friends Service Committee v. Denver*—have been major forces in shaping law enforcement policy on information collection for the intelligence process, notably related to First Amendment expressive activity and the inferred right to privacy.

As a matter of U.S. public policy today, the prevention of terrorist attacks against the U.S. homeland and other U.S. interests is the primary goal of the intelligence community and of federal law enforcement agencies. Prevention of terrorist attacks is necessarily a proactive and ongoing role, and thus it is not necessarily carried out in response to any particular external event. Countercrime activities are usually focused on investigation and developing the information basis for criminal prosecution. As a practical matter, most such investigations are reactive—that is, they are initiated in response to a specific occurrence of criminal activity.

These comments are not intended to imply that there is no overlap between the counterterrorist and countercrime missions. For example, law enforcement authorities are also concerned about the prevention of crimes through the perhaps difficult-to-determine deterrent effect of postattack prosecution of terrorists and their collaborators. In addition, preparation for future criminal acts can themselves be a current criminal violation under the conspiracy or attempt provisions of federal criminal law or other provisions defining preparatory crimes, such as solicitation of a crime of violence or provision of material support in preparation for a terrorist crime. The standard for opening an investigation—and thus for collecting personally identifiable information—is satisfied when there is not yet a current substantive or preparatory crime but facts or circumstances reasonably indicate that such a crime will occur in the future (i.e., when there is a valid criminal predicate).[9]

Although most crimes do not have a direct terrorism nexus, it is not uncommon to find that terrorists engage in criminal activities that are on the surface unrelated to terrorism. For example, a terrorist group with-

---

[9]Information on investigations and inquiries is derived from *The Attorney General's Guidelines on General Crimes, Racketeering Enterprise and Terrorism Enterprise Investigations*, Attorney General John Ashcroft, U.S. Department of Justice, Washington, D.C., May 30, 2002, available at http://www.usdoj.gov/olp/generalcrimes2.pdf.

out financial resources provided from an external source may engage in fraud. One well-known case in 2002 involved cigarette smuggling in support of Hezbollah.[10]

In addition, law enforcement agencies are often unable to deploy personnel and other resources if they are not being used to further active criminal investigations, so counterterrorism investigations are often part of an "all crimes" approach—that is, law enforcement agencies focus on an overall goal of public safety and stay alert for any threats to the public safety, including but not limited to terrorism.

Finally, both criminals and terrorists (foreign or domestic) operating in the United States are likely to blend very well and easily into the society in which they are embedded. That is, ordinary criminals are likely to be similar in profile to decentralized terrorist groups that also would draw their members from the ranks of disaffected Americans (or from individuals who are already familiar with each other or trusted, such as family members). Thus, both counterterrorist and countercrime efforts are likely to depend a great deal on information originating in the relevant neighborhoods or communities or observations made by local law enforcement authorities.

## 1.6  TECHNOLOGIES OF INTEREST IN THIS REPORT

The counterterrorist activities of the U.S. government depend heavily on many different kinds of technology. A comprehensive assessment of all technologies relevant to these efforts would be extensive and resource-intensive, not to mention highly classified at least in part, and indeed the committee was not charged with conducting such an assessment. Instead, the charge directed the committee to focus primarily on two important technologies—data mining and behavioral and physiological surveillance—and their relationship to and impact on privacy.[11]

The focus of the committee's charge does not negate the value of other technologies or programs that generate information relevant to the counterterrorist mission, such as technologies for tagging and track-

---

[10]A North Carolina-based Hezbollah cell smuggled untaxed cigarettes into North Carolina and Michigan and used the proceeds to provide financial support to terrorists in Beirut. See D.E. Kaplan, "Homegrown terrorists: How a Hezbollah cell made millions in sleepy Charlotte, N.C.," *U.S. News and World Report*, March 2, 2003, available at http://www.usnews.com/usnews/news/articles/030310/10hez.htm.

[11]Despite the focus on data mining and behavioral surveillance, the committee does recognize that most of the issues related to privacy and these technologies also apply more broadly to other information technologies as they might be used for counterterrorism. Nevertheless, this is mostly a report about privacy as it relates to these two specific technologies of interest.

ing for identity management, or even for the admittedly controversial use of so-called "national security letters" for information gathering. (A national security letter (NSL) is a demand for information from a third party issued by the FBI or by other government agencies with authority to conduct national security investigations. No judicial approval is needed for the issuance of an NSL, and many NSLs have been issued pursuant to statutory nondisclosure provisions that prevent the issuance from being made known publicly. Both of these provisions have created controversy.) Indeed, regardless of whether a given information-generating program or technology is or is not classified, it can be said openly that the purpose of the program or technology is to generate information. Mission-directed intelligence analysis is an all-source enterprise—that is, the purpose of the analytical mission is to make sense out of information coming from multiple sources, classified and unclassified. Data mining and information fusion are technologies of analysis rather than collection, and thus they are intended to help analysts find patterns of interest in all of the available data.

### 1.6.1  Data Mining

Under the rubric of data mining techniques fall a diverse set of tools for mathematical modeling, including machine learning, pattern recognition, and information fusion.

- Machine learning is the study of computer algorithms that improve automatically through experience.
- Pattern recognition addresses a broad class of problems in which a feature extractor is applied to untreated (usually image) input data to produce a reduced data set for use as an input to a classification model, which then classifies the treated input data into one of several categories.
- Information fusion refers to a class of methods for combining information from different sources in order to make inferences that may not be possible from a single source alone. Some information fusion methods use formal probabilistic models, and some include ways of assessing rates of linkage error; others include only one or none of these things.

There is a continuum of sophistication in techniques that have been referred to as data mining that may provide assistance in counterterrorism. On the more routine end of the spectrum (sometimes called subject-based data mining and often so routine as to not be included in the portfolio of techniques referred to as data mining) lies the automation of typical investigative techniques, especially the searching of large databases for characteristics that have been associated with individuals of interest, that is, people who are worthy of further investigation. Through the benefits of automation, the investigative power of these traditional techniques can be greatly expedited and broadened in comparison to former practices, and therefore they can provide important assistance in the fight against terrorism.

Subject-based data mining can include, for example, people who own cars with license plates that are discovered at the scene of a terrorist act or whose fingerprints match those of people known to be involved in terrorist activity. Subject-based data mining might also include people who have been in communication with other persons of interest, people who have traveled to various places recently, and people who have transferred large sums of money to others of interest. When several disparate pieces of information of this type are obtained that are associated with terrorist activity, identifying a subset of a database that matches one or more of these various pieces of information can be referred to as "drilling down." This is a data mining technique that simply expands and automates what a police detective or intelligence analyst would carry out with sufficient time.

There are two key requisites for this use of data mining. One is the development of linkages relating data and information in the relevant databases, which facilitates response to these types of queries—for example, being able to identify all numbers that have recently called or been called by a given telephone number. Of course, attestations regarding the accuracy and provenance of such identification are also necessary for confidence in the ultimate results. The second requisite is the quality of the information collected. Individuals claimed by law enforcement officials to match prints found at a crime scene have sometimes turned out not to match upon further investigation.[12] Also, matching names or other forms of record linkage are error-prone operations, generally because of data quality issues.

Similarly, so-called rule-based techniques collect joint characteristics or data for individuals (or other units of analysis, such as networks of individuals) whom detectives or intelligence analysts view as being potentially associated with terrorist activity. This activity can include, for example, the recent purchase, possibly as a member of a group, of chemicals or biological agents that can be used to create explosives or toxins. Again, this is a simple extension of what analysts would do with sufficient resources and represents a relatively unsophisticated application of data mining. The key element is the use of analysts to identify the important

---

[12]S. Kershaw, "Spain and U.S. at odds on mistaken terror arrest," *New York Times*, June 5, 2004, available at http://query.nytimes.com/gst/fullpage.html?res=9800EFDB1031 F936A35755C0A9629C8B63.

rules or patterns that are indicative of or associated with terrorist activity. Given that terrorists often operate in groups, network-based methods have particular importance and should be used in concert with rule-based methods when possible. As above, the use of rule-based techniques can be greatly compromised by poor-quality data.

Pattern-based data mining techniques either require a feedback mechanism to generate learning over time or are more assumption-dependent than subject-based techniques. Machine learning is one such technique: in situations in which the truth of a decision process can often be made known, the feedback of knowing which results were decided correctly and incorrectly can be used to improve the decision process, which "learns" over time to become a better discriminator. For example, in scanning carry-on luggage to decide which contents are of concern and which are not, the process of simultaneously and individually searching a large number of the bags identified both of concern and not of concern and feeding back this information into the decision algorithm, can be used to improve the algorithm. Over time the algorithm can learn which patterns are associated with bags of concern. These situations in which cases of interest and cases not of interest become known for a large number of instances, referred to as a training set, permit machine learning to operate. This represents a collection of techniques that might have important applicability to specific, limited components of the counterterrorism problem.

There are also a number of situations in which the identification of anomalous patterns, in comparison to a long historical pattern of behavior or use, might make it possible to ultimately discriminate between activities of interest and activities not of interest to intelligence analysts. Referred to as signature-based analysis, current successful applications of data mining in these situations include the identification of anomalous patterns of credit card use or the fraudulent use of a telephone billing account. However, in those applications, a training set is available to help evaluate the extent to which the pattern of interest is useful in discriminating the behavior of interest from that not of interest.

When a training set or some formal means for assessing predictive validity is not available (i.e., if there is no way to test predictions against some kind of ground truth), these techniques are unlikely to provide useful information for counterterrorism. Nevertheless, it may be possible to use subject-matter experts to identify discriminating patterns, and one cannot reject a priori the possibility that anomalous patterns might be identified that intelligence analysts would also view as very likely to be associated with terrorist activity. Working collaboratively, signature-based data mining techniques might be developed that could effectively discriminate in counterterrorism between patterns of interest and those not of interest. Such patterns might then provide leads for further investigation through traditional law enforcement or national security means.

This rough partitioning of data mining techniques into pattern-based and subject-based approaches is meant to describe two relatively broad classes of techniques representing two "pure types" of methods used. However, many of the approaches used in practice can be considered combinations of these pure types, and therefore the examples included here of these two approaches do not fully explore the richness of techniques that is possible. Indeed, the data mining components of a real system are likely to reflect aspects of both subject-based and pattern-based data mining algorithms, through joint use of several perspectives using different units of analysis, combining evidence in several ways.

In many cases, the unit of analysis is the individual, and the objective is discriminating between the people who are and are not of interest. However, rather than using the individual as the basic unit of analysis, many techniques may use other constructs, such as the relevant group of close associates, the date of some possible terrorist activity, or the intended target, and then tailor the information retrieval and the analysis using that perspective. To best address a given problem, it may be beneficial at times to use more than one unit of analysis (such as a group), and to combine such analyses so that mutually consistent information can be recognized and used. The unit of analysis selected has implications for the rule-based techniques that might be used, or what patterns or signatures might be seen to be anomalous and therefore of interest.[13]

In addition, the use of data mining procedures may occur as component parts of a counterterrorism system, in which data mining tools address specific needs, such as identifying all the financial dealings, contacts, events, travels, etc., corresponding to a person of interest. The overall system would be managed by intelligence agents, who would also have impacts on both the design of the data mining components and on the remaining components, which might involve skills that could not be automated. The precise form of such a system is only hinted at here, and both system development and deployment are likely to require a substantial investment of time and resources as well as collaboration with those with state-of-the-art expertise in data mining, database management, and counterterrorism.

Finally, no single operational system has access to all of the relevant data at the same time. In practice, the results of an analysis from any given system will often result in queries being made of other systems exploit-

---

[13]Additional examples in the fraud detection context are provided, for example, in R.J. Bolton and D.J. Hand, "Statistical fraud detection, a review," *Statistical Science* 17(3):235-255, 2002.

ing different analytical techniques with access to different databases. For example, an intensive analysis on one system may be made using a limited set of records to identify a set of initial leads. In subsequent stages using different systems, progressively more extensive sets of data may be analyzed to winnow the set of initial leads. Such a practice—often known as multistage inference—may help to improve efficiency and to reduce privacy impacts.

In general, there is little doubt about the utility of subject-based data mining as an investigative technique. However, the utility of pattern-based data mining and information fusion depends on the availability of a training set and the application to which the techniques are applied. Pattern-based data mining is most likely to be useful when training sets are available; there are supplementary tasks for which data mining tools might be helpful that do not require a training set. At the same time, the utility of pattern-based data mining, without a training set, to identify patterns indicative of individuals and events worth additional investigation, is very unclear. Although there is no a priori argument that categorically eliminates pattern-based data mining as useful tools for counterterrorism applications, considerable basic research will be necessary to develop operational systems that help to provide a prioritization of cases for experts to examine in more depth. Such research would examine the feasibility and utility of pattern-based data mining and information fusion for counterterrorism applications and subsequent development into specific applications components. That approach to the problem in question might not succeed but the potential gains are large, and for this reason such a modest program investment, structured in accordance with the framework proposed in Chapter 2, may be well worth making.

### 1.6.2  Behavioral Surveillance

Behavioral surveillance seeks to detect physiological behaviors, conditions, or responses and the attendant biological activity that indicate that an individual is about to commit an act of terrorism. Specifically, behavioral surveillance seeks to detect patterns of behavior thought to be precursors or correlates of wrongdoing (e.g., deception, expressing hostile emotions) or that are anomalous in certain situations (e.g., identifying a person who shows much greater fidgeting and much more facial reddening than others in a security line).

If people were incapable of lying, the easiest and most accurate way to determine past, current, and future behavior would be to ask them what they have been doing, what they are doing, and what they plan to do. But people are highly capable of lying, and it is currently very difficult to detect lying with great degrees of accuracy (especially through automated means). Thus, the terrorist's desire to avoid detection makes this verbal channel of information highly unreliable.

For this reason, behavioral surveillance focuses on biological or physiological indicators that are relatively involuntary (i.e., whose presence or absence is not subject to voluntary control) or provide detectible signs when they are being manipulated. For example, physiological indicators, such as cardiac activity, facial expressions, and voice tone, can be monitored and the readings used to make inferences about internal psychological states (e.g., "based on this pattern of physiological activity, this person is likely to be engaged in deception"). However, such indicators do not provide direct evidence of deception of any sort, let alone terrorist behavior (e.g., the deception if present at all may not relate to terrorist behavior but rather to cheating on one's income tax or spouse), and thus the problem becomes one of inferring the specific (i.e., terrorist behavior) from more general indicators.

To illustrate the government interest in behavioral surveillance, consider Project Hostile Intent, conducted under the auspices of the U.S. Department of Homeland Security's Human Factors Division in the Science and Technology Directorate. This project seeks to develop models of hostile intent and deception, focusing on behavioral and speech cues. These cues would be determined from experiments and derived from operationally based scenarios that reflect the screening and interviewing objectives of the department. In addition, the project seeks to develop an automated suite of noninvasive sensors and algorithms that can automatically detect and track the input cues to the models. If successful, the resulting technologies would afford capabilities to identify deception and hostile intent in real time, on the spot, using noninvasive sensors, with the goal of being able to screen travelers in an automated fashion with equal or greater effectiveness than the methods used today without impeding their flow.[14]

Although behavioral methods are useful under some circumstances (such as real-life circumstances that closely approximate laboratory conditions), they are intrinsically subject to three limitations:

- *Many-to-one.* Any given pattern of physiological activity can result from or be correlated with a number of quite different psychological or physical states.
- *Probabilistic.* Any detected sign or pattern conveys at best a change

---

[14]U.S. Department of Homeland Security, "Deception detection: Identifying hostile intent," *S&T Snapshots: Science Stories for the Homeland Security Enterprise* 1(1), May 2007, available at http://www.homelandsecurity.org/snapshots/newsletter/2007-05.htm#deception.

in the likelihood of the behavior, intent, or attitude of interest and are far from an absolute certainty.

  • *Errors.* In addition to the highly desirable true positives and true negatives that are produced, there will be the very troublesome false positives (i.e., a person telling the truth is thought to be lying) and false negatives (i.e., a person lying is thought to be telling the truth). Such errors are linked to the probabilistic nature of behavioral signals and a lack of knowledge today about how to interpret such signals.

Privacy issues associated with behavioral surveillance are regarded by the committee to be far more significant and far-ranging than those associated with the collection and use of electronic databases, in part because of their potential for abuse, in part because of what they may later reveal about an individual that is potentially unconnected to terrorist activities, in part because of a sense that the intrusion is greater if mental state is being probed, in part because people expect to be allowed to keep their thoughts to themselves, and in part because there is often much more ambiguity regarding interpretation of the results.

## 1.7  THE SOCIAL AND ORGANIZATIONAL CONTEXT

Technology is always embedded in a social and organizational context. People operate machines and devices and make decisions based on what these machines and devices tell them. In turn, these decisions are based on certain criteria that are organizationally specified. For example, a metal detector is placed at the entrance to a building. At the request of the security guard, a visitor walks through the detector. If the detector buzzes, the guard searches the visitor more closely. If the guard finds a weapon, the guard confiscates it and calls his superior to take the visitor for additional questioning. The guard carries out these procedures because they are required by the organization responsible for building security—and if the guard does not carry out these procedures, security may be compromised despite the presence of the metal detector.

Nor can the presence of the relevant machines and devices be taken for granted. There are many steps that must be taken before the relevant machine or device is actually deployed and put into use at a security checkpoint, and even when the science underpinning the relevant machines and devices is known, the science must be instantiated into artifacts. For example, a functional metal detector depends on some understanding of the science of metal detection, even if the theory is not completely known. Prototypes must be built and problems in the manufacturing process overcome. Budgets must be available to acquire the necessary devices and to train security guards in their operation. Oversight must be exercised to ensure that the processes and procedures necessary to decide on and implement a program are correctly followed.

Protecting privacy often depends on social and organizational factors as well. For example, the effectiveness of rules that prohibit agents or analysts from disclosing certain kinds of personal information about the targets of their investigations is based on the willingness and ability of those agents or analysts to follow the rules and the organization's willingness and ability to enforce them. While encryption may provide the technical capability to protect data from being viewed by someone without the decryption key, policies and practices determine whether encryption capabilities are actually used in the proper circumstances.

The social and organizational context in which technology is embedded is important from a policy standpoint because the best technology embedded in a dysfunctional organization or operated with poorly trained human beings is often ineffective. This point goes beyond the usual concerns about a technology that is promising in the laboratory being found too unwieldy or impractical for widespread use in the field.

## 1.8 KEY CONCEPTS

### 1.8.1 The Meaning of Privacy[15]

In both everyday discourse and the scholarly literature, a commonly agreed-on abstract definition of privacy is elusive. For example, privacy may refer to protecting the confidentiality of information; enabling a sense of autonomy, independence, and freedom to foster creativity; wanting to be left alone; or establishing enough trust that individuals in a given community are willing to disclose data under the assumption that they will not be misused. For purposes of this report, the term "privacy" is generally used in a broad and colloquial sense that encompasses the technical definitions of privacy and confidentiality commonly used in the statistical literature. That is, the statistical community's definition of privacy is an individual's freedom from excessive intrusion in the quest for information and an individual's ability to choose the extent and circumstances under which his or her beliefs, behaviors, opinions, and attitudes will be shared with or withheld from others. Confidentiality is the care in dissemination of data in a manner that does not permit identification of the respondent or would in any way be harmful to him or her and that

---

[15]An extended discussion of the meaning of privacy can be found in National Research Council, *Engaging Privacy and Information Technology in a Digital Age*, The National Academies Press, Washington, D.C., 2007.

the data are immune from legal process.[16] Put differently, privacy relates to the ability to withhold personal data, whereas confidentiality relates to the activities of an agency that has collected such data from others. Yet another sense of privacy to keep in mind is that of a set of restrictions on how or for how long personal information can be used. In this report, when these distinctions are important, these different senses of meaning will be explicitly addressed, but in the less technical sections, the term "privacy" will be used in a more generic fashion.

In its starkest terms, privacy is about what personal information is being kept private and which parties the information is being kept from. For example, one notion of privacy involves confidentiality or secrecy of some specific information, such as preventing disclosure of an individual's library records to the government. A second notion of privacy involves anonymity, as reflected in, for example, an unattributable chat room discussion that threatens the use of violence against innocent parties.

These two simple examples illustrate two key points regarding privacy. First, the party against which privacy is being invoked may have a legitimate reason for wanting access to the information being denied—a government conducting a terrorist investigation may want to know what a potential suspect is reading, or a law enforcement official may need the identity of the person threatening violence in order to protect innocent people. Second, some kind of balancing of competing interests may be necessary—thus raising the question of the circumstances under which the government should have access to such information.

In practice, three other issues are also critical to understanding the privacy interests in any given situation: what the information will be used for, where the information comes from, and what the consequences are for the individual whose information is at issue. Regarding purpose, divulging personal information for one purpose may not be regarded as a violation of privacy, whereas divulging the same information for a different purpose may be regarded as a clear violation of privacy. (In other words, a "justified" violation of an individual's privacy—that is, for a reason that is good and valid to the individual in question—is generally not viewed as a violation of his or her privacy interests by that individual.) Regarding source, government collection of personal information is often regarded as different in kind from private collection of personal information, although government is increasingly making use of personal data gathered by private parties. This point is especially significant because laws that restrict government collection of personal information often do

---

[16]National Research Council, *Private Lives and Public Policies: Confidentiality and Accessibility of Government Statistics*, National Academy Press, Washington, D.C., 1993.

not apply to private collection, and the government breaks no law in purchasing such information from private parties. Regarding consequences, for many people, a primary consideration in privacy is the adverse consequences they may experience if their privacy is compromised—denial of financial benefits, personal embarrassment or shame, and so on.

The notion of trust is intimately related to the meaning of privacy. Briefly put, people tend to invoke rights to privacy much more strongly when they fear the motivations or intent of the entity that is to receive their data. That is, a lack of trust in these data-receiving entities drives both the strength of people's desires for privacy and their conceptions of privacy. This is especially true when the data-receiving entity is capable of imposing an adverse consequence on them. Box 1.1 addresses this point further.

Privacy also has a variety of more technical meanings, some of which are elaborated in Appendix L ("The Science and Technology of Privacy Protection"). The most well-defined of these meanings for scientific study is based on the intuitive notion that a system containing an individual's information protects his or her privacy if all events, such as being singled out for additional attention at airport security, being denied medical insurance coverage, or gaining entrance to the college of his or her choice, are no more likely than if the system did not contain that information. This meaning can be formalized, as described in Appendix L (Section L.2).

### 1.8.2 Effectiveness

The effectiveness of a technology, a system, or a program is judged by the extent to which it directly furthers the objective being sought. Effectiveness is a measure of technical performance, and policy makers and government officials responsible for developing, purchasing, deploying, and using information-based programs must make judgments regarding whether a given level of effectiveness is sufficient to proceed with the use or deployment of a given technology, system, or program. Section 1.8.4 addresses false positives and false negatives as essential elements of judging the effectiveness of a program.

The qualification of "directly" furthering the objective being sought is an important one. From time to time, technologies, systems, or programs are admittedly ineffective from a technological point of view and yet are justified on the basis of their alleged deterrent value. That is, their mere presence and the adversary's concern that they might work are said to help deter the adversary from taking such an action.[17] The desirability of

---

[17]In the example of the metal detector, one might imagine that the device consisted only of a buzzer activated randomly on 30 percent of the individuals passing through and with no

---

**BOX 1.1**
**A Relationship Between Privacy and Trust**

The National Research Council report *Engaging Privacy and Information Technology in a Digital Age* (2007) explicitly addresses the relationship between privacy and trust. Specifically, that committee found (in Finding 4) that "privacy is particularly important to people when they believe that the entity receiving their personal information is not trustworthy and that they may be harmed by sharing that information."
That report goes on to explain (pp. 311-312):

> Trust is an important issue in framing concerns regarding privacy. In the context of an individual providing personal information to another, the sensitivities involved will depend on the degree to which the individual trusts that party to refrain from acting in a manner that is contrary to his or her interests (e.g., to pass it along to someone else, to use it as the basis for a decision with inappropriately adverse consequences). As an extreme case, consider the act of providing a complete dossier of personal information on a stack of paper—to a person who will destroy it. If the destruction is verifiable to the person providing the dossier (and if there is no way for the destroyer to read the dossier), it would be hard to assert the existence of any privacy concern at all.
> But for most situations in which one provides personal information, the basis for trust is less clear. Children routinely assert privacy rights to their personal information against their parents when they do not trust that parents will not criticize

---

admittedly ineffective systems that might help to deter adversaries is not considered in this report.

### 1.8.3 Law and Consistency with Values

Measures of effectiveness deal with issues of feasibility. Legality and ethicality, in contrast, address issues of desirability. Not all technically feasible technologies, systems, or programs are desirable. Law provides one codification of national values that prescribes required actions and

---

metal detection circuitry at all. Such a device would be useless in detecting metal knives and guns, but assuming its internal operation were kept secret, its presence could still arguably provide deterrent value—would-be carriers of guns and knives would be deterred by the possibility that they faced some chance that they would be physically searched. In practice, the physical search would be the mechanism for detecting 30 percent of the individuals carrying guns or knives, but the detector as such would have no value. Note also that a policy of sampling 30 percent of individuals at random would have the same effect in practice, and both would arguably have a comparable deterrent effect.

them or punish them or think ill of them as a result of accessing that information. (They also assert privacy rights in many other situations.) Adults who purchase health insurance often assert privacy rights in their medical information because they are concerned that insurers might not insure them or might charge high prices on the basis of some information in their medical record. Many citizens assert privacy rights against government, although few would object to the gathering of personal information within the borders of the United States and about U.S. citizens if they could be assured that such information was being used only for genuine national security purposes and that any information that had been gathered about them was accurate and appropriately interpreted and treated . . . . Perversely, many people hold contradictory views about their own privacy and other people's privacy—that is, they support curtailing the privacy of some demographic groups at the same time that they believe that their own should not be similarly curtailed. This dichotomy almost certainly reflects their views about the trustworthiness of certain groups versus their own.

In short, the act of providing personal information is almost always accompanied to varying degrees by a perceived risk of negative consequences flowing from an abuse of trust. The perception may or may not be justified by the objective facts of the situation, but trust has an important subjective element. If the entity receiving the information is not seen as trustworthy, it is likely that the individuals involved will be much more hesitant to provide that information (or to provide it accurately) than they would be under other circumstances involving a greater degree of trust.

prohibits other actions. Although society expects its government to obey the law, it is also true that technologies and events outpace the rate at which law changes. Such rapid changes often leave policy makers with a difficult gray area in which certain actions are not explicitly prohibited but that nevertheless may be inconsistent with a broad notion of American values.

A good example of the impact of technological change on the law is the interpretation of the Supreme Court in 1976 in *United States v. Miller*[18] that there can be no reasonable expectation of privacy in information held by a third party. The case involved cancelled checks, to which, the Court noted, "respondent can assert neither ownership nor possession."[19] Such documents "contain only information voluntarily conveyed to the banks and exposed to their employees in the ordinary course of business,"[20] and

---

[18]*United States v. Miller*, 425 U.S. 435 (1976).
[19]Id. at 440.
[20]Id. at 442.

therefore the Court found that the Fourth Amendment is not implicated when the government sought access to them:

> The depositor takes the risk, in revealing his affairs to another, that the information will be conveyed by that person to the Government. This Court has held repeatedly that the Fourth Amendment does not prohibit the obtaining of information revealed to a third party and conveyed by him to Government authorities, even if the information is revealed on the assumption that it will be used only for a limited purpose and the confidence placed in the third party will not be betrayed.[21]

Congress reacted to the decision by enacting modest statutory protection for customer financial records held by financial institutions,[22] but there is no constitutional protection for financial records or for any other personal information that has been disclosed to third parties. As a result, the government can collect even the most sensitive information from a third party without a warrant and without risk that the search may be found unreasonable under the Fourth Amendment.

The Court reinforced its holding in *Miller* in the 1979 case of *Smith v. Maryland*, involving information about (as opposed to the content of) telephone calls.[23] The Court found that the Fourth Amendment is inapplicable to telecommunications "attributes" (the number dialed, the time the call was placed, the duration of the call, etc.), because that information is necessarily conveyed to, or observable by, third parties involved in connecting the call.[24] "[T]elephone users, in sum, typically know that they must convey numerical information to the phone company; that the phone company has facilities for recording this information; and that the phone company does in fact record this information for a variety of legitimate business purposes."[25] As with information disclosed to financial institutions, Congress reacted to the Supreme Court's decision by creating a statutory warrant requirement for pen registers,[26] but the Constitution does not restrict government action in this area.

Some legal analysts believe that this interpretation regarding the categorical exclusion of records held by third parties from Fourth Amendment protection makes less sense today because of the extraordinary increase in both the volume and the sensitivity of information about individuals so often held by third parties. In this view, the digital transactions of daily

---

[21]Id. at 443 (citation omitted).

[22]*Right to Financial Privacy Act of 1978*, 12 U.S.C. §§ 3401-3422.

[23]442 U.S. 735 (1979).

[24]442 U.S. 735 (1979).

[25]Id. at 743.

[26]18 U.S.C. §§ 3121, 1841. A pen register is a device that records all numbers dialed from a particular telephone line.

life have become ubiquitous.[27] Such transactions include detailed information about individuals' behavior, communications, and relationships.

At the same time, people who live in modern society do not have a real choice to refrain from leaving behind such trails. Even in the 1970s when *Miller* and *Smith* were decided, individuals who wrote checks and made telephone calls did not voluntarily convey information to third parties—they had no choice but to convey the information if they wanted to make large-value payments or communicate over physical distances. And in those cases, the third parties did not voluntarily supply the records to the government. Financial institutions are required to keep records (ironically, this requirement is found in the Right to Financial Privacy Act), and telephone companies are subject to a similar requirement about billing records. In both cases, the government demanded the records. And, at the same time, the information collected and stored by banks and telephone companies is subject to explicit or implicit promises that it will not be further disclosed. Most customers would be astonished to find their checks or telephone billing records printed in the newspaper.

Today, such transactional records may be held by more private parties than ever before. For example, a handful of service providers already process, or have access to, the large majority of credit and debit card transactions, automated teller machine (ATM) withdrawals, airline and rental car reservations, and Internet access, and the everyday use of a credit card or ATM card involves the disclosure of personal financial information to multiple entities. In addition, digital networks have facilitated the growth of vigorous outsourcing markets, so information provided to one company is increasingly likely to be processed by a separate institution, and customer service may be provided by another. And all of those entities may store their data with still another. Moreover, there are information aggregation businesses in the private sector that already combine personal data from thousands of private-sector sources and public records. They maintain rich repositories of information about virtually every adult in the country, which are updated daily by a steady stream of incoming data.[28]

Finally, in this view, the fact that all of the data in question are in digital form means that increasingly powerful tools—such as automated data mining—can be used to analyze it, thereby reducing or eliminating privacy protections that were previously based on obscurity and difficulty

---

[27]K.M. Sullivan, "Under a watchful eye: Incursions on personal privacy," pp. 128-146 in *The War on Our Freedoms: Civil Liberties in an Age of Terrorism* (R.C. Leone and G. Anrig Jr., eds.), The Century Foundation, New York, N.Y., 2003.

[28]See, generally, U.S. Government Accountability Office, *Personal Information: Agency and Reseller Adherence to Key Privacy Principles*, GAO 06-421, Washington, D.C., 2006.

of access to the data. The impact of *Miller* in 1976 was limited primarily to government requests for specific records about identified individuals who had already done something to warrant the government's attention, whether or not the suspicious activity amounted to probable cause. Today, the *Miller* and *Smith* decisions allow the government to obtain the raw material on millions of individuals without any reason for identifying anyone in particular.

Thus, in this view, the argument suggests that by removing the protection of the Fourth Amendment from all of these records solely because they are held by third parties, there is a significant reduction in the constitutional protection for personal privacy—not as the result of a conscious legal decision, but through the proliferation of digital technologies. In short, under current Fourth Amendment jurisprudence, all personal information in third-party custody, no matter how sensitive or how revealing of a person's health, finances, tastes, or convictions, is available to the government without constitutional limit. The government's demand need not be reasonable, no warrant is necessary, no judicial authorization or oversight is required, and it does not matter if the consumer has been promised by the third party that his or her data would be kept confidential as a condition of providing the information.

A contrary view is that *Miller* and *Smith* are important parts of the modern Fourth Amendment and that additional privacy protections in this context should come from Congress rather than the courts. According to this view, *Miller* and *Smith* ensure that there are some kinds of surveillance that the government can conduct without a warrant. Fourth Amendment doctrine has always left a great deal of room for unprotected activity, such as what happens in public: the fact that the police can watch *in public areas* for criminal activity without being constrained by the Fourth Amendment is critical to the balance of the Fourth Amendment's rule structure. In switching from physical activity to digital activity, everything becomes a record. If all records receive Fourth Amendment protection, treating every record as private, the equivalent of something inside the home, then the government will have considerable difficulty monitoring criminal activity without a warrant. In effect, under this interpretation, the Fourth Amendment would apply much more broadly to records-based and digital crimes than it does to physical crimes, and all in a way that would make it very difficult for law enforcement to conduct successful investigations. In this view, the best way forward is for the Supreme Court to retain *Smith* and *Miller* and for Congress to provide statutory protections when needed, much as it has done with the enactment of privacy laws, such as the Electronic Communications Privacy Act.

Given these contrasting perspectives and the important issues they raise, the constitutional and policy challenges for the future are to decide—explicitly and in light of new technological developments—the appropriate boundaries of Fourth Amendment jurisprudence regarding the disposition of data held by third parties. The courts are currently hearing cases that help get to this question; so far they have indicated that noncontent information is covered by *Miller* but that content information receives full Fourth Amendment protection. But these cases are new and may be overturned, and it will be some years before clearer boundaries emerge definitively.

### 1.8.4 False Positives, False Negatives, and Data Quality[29]

False positives and false negatives arise in any kind of classification exercise.[30] For example, consider a counterterrorism exercise in which it is desirable to classify each individual in a set of people as "not worthy of further investigation/does not warrant obtaining more information on these people" or "worthy of further investigation/does warrant obtaining more information on these people," based on an examination of data associated with each individual. A false positive is someone placed in the latter category who has no terrorist connection. A false negative is someone placed in the former category who has a terrorist connection.

Consider a naïve notional system in which a computer program or a human analyst examines the data associated with each individual, searching for possible indications of terrorist attack planning. This examination results in a score for each individual that indicates the relative likelihood of him or her being "worthy of further investigation" relative to all of the others being examined.[31] When all of the individuals are examined, they are sorted according to this score.

This rank ordering does not, in itself, determine the classification—in addition, a threshold must be established to determine what scores will correspond to each category. The critical point here is that setting this threshold is the responsibility of a human analyst—technology does not,

---

[29]This section is adapted largely from National Research Council, *Engaging Privacy and Information Technology in a Digital Age*, The National Academies Press, Washington, D.C., 2007, Chapter 1.

[30]An extensive treatment of false positives and false negatives (and the trade-offs thereby implied) can be found in National Research Council, *The Polygraph and Lie Detection*, The National Academies Press, Washington, D.C., 2003.

[31]The score calculated by any given system may simply be an index with only ordinal (rank-ordering) properties. If more information is available and a more sophisticated analytical approach is possible, the score may be an actual Bayesian probability or likelihood that could be manipulated quantitatively in accordance with the mathematics of probability and statistics.

indeed cannot, set this threshold. Moreover, it is likely that the appropriate setting of a threshold depends on the consequences for the individual being miscategorized. If the real-world consequence of a false positive for a given individual is being denied boarding of an airplane compared with looking at more records relevant to that individual, one may wish greater certainty to reduce the likelihood of a false positive—this desire would tend to drive the threshold higher in the first instance than in the second. In addition, any real analyst will not be satisfied with a system that impedes the further investigation of someone whose score is below the threshold. That is, an analyst will want to reserve the right (have the ability) to designate for further examination an individual who may have been categorized as below threshold—to say, in effect, "That guy has a lower score than most of the others, but there's something strange about him anyway, and I want to look at him more closely even if he is below threshold."

Because the above approach is focused on individuals, any realistic setting of a threshold is likely to result in enormous numbers of false positives. One way to reduce the number of false positives significantly is to exploit the fact that terrorists—especially those with big plans in mind—are most likely to operate in small groups (also known as cells). Thus, a more sophisticated system could consider a different unit of analysis—groups of individuals rather than individuals—that might be worth further investigation. This approach, known as collective inference, focuses on analyzing large collections of records simultaneously (e.g., people, places, organizations, events, and other entities).[32] Conceptually, the output of this system could be a rank ordering of all possible groups (combinations) of two individuals, another rank ordering of all possible groups of three individuals, and so on. Once again, thresholds would be set to determine groups that were worth further investigation. The rank orderings resulting from a group-oriented analysis could also be used to rule out individuals who might otherwise be classified as worthy of further investigation—if an individual with an above-threshold score was not found among the groups with above-threshold scores, that individual would be either a lone wolf or clearly seen to be a false positive and thus eliminated before the investigation went any further.

A "brute-force" search of all possible groups of two, of three, and so on when the population in question is that of the United States is daunting, to say the least. But in practice, most of those groups will be individuals with no plausible connections among them, and thus the

---

[32]More detail on these ideas can be found in D. Jensen, M. Rattigan, and H. Blau, "Information awareness: A prospective technical assessment," *Proceedings of the 9th ACM SIGKDD International Conference on Knowledge Discovery and Data Mining*, 2003, available at http://kdl. cs.umass.edu/papers/jensen-et-al-kdd2003.pdf.

records containing information about those groups need not be examined. Identifying such groups is a problem, but other techniques may be useful in eliminating some groups at a fairly early stage—for example, if a group does not contain individuals who have communicated with each other, that group might be eliminated from further consideration. All such criteria also run the risk of incurring false negatives, and it remains to be seen how useful such pruning efforts are in practice.

False positives and false negatives arise from two other sources. One is the validity of the model used to distinguish between terrorists and innocent individuals. A perfectly valid model of a terrorist is one in which a set of specific measurable characteristics, if correctly associated with a given individual, would correctly identify that individual as a terrorist with 100 percent accuracy, and other individuals lacking one or more of those characteristics would be correctly identified as an innocent individual. Of course, in the real world, no model is perfect, and so false positives and false negatives are inevitable from the imperfection of models.

The second and independent source of false positives and false negatives is imperfect data. That is, even if a model were perfect, in the real world, the data asserted to be associated with a given individual is not in fact associated with that individual. For example, an individual's height may be recorded as 6.1 meters, whereas his height may in fact be 1.6 meters. Her religion may be recorded as Protestant, but in fact she may be a practicing Catholic. Such data errors arise for a wide range of reasons, including keyboarding errors, faulty intelligence, errors of translation, and so on. Improving data quality can thus reduce the rate of false positives and false negatives, but only up to the limits inherent in the imperfections of the model. Since models, for computability, abstract only some of the variables and behaviors of reality, they are by design imperfect. Model imperfections are a built-in source of error, and better data cannot compensate for a model's inadequacies.

Model inadequacies stem from several possible sources: (1) the required data for various characteristics in the assumed causal model may not be available, (2) some variables may be left out to simplify computations, (3) some variables that are causal may be available but unknown, (4) the precise form of the relationship between the predictor variables and the assessment of degree of interest is unknown, (5) the form of the relationship may be simplified to expedite computation, and (6) the phenomenon may be dynamic in nature and therefore any datedness in the inputs could cause erroneous improper predictions.

Data quality is the property of data that allows them to be used effectively and rapidly to inform and evaluate decisions.[33] Ideally, data should

---

[33]A.F. Karr, A.P. Sanil, and D.L. Banks, "Data quality: A statistical perspective," *Statistical Methodology* 3:137-173, 2006; T.C. Redman, "Data: An unfolding quality disaster,"

be correct, current, complete, and relevant. Data quality is intimately related to false positives and false negatives, in that it is intuitively obvious that using data of poor quality is likely to result in larger numbers of false positive and false negatives than would be the case if the data were of high quality.

Data quality is a multidimensional concept. Measurement error and survey uncertainty contribute (negatively) to data quality, as do issues related to measurement bias. Many issues arise as the result of missing data fields; inconsistent data fields in a given record, such as recording a pregnancy for a 9-year-old boy; data incorrectly entered into the database, such as that which might result from a typographical error; measurement error; sampling error and uncertainty; timeliness (or lack thereof); coverage or comprehensiveness (or lack thereof); improperly duplicated records; data conversion errors, as might occur when a database of vendor X is converted to a comparable database using technology from vendor Y; use of inconsistent definitions over time; and definitions that become irrelevant over time.

All of the forgoing discussion relates to the implications of measurement error that could easily arise in a given environment or database. However, when data come from multiple databases, they must be linked, and the methodology for performing data linkages in the absence of clear, unique identifiers is probabilistic in nature. Even in well-designed record linkage studies, such as those developed by the Census Bureau, automated matching is capable of reliably matching only about 75 percent of the people (although some appreciable fraction of the remainder are not matchable), and hand-matching of records is required to reduce the remaining number of unresolved cases.[34] The difficulty of reliable matching, superimposed on measurement error, will inevitably produce much more substantial problems of false positives and false negatives than most analysts recognize.

Data issues also arise as the result of combining databases—syntactic inconsistencies (one database records phone numbers in the form 202-555-1212 and another in the form 2025551212); semantic inconsistencies (weight measured in pounds vs. weight measured in kilograms); different

---

*DM Review Magazine*, August 2004, available at http://www.dmreview.com/article_sub.cfm?articleId=1007211; W.W. Eckerson, "Data warehousing special report: Data quality and the bottom line," *Application Development Trends Magazine*, May 1, 2002, available at http://www.adtmag.com/article.aspx?id=6303; Y. Wand and R. Wang, "Anchoring data quality dimensions in ontological foundations," *Communications of the ACM* 39(11):86-95, November 1996; and R. Wang, H. Kon, and S. Madnick, "Data quality requirements analysis and modelling," *Ninth International Conference of Data Engineering*, Vienna, Austria, 1993.

[34]M.J. Anderson and S.E. Fienberg, *Who Counts? The Politics of Census-Taking in Contemporary America*, Russell Sage Foundation, New York, 1999, p. 70.

provenance for different databases; inconsistent data fields for records contained in different databases on a given data subject; and lack of universal identifiers to specify data subjects.

Missing data are a major cause of reduction in data quality. In the situation in which network linkages are of interest and are directly represented in a database, the problem of missing data can sometimes be easier and sometimes more challenging than in the case of a rectangular file. A rectangular file usually consists of a list of individuals with their associated characteristics. In this situation, missing data can be of three general types: item nonresponse, unit nonresponse, and undercoverage. Item and unit nonresponse, while certainly problematic in the current context, are limited in impact and can sometimes be addressed using such techniques as imputation. Even undercoverage, while troubling, is at least limited to the data for the individual in question. (If such an individual is represented on another database to which one has access, merging and unduplicating operations can be helpful to identification, and estimates of the number of omissions can be developed using dual-systems estimation.)

On one hand, when the appropriate unit of analysis is networks of individuals (i.e., the individuals and their characteristics along with the various linkages between them are represented as being present or absent), the treatment of missing data can be easier when linkages from other individuals present in a database, such as phone calls, e-mails, or the joint issuance of plane tickets, etc., can help inform the analyst of another individual's existence for whom no direct information was collected.

On the other hand, treating missing data can also be a challenging problem. If the data for a person in a network is missed, not only is the information on that individual unavailable, but also the linkages between that person and others may be missing. This can have a substantial impact on the data for the missing individual, as well as the data for the other members of the group in the network and even the structure of the network, since in an extreme case it may be that the missing individual is the sole linkage between two otherwise separate groups. It is likely that existing missing data techniques can be adapted to provide some assistance in the less extreme cases, but at this point this is an area in which additional research may be warranted.

False positives and false negatives are in some sense complementary for any given database and given analytical approach. More precisely, for a given database and analytical approach, one can drive the rate of false positives to zero or the rate of false negatives to zero, but not simultaneously. Decreases in the false positive rate are inevitably accompanied by increases in the false negative rate and vice versa, although not necessarily in the same proportion. However, as the quality of the data is improved or if the classification technique is improved, it is possible to reduce both the false positive rate and the false negative rate, provided an accurate model for true positives and negatives is used.

Both false positives and false negatives pose problems for counterterrorist efforts. In the case of false positives, a counterterrorism analyst searching for evidence of terrorist attack planning may obtain personal information on a number of individuals. All of these individuals surrender some privacy, and those who have not been involved in terrorist activity (the false positives) have had their privacy violated or their rights compromised despite the lack of such involvement. Moreover, the use of purloined identities—identity theft—has enabled various kinds of fraud and evasion of law enforcement already. If terrorists are able to assume other identities, not only will that capability enable them to evade some detection and obfuscate the data used in the models—that is, deliberately manipulate the system, resulting in the generation of false positives against innocent individuals—but also it also might result in extreme measures being taken against the innocent individuals whose identities have been stolen.

Every false positive also has an opportunity cost; that is, it is associated with a waste of resources—precious investigative or analytical resources that are expended in the investigation of a innocent individual. In addition, false positives put pressure on officials to justify the expenditure of such resources, and such pressures may also lead to abuses against innocent individuals. From an operational standpoint, the key question is how many false alarms are acceptable. If one has infinite resources, it is easy to investigate every false alarm that may emerge from any system, no matter how poor its performance. But in the real world of constrained resources, it is necessary to balance the number of false alarms against the resources available to investigate them as well as the severity of the perceived threat. Furthermore, it is also important to consider other approaches that might be profitably applied to the problem, as well as other security issues in need of additional effort.

False negatives are also a problem and the nightmare of the intelligence analyst. A false negative is someone who should be under suspicion and is not. That is, the analyst simply misses the terrorist. From a political standpoint, the only truly acceptable number for false negatives is zero—but this political requirement belies the technical reality that the number of false negatives can never be zero. Moreover, identifying false negatives in any given instance may be problematic. In the case of the terrorist investigation, it is essentially impossible to know with certainty if a person is a false negative until he or she is known to have committed a terrorist act.

False positives and false negatives (and data quality, because it affects both false positives and false negatives) are important in a discussion of privacy because they are the language in which the trade-offs between privacy and other needs are often cast. One might argue that the consequences of a false negative (a terrorist plan is not detected and many people die) are in some sense much larger than the consequences of a false positive (an innocent person loses privacy or is detained). For this reason, many decision makers assert that it is better to be safe than sorry. But this argument is fallacious. There is no reason to expect that false negatives and false positives trade off against one another in a one-for-one manner. In practice, the trade-off will almost certainly entail one false negative against an enormous number of false positives, and a society that tolerates too much harm to innocent people based on large a number of false positives is no longer a society that respects civil liberties.

### 1.8.5 Oversight and Prevention of Abuse

Administrators of government agencies face enormous challenges in ensuring that policies and practices established by higher authorities (e.g., Congress, the Executive Office of the President, the relevant agency secretary or director) are actually followed in the field by those who do the day-to-day work of the agency. In the counterterrorism context, one especially important oversight responsibility is to ensure that the policies and practices meant to protect citizen privacy are followed in a mission environment that is focused on ensuring transportation safety, protecting borders, and pursuing counterterrorism. Challenges in this domain arise not only from external pressures based on public concern over privacy but also from internal struggles about how to motivate high performance while adhering to legal requirements and staying within budget.

Preventing privacy abuses from occurring is particularly important in a counterterrorism context, since privacy abuses can erode support for efforts that might in fact have some effectiveness in or utility for the counterterrorist mission. In this context, abuse refers to practices that result in a dissemination of personally identifiable information and thereby violate promised, implied, or legally guaranteed confidentiality or civil liberties.[35] This point implies that oversight must go beyond the enforcement

---

[35]Personally identifiable information (PII) refers to any information that identifies or can be used to identify, contact, or locate the person to whom such information pertains. This includes information that is used in a way that is personally identifiable, including linking it with identifiable information from other sources, or from which other personally identifiable information can easily be derived, including, but not limited to, name, address, phone number, fax number, e-mail address, financial profiles, Social Security number, credit card information, and in some cases Internet IP address. Although PII is also said to not include information collected anonymously, the discussion above suggests that the ability to make

of rules and procedures established to cover known and anticipated situations, to be concerned with unanticipated situations and circumstances.

Oversight can occur at the planning stage to approve intended operations, during execution to monitor performance, and retrospectively to assess previous performance so as to guide future improvements. Effective oversight may help to improve trust in government agencies and enhance compliance with stated policy.

## 1.9 THE NEED FOR A RATIONAL ASSESSMENT PROCESS

In the years since the September 11, 2001, attacks, the U.S. government has initiated a variety of information-based counterterrorist programs that involved data mining as an important component. It is fair to say that a number of these programs, including the Total Information Awareness program and the Computer-Assisted Passenger Prescreening System II (CAPPS II), generated significant controversy and did not meet the test of public acceptability, leaving aside issues of technical feasibility and effectiveness.

Such outcomes raise the question of whether the nature and character of the debate over these and similar programs could have been any different if policy makers had addressed in advance some of the difficult questions raised by a program. Although careful consideration of the privacy impact of new technologies is necessary even before a program seriously enters the research stage, it is interesting and important to consider questions in two categories: effectiveness and consistency with U.S. laws and values.

The threshold consideration of any privacy-sensitive technology is whether it is effective toward a clearly defined law enforcement or national security purpose. The question of effectiveness must be assessed through rigorous testing guided by scientific standards. Research on the question of how large-scale data analytical techniques, including data mining, could help the intelligence community identify potential terrorists is certainly a reasonable endeavor. Assuming that the initial scientific research justifies additional effort based on the scientific community's standards of success, that work should continue, but it must be accompanied by a clear method for assessing the reliability of the results.

---

an identification may depend both on the specific values of the PII in question and on the ability to aggregate data in ways that reduce significantly or even eliminate the anonymity originally promised or implied. Thus, information that previously was not PII may at a later date become PII as new techniques are developed or as other non-PII information becomes available. In short, the definition of PII can easily vary with context. For more discussion, see National Research Council, *Engaging Privacy and Information Technology in a Digital Age*, The National Academies Press, Washington, D.C., 2007.

Even if a proposed technology is effective, it must also be consistent with existing U.S. law and democratic values. Addressing this issue may involve a two-part inquiry. One must assess whether the new technique and objective comply with existing law, yet the inquiry cannot end there. Inasmuch as some programs seek to enable the deployment of very large-scale data mining over a larger universe of data than the U.S. government has previously analyzed, the fact that a given program complies with existing law does not establish that such surveillance practice is consistent with democratic values.

A framework for decision making about information-based programs couched in terms of questions in these two categories is presented in Chapter 2.

# 2

# A Framework for Evaluating Information-Based Programs to Fight Terrorism or Serve Other Important National Goals

The government increasingly uses technologies, programs, and systems that involve the acquisition, use, retention, or sharing of information about individuals to fight terrorism or serve other important national goals. These systems are very diverse and in the counterterrorism context range from requiring identification to board airplanes or enter government buildings to telephone and e-mail surveillance and intensive mining of commercial records. For purposes of this framework, this chapter describes all of these, together with the people who operate them, as information-based programs because they have in common their reliance on information about individuals.

This chapter proposes a framework for evaluating and deploying technologies, programs, and systems that rely on personal data to prevent terrorism or to serve other important national goals. This framework establishes sets of criteria to address the likely effectiveness and the lawfulness and consistency with U.S. values of any proposed information-based program.

## 2.1 THE NEED FOR A FRAMEWORK FOR EVALUATING INFORMATION-BASED PROGRAMS

Although information-based programs are not new, advances in digital technology and the proliferation of digital information about individuals have expanded their variety, the interest in their use, and potentially their impact. As a result, information-based programs often raise difficult questions about privacy and other civil liberties, cost, effectiveness, legality, and consistency with societal values.

These issues and the lack of consensus about how they should be evaluated have contributed to limiting the ability of public officials to make rational and informed choices about information-based programs for counterterrorism, research on potentially promising systems, and the availability of information about such systems and their use.

Many groups and individuals have considered how information-based programs should be evaluated and under what conditions they should be deployed. The U.S. Department of Defense Technology and Privacy Advisory Committee,[1] the U.S. Department of Homeland Security Privacy and Integrity Advisory Committee,[2] the Markle Foundation Task Force on National Security in the Information Age,[3] and the McCormick Tribune Foundation's Cantigny Conference on Counterterrorism Technology and Privacy[4] are among the many groups—inside and outside government—to address these vital issues. There is a striking degree of consistency among their recommendations and also in the extent to which they have not been implemented.

Building on the work of these prior efforts and informed by the members' experiences and research, the committee designed a framework to guide public officials charged with making decisions about the development, procurement, and use of information-based programs. Its purpose is not to impose bureaucratic compliance requirements, but rather to assist well-meaning people at every level of government to do their jobs better, to enhance their effectiveness in countering terrorist threats, to facilitate the wise and timely implementation of new programs, to invest limited government resources wisely, and to ensure that basic American values are not compromised when doing so. The committee also intends the framework to assist judges and policy makers responsible for approving or evaluating those decisions, legislators in crafting the law that governs these programs, and the press and the public in their broad and critical oversight of government activities.

This framework not only shares much in common with the recommendations of prior groups, but it is also consistent with many of the widely recognized standards that already guide information technology procurement, deployment, and use decisions in industry and other areas

---

[1]See Technology and Privacy Advisory Committee, *Safeguarding Privacy in the Fight against Terrorism*, Department of Defense, Washington, D.C., March 2004, available at http://www.cdt.org/security/usapatriot/20040300tapac.pdf.

[2]See http://www.dhs.gov/xinfoshare/committees/editorial_0512.shtm.

[3]For more information, see http://www.markletaskforce.org/.

[4]See "The Cantigny principles on technology, terrorism, and privacy," *National Security Law Report* 27(1):14-16, February 2005.

of government. Although this framework is necessarily broader, since it reaches far beyond information technology, it mirrors many of the best practices reflected in the Control Objectives for Information and Related Technologies (COBIT), the IT Infrastructure Library (ITIL), International Organization for Standards (ISO) 17799, and the standards promulgated by the National Institute of Standards and Technology (NIST), among others.

In short, the individual elements of what the committee proposes are not wholly new. They reflect much of the wise advice that the government has received—and largely failed to implement—many times before, advice that both it and the private sector do follow in other areas. It is the committee's hope that by adding to this prior work the breadth of experience, knowledge, and expertise reflected in its membership, it can offer a comprehensive framework that policy makers will, in fact, implement. It is the integration of the individual elements that the committee does think is new.

At the heart of this framework are two sets of questions: First, is an information-based program effective or likely to be effective in achieving its intended goal—in short, does it work? Second, does the program comply with the law and reflect the values of society, especially concerning the protection of data subjects' civil liberties?

Although these questions are posed as having yes-no answers, any serious application of the framework will almost certainly result in information on *how* effective and *how* protective of civil liberties any given information-based program is. This is critical knowledge when determining which of many competing systems, if any, should be developed, acquired, or deployed, and how they might be used or improved. For any potential program, policy makers will have to exercise sound judgment in deciding whether the program is sufficiently effective and sufficiently protective of privacy to warrant proceeding with it, although such judgment should be undertaken after the framework has been applied rather than before.

The questions posed by this framework should be asked not only of all new information-based programs, but also of existing programs today, at regular intervals in the future, and any time that a program is to be altered or put to a different use, to ensure that scarce resources are invested wisely; tools are used appropriately, lawfully, and consistently with societal values; and the best protection is pursued for national security and civil liberties. As discussed in greater detail below, achieving such goals requires routine monitoring, ongoing auditing, and clear, competent oversight. In short, the application of the framework is an ongoing process that should last throughout the operational lifetime of a program.

Technology can aid considerably in the application of the framework, and the effectiveness with which the framework addresses many issues can be enhanced through the use of technology—for example, the creation of immutable audit records and the continuous, automated analysis of those records. But technology alone is not sufficient. What is most critical is that the tools necessary to ensure compliance with the framework—whether or not they are technological—be built into information-based programs to the greatest extent possible and internalized into the processes by which they are developed, acquired, deployed, and used.

The framework is deliberately and necessarily broad because it is designed to apply to all information-based programs. As a result, not all of the points addressed by the framework may be applicable to all programs. Points that are inapplicable should be noted explicitly, along with a clear explanation of why they are inapplicable. The fact that a point is difficult to address should not be a justification for ignoring it. Honest, well-reasoned responses are far more useful to system developers, users, and overseers than none at all, and incomplete or erroneous responses can be supplemented or corrected as additional experience with a program is gained.

The framework and the processes by which it is implemented need to be evaluated regularly and revised as necessary to ensure that it is achieving these objectives. The fact that the framework is undoubtedly imperfect is no reason for avoiding it. Too frequently the argument is heard that national security is too important and the terrorist threat too great to pause to ask hard questions of the systems to be deployed to protect the nation. In the committee's view, that is the wrong approach. It is precisely because national security is important and the threats to it are great that it is so important to ensure that the systems to be deployed to protect the nation are effective and are consistent with U.S. values.

## 2.2  EVALUATING EFFECTIVENESS

The first inquiry about an information-based program is concerned with effectiveness: whether a program achieves its intended purpose (i.e., Does it work?), with what precision it does so (i.e., How well does it work?), how it might be made to work better in the future, and how its effectiveness compares with that of other available alternatives. For example, grounding all airplanes would be a highly effective technique for preventing terrorist bombings of airplanes in flight, but it would not be a workable solution because it would also keep millions of law-abiding passengers from flying. As this example suggests, ineffective or overly broad programs often create significant side effects that extend far beyond the immediate impact on the data subjects.

It is impossible in the abstract to establish acceptable levels of effectiveness because the level that society demands of any given program is likely to depend on the severity and likelihood of the consequences it is designed to guard against and the burden on individuals and overall cost of the program designed to prevent those consequences.

What matters is that policy makers and government officials responsible for developing, purchasing, deploying, and using information-based programs systematically evaluate the effectiveness of those programs and assess whether they are warranted in light of their likely effectiveness. This is seldom easy, and it is made more difficult by four factors: the rapid change in technologies and applications, the evolving nature of terrorist threats, the fact that so much of the information about terrorist threats and countermeasures is classified, and the reality that dealing with broad-based terrorist threats will require many programs to be scalable to a level far beyond what is typically required in industry or academic settings.

The following criteria are designed to assess and enhance effectiveness in light of these challenges. They are intended to ensure that the nation invests its human, technological, and financial resources wisely. They should be addressed before a new information-based program is procured or deployed and, as appropriate, at regular intervals during the development and use of such a program.

1. There should be a *clearly stated purpose* for the information-based program. It is impossible to assess a program's effectiveness without knowing what it was intended to accomplish. A clear, precise objective is the foundation for any system.

    a. Is that objective worthwhile?

    b. Is it legally appropriate?

    c. Is there a demand or need for it?

    d. Is it already being accomplished or could it be accomplished through less intrusive or less costly means?

A system's purpose should be the basis for judging if the system is appropriate, and thereafter a basis for assessment of the system and for audits of its use. The purpose may be updated in response to changed circumstances or new experience with the system, but changes to the purpose should be explicit.

2. There should be a *sound rational basis* for the information-based program and each of its components. Is there a scientific foundation for the system? For most information-based programs, the rational basis will have to take into account not only how individual components work in a laboratory, but also how they will work together and in connection with other systems in the field. This inquiry is likely to involve not only computer science, statistics, and related fields, but also a range of other social and behavioral sciences.

3. There should be a *sound experimental basis* for the information-based program and each of its components. Experimental science, and much of engineering as well, generally involves a logical progression from theory to simulations to laboratory tests, to small-scale field tests, to larger scale tests. In the rush to find quick responses to pressing national security concerns, there is a natural tendency to want to skip one or more of these phases, but the hundreds of millions of dollars wasted on systems that did not go through appropriate experimentation and subsequently did not work suggest that such omissions seldom pay off.

a.  Does the system work to achieve its stated purpose?

b.  Has the new system been shown to work in simulations or laboratory settings or has it been field-tested?

c.  Did the test conditions take into account real-world conditions?

d. Has it been applied to historical data to determine if it accurately accomplished its objective?

e.  Have experimental successes been replicated to demonstrate that they were not coincidence?

f.  Has the system been subjected to critical analysis, challenge, and likely countermeasures (for example, through "red-teaming")?[5]

4. The information-based program should be *scalable*. A system for enhancing security that appears promising in the laboratory may well fail in the field if it cannot be scaled up to deal with the real-world flood of data (or even the physical demands of conducting background checks or security scans at airports). Testing scalability has been a special challenge in this area because of the difficulty of obtaining data sets for testing of appropriate size and complexity. In some instances, Congress has proven too quick to rush to judgment on potential systems that were being tested but not deployed, and administration officials have been insufficiently frank about the need for data for testing. Testing on a data set of adequate size is essential to predicting the scalability and therefore the effectiveness of any information-based program.

5. There should be a clearly stated set of *operational or business processes* that comprehensively specify how the information-based program should operate in the organization, including who interacts with the program, whether programmatically for input, analysis, or obtaining results, or operationally for maintenance and modification, and with what authority; the information sources and how they are processed; and how the operations defined by the processes contributes to achieving

---

[5]"Red-teaming" refers to the practice of conducting realistic "blind" tests against a system. Such tests are blind in the sense that the operators of the system do not know that they are being tested, and realistic in the sense that the testers are free to do most or all of the things that actual terrorists might or could do in challenging the system.

the stated purpose. This criterion addresses issues related to operational integration of the program with the organization.

6. The information-based program should be capable of being *integrated* in practice with relevant systems and tools inside and outside the organization. For example:

a. Does the system interact effectively with the sources of information on which it relies?

b. If it requires combining data, can it do so in practice to yield meaningful results, at the necessary speed, while maintaining an appropriate level of information integrity?

c. Can the end product of the system be acted on meaningfully by people or other systems?

7. Information-based programs should be *robust*. This requires not only that the program work reliably in the field, but also that it not easily be compromised by user errors or circumvented by countermeasures. Investments in programs that are easily undercut or avoided are rarely sound.

8. There should be adequate guarantees that the data on which the information-based program depends are *appropriate and reliable*. Data should be stored as long as necessary, but they should be deleted when appropriate and regularly updated if they are needed by the system on an ongoing basis.

a. Are there adequate guarantees of the information's validity, provenance, availability, and integrity? Such guarantees are particularly important if a failure to meet the guarantees might adversely affect an individual.

b. Are the data easily compromised or manipulated so that the system can be defeated?

An information-based program is no better than the data on which it relies, and too many proposals for systems that initially appeared promising foundered when questions were raised about the adequacy and reliability of the source data.

9. The information-based program should provide for appropriate *data stewardship*, a term that refers to accountability for program resources being used and protected appropriately according to the defined and authorized purpose. The data must be protected from unlawful or unauthorized disclosure, manipulation, or destruction. In addition, there should be technologies and/or procedures built into the system to ensure that privacy, security, and other data stewardship and governance policies are followed.

10. There should be adequate guarantees of *objectivity* in the testing and assessment of the information-based program. In the race for success stories and government contracts in the fight against terrorism, there is

a clear tendency to promote systems that lack appropriate guarantees of objectivity in the testing of their effectiveness. This is unacceptable when spending public money, especially when the stakes are so high. No agency or vendor should do all of the testing on the information-based programs it is promoting. Academics typically depend on peer review. That may be more difficult when the systems involved are classified, but it is the standard that the government should be seeking to achieve through appropriate measures. Often scientists or other experts with clearances can help test and evaluate the test results on systems they have not been involved in developing. Technical advisory committees, with members with appropriate clearances, are useful. Third-party assessment even within the government, so that one agency tests another's systems, would help bring independence to the development and evaluation process. The government should assess independently the effectiveness of any system that it is considering purchasing or deploying. To the extent possible, testing should be blind—to both researchers and research subjects—so that the risk of biasing the outcome is diminished. The causes of failures should be documented so that they can be avoided in developing future systems, or reexplored as technologies and data sources evolve. Failures, as well as successes, should be reported together with what the agency has learned about the cause of those failures.

11. There should be *ongoing assessment* of the information-based program. No system, no matter how well designed or tested, will be perfect. There will always be not only unforeseen issues, but also entirely foreseeable ones, such as erroneous or mismatched data, false positives, and false negatives. Assessment is critical to detecting errors, correcting them, and improving systems to reduce errors in the future. Assessment is also essential to ensuring that the system is used properly and only for appropriate purposes. Are there mechanisms for detecting, reporting, and correcting errors? Are there monitoring tools and regular audits to assess system and operator performance?

12. The effectiveness of the information-based program and its compliance with these key requirements should be *documented*. Documentation is necessary to ensure that these critical issues are addressed during the development of new information-based programs, and also to respond to subsequent inquiries about their effectiveness. Satisfactory documentation should be required before any information-based program is procured or deployed. When such a system uses personally identifiable information or otherwise affects privacy, the documentation should be examined by an entity, such as an independent scientific review committee, that is capable of evaluating the scientific evidence of effectiveness outside the agency promoting the new system.

## 2.3  EVALUATING CONSISTENCY WITH U.S. LAW AND VALUES

The second inquiry is concerned with whether a proposed (or existing) information-based program is consistent with U.S. law and values. Lawfulness is more likely to be binary: a proposed action either is or is not against the law. U.S. society expects its government to obey the law, and it is required by the Constitution to do so. In addition, because technologies and events usually outpace law, it is necessary to constantly consider what types of information-based programs *should be* lawful. In short, are they consistent with the values of U.S. society?

The values inquiry is always difficult, especially in the context of a diverse and pluralistic society like that of the United States. But it is essential in order to respect the values that undergird the system of government and bind people together. Evaluating information-based programs in light of values is also essential because the Supreme Court has limited the Fourth Amendment to protect only "reasonable expectations" of privacy, and it has found that reasonableness is measured in part by what society is willing to accept as reasonable and in part by what individuals' subjective expectations are. An awareness of society's values and individual expectations is therefore critical for understanding what expectations of privacy the law is likely to regard as reasonable and therefore afford legal protection. In addition, paying attention to core values is necessary to avoid creating a race to the bottom—in which the public begins to accept uses of personal data only because the law permits them.

There are also practical, utilitarian reasons for concern about values. Promising antiterrorism systems may be derailed, even ones well within existing law, because they so offend popular and political understandings of privacy that go beyond existing legal requirements.

The determination as to whether a proposed system is lawful, or should be lawful, often requires evaluating the effectiveness of the system in light of its purpose, cost, and the consequences if it fails. As a result, while clear and unambiguous (bright-line) legal rules are desirable, they inevitably rely on subjective judgments that overlap with the effectiveness criteria described above. For example, the precision and accuracy of a system are key aspects of any determination of legality in which individual rights are involved. If the government obtains a warrant to tap a specified phone line but taps another line instead, it has probably broken the law. Or if a surveillance order from a court requires the government to delete nonrelevant communications but it fails to do so, the entire court order and all of the evidence obtained through it can be thrown out. Understanding a program's effectiveness is also often necessary because the law requires the government and courts to assess whether there are any equally effective but less intrusive means of accomplishing the purpose.

In the absence of an assessment of effectiveness, such a requirement is impossible to satisfy.

Effectiveness also matters from the standpoint of values, not so much as a requirement of a specific law, but as a commonsense or even an ethical requirement. Any intrusion on privacy would be entirely unjustified if it were not accompanied by some reasonable chance of accomplishing a worthwhile purpose. If an intrusion is perforce ineffective, it would seem by its very nature unwarranted. (Of course, the converse is not necessarily true—it may be that even effective programs should not be deployed because they *do* offend the ethical sensibilities of the citizenry.)

The following criteria are therefore designed not only to ensure that a proposed system is lawful in the face of existing laws, but also to reduce the impact on privacy that might otherwise render the system either unlawful in the future or politically impractical. They should be addressed by agency officials before a new information-based program is procured or deployed and, as appropriate, at regular intervals during the development and use of such a system. The committee also believes that the criteria should be useful to judicial and congressional officials as they evaluate new and existing programs and determine the boundaries of the nation's laws protecting privacy and other civil liberties. The criteria are divided into three categories to facilitate their application.

### 2.3.1 Data

1. *Need for personal data.* The need for personal data to accomplish the stated purpose and the specific uses for personal data should be clearly identified. Personal data should not be used unless they are reasonably necessary to achieve the stated objective and effective in doing so. Alternatives should be explicitly considered to determine whether there are equally effective means of achieving the same purpose that rely less on personal data (or on less personal data). Such alternatives are usually preferable.

2. *Sources of data.* The sources of those personal data should be clearly identified. It must be lawful for the source to supply the data and for the agency to obtain them.

3. *Appropriateness of data.* The personal data should be determined to be appropriate for the intended use, taking into account the purpose(s) for which the data were collected, their age, and the conditions under which they have been stored and protected. Data quality, integrity, and provenance should be assessed explicitly and determined to be appropriate for the intended use and objective. In addition, information-based programs should not rely exclusively on data that relate to the exercise of

rights protected by the First Amendment (i.e., freedom of expression, the press, assembly, religion, and petition).

4. *Third-party data.* Because using personal data from other government agencies or from private industry may present special risks, such third-party data should be subject to additional protections:

a. The agency should take into account the purpose for which the data were collected, their age, and the conditions under which they have been stored and protected when determining whether the proposed information-based program is appropriate.

b. If data are to be used for purposes that are inconsistent with those for which they were originally collected, the agency should specifically evaluate whether the inconsistent use is justified and whether the data are appropriate for such use.

c. Because of the difficulty of updating, overseeing, and maintaining the accuracy and context of data that have been copied from place to place, data should be left in place whenever possible (i.e., in the hands of the third parties that originally controlled those data). If this is impossible, they should be returned or destroyed as soon as practicable.

d. Private entities that provide data to the government on request or subject to judicial process should be reasonably compensated for the costs they incur in complying with the government's request or order.

### 2.3.2 Programs

5. *Objective.* The objective of the information-based program should be clearly stated. That objective must be lawful to pursue by the agency developing, procuring, or deploying the program.

6. *Compliance with existing law.* The information-based program should comply with applicable existing law.

7. *Effectiveness.* Using scientifically valid criteria, the information-based program should be demonstrated to be effective in achieving the intended objective.

8. *Frequency and impact of false positives.* The information-based program should be demonstrated to yield a rate of false positives that is acceptable in view of the purpose of the search, the severity of the effect of being identified, and the likelihood of further investigation.

9. *Reporting and redress of false positives.* There must be in place a process for identifying the frequency and effects of false positives and for dealing with them (e.g., reporting false positives to developers to improve the system, correcting incorrect information if possible, remedying the effects of false positives as quickly as practicable), as well as a specific locus of responsibility for carrying out this process.

10. *Impact on individuals.* The likely effects on individuals identified through the information-based program should be defined clearly (e.g., they will be the subject of further investigation for which a warrant will be sought, they will be subject to additional scrutiny before being allowed to board an aircraft, and so on).

11. *Data minimization.* The information-based program should operate with the least personal data consistent with its objective. Only the minimally necessary data should be accessed, disseminated, or retained. This has long been a requirement of U.S. surveillance law, although it has been rendered largely irrelevant in recent years as technology and applications have evolved so that vast streams of data are recorded and stored, rather than just limited, relevant elements. Moreover, the proliferation of digital data and dramatic reductions in the costs associated with sharing and storing data mean that even irrelevant data are routinely retained by the government indefinitely. Giving new force to minimization requirements is essential to avoiding the situation of government maintaining ubiquitous data records that threaten to invade personal privacy and overwhelm efforts to use data effectively to enhance security. Whenever practicable, the information-based program should rely on personal data from which information by which specific individuals can be commonly identified (e.g., name, address, telephone number, Social Security number, unique title) has been removed, encrypted, or otherwise obscured.

12. *Audit trail.* The information-based program should create a permanent, tamper-resistant record of when data have been accessed and by whom. Continuous, automated analysis of audit records can help ensure compliance with applicable laws and policies. This is especially important when sensitive or potentially sensitive data are involved.

13. *Security and access.* The information-based program should be secured against accidental or deliberate unauthorized access, use, alteration, or destruction. Access to such an information-based program should be restricted to persons with a legitimate need and protected by appropriate access controls, taking into account the sensitivity of the data.

14. *Transparency.* The information-based program should be developed, deployed, and operated with the greatest transparency possible, consistent with its objective. Persons affected by the program and the public generally should be informed as fully as practicable of the existence of the program, its purpose, cost, the laws and regulations under which it operates, the measures in place for assessing its effectiveness and protecting privacy, and the process for reporting and obtaining redress of grievances concerning its operation.

### 2.3.3 Administration and Oversight

15. *Training.* All persons engaged in developing or using information-based programs should be trained in their appropriate use and the laws and regulations applicable to their use.

16. *Agency authorization.* No information-based program that involves the acquisition, use, retention, or sharing of personally identifiable information should be developed, procured, or deployed until a senior agency official, preferably one subject to Senate confirmation, has certified in writing that it complies with the requirements of this framework.

17. *External authorization.* The deployment or use of any information-based program that relies on sensitive personally identifiable information, personally identifiable information collected surreptitiously, personally identifiable information that has been obtained from a third party without individual consent, or personally identifiable information that is being used for a purpose that is incompatible with that for which it was originally collected should be conditioned on an appropriately specific authorization from a source external to the information-based program.[6] Typically, this would be authorization by an appropriate court (federal Article III, Foreign Intelligence Surveillance, or state), but Congress may provide for other forms of external authorization.

18. *Auditing for compliance.* Information-based programs should be audited not less than annually to ensure compliance with the provisions of this framework and other applicable laws and regulations. The party conducting such audits may or may not be in the department responsible for the program but should operate and report independently of the program in question.

19. *Privacy officer.* Before an agency develops, procures, or deploys an information-based program, it should have in place a policy-level privacy officer. The privacy officer would be responsible for ensuring the training of appropriate agency personnel on privacy issues; assisting in the design and implementation of systems to protect privacy; working with the general counsel, inspector general, other appropriate officials in

---

[6]The specificity of the authorization required in any given instance is an issue that changing technologies have highlighted in the context of the wiretapping of voice calls. For example, for criminals who use throwaway cell phones, authorizations that grant wiretap authority to law enforcement agencies only for specific phone numbers are obviously much less useful than authorizations that grant wiretap authority for all phones that a specific individual might use. Furthermore, the committee expects that the issue of specificity will become more important as the scope of information sought becomes broader. Because the nature of the appropriate specificity depends on the particular information needs of a given program, it is impossible for the committee to specify in advance in its broad framework the appropriate level of specificity. However, it does note that policy makers should make explicit decisions regarding the appropriate level of specificity.

the agencies to ensure compliance with such systems; providing advice and information on privacy issues and tools for protecting privacy; and advising agency leaders and personnel on privacy matters and the implementation of this framework.

20. *Reporting.* An agency that develops, procures, or deploys an information-based program should report to Congress not less than annually, or more frequently as required by law, on the use of the system; its effectiveness; the nature, use, and timeliness of redress mechanisms; and the integrity of the system and the data on which it relies. The report should be made public to the greatest extent possible.

## 2.4  A NOTE FOR POLICY MAKERS: APPLYING THE FRAMEWORK IN THE FUTURE

In times of crisis, policy makers are often pressured into making important decisions with inadequate information and too little time for consultation and deliberation. When those decisions involve laws concerning information-based programs, the consequences can be especially significant and long-lasting. Law inevitably tends to lag behind technology, yet dramatic technological changes can alter the scope of laws overnight. So, for example, when the Supreme Court excluded records maintained by third parties from the scope of the Fourth Amendment in 1976, it created a situation in which, 30 years later, because of the proliferation of digital records maintained by third parties, almost all information about individuals would be accessible to the government without judicial authorization.

The committee intends the entire framework proposed in this chapter to be useful to policy makers in outlining issues to be addressed through legislation or regulatory policy, as well as in proposing specific steps for ensuring that the nation fights terrorism effectively and consistently in accord with its core values. However, the breadth and variety of information-based programs, as well as the constantly changing capacity of technology, make crafting legislation governing those programs and protecting civil liberties a difficult task. To further facilitate effective legislation to achieve these critical goals, the committee presents this additional brief discussion of how the framework might be applied in the legislative context.

In the committee's view, all such legislation should specifically address the following eight areas (many specific elements of which have already been described above):

1. *Agency competency.* Is the agency being authorized to operate or use the information-based program competent to do so? Is the program

consistent with its mission? Is it staffed appropriately? Are its staff trained appropriately? Does it have a policy-level chief privacy officer? Does it have a culture of respecting the law and civil liberties?

2. *Purpose.* Does the information-based program have a clearly articulated purpose against which its effectiveness and impact on civil liberties can be assessed? Are there appropriate protections to guard against mission creep or repurposing of the program without careful deliberation? Will that purpose remain valid in the face of countermeasures or likely technological changes? Are there procedures in place for reevaluating that purpose?

3. *Effectiveness.* Are there appropriate guarantees that the information-based program and each of its components are effective? Are credible processes in place to measure effectiveness and to ensure continual assessment of effectiveness and efforts to improve effectiveness? Are measures of effectiveness documented?

4. *Authorization.* Are requirements in place for authorization by an identified, accountable official both before an information-based programs is created, procured, or deployed and before such programs are applied to personal data about a specific individual? Does the authorization for applying the program to a specific individual come from a court or other source external to the agency operating the program, especially if the data gathering or use is covert?

5. *Data.* Are there reasonable guarantees that the personal data to be used by an information-based program are appropriate, sufficiently accurate for the stated purpose, and reliably available on a timely basis? Are there protections to ensure that only necessary personal data are used, retained no longer than necessary, and protected against accidental or deliberate misuse? Are the data and the manner in which they are obtained consistent with U.S. values? Does their use deter the exercise of constitutionally protected rights?

6. *Redress.* Are there robust systems in place to identify errors, such as false positives, use them systematically to improve information-based programs, and provide rapid, effective redress to affected individuals?

7. *Assessment.* Are there reliable tools for assessing the performance of information-based programs and their compliance with applicable laws and regulations, as well as for acting on those assessments? Are the results of ongoing assessment documented?

8. *Oversight.* Is the information-based program subject to meaningful oversight from both inside and outside the agency, including from Congress? Are the program and its oversight mechanism transparent to the public and the press to the greatest extent possible? If transparency is impossible, are there reliable means for heightened independent agency, judicial, and/or congressional oversight?

## 2.5 SUMMARY OF FRAMEWORK CRITERIA

### 2.5.1 For Evaluating Effectiveness

1. Is there a clearly stated purpose for the information-based program?
   - Is that objective worthwhile?
   - Is it legally appropriate?
   - Is there a demand or need for it?
   - Is it already being accomplished or could it be accomplished through less intrusive or less costly means?

2. Is there a sound rational basis for the information-based program and each of its components?
   - Is there a scientific foundation for the system?

3. Is there a sound experimental basis for the information-based program and each of its components?
   - Does the system work to achieve its stated purpose?
   - Has the new system been shown to work in simulations or laboratory settings or has it been field-tested?
   - Did the test conditions take into account real-world conditions?
   - Has it been applied to historical data to determine if it accurately accomplished its objective?
   - Have experimental successes been replicated to demonstrate that they were not coincidence?
   - Has the system been subjected to critical analysis, challenge, and likely countermeasures (for example, through "red-teaming")?

4. Is the information-based program scalable?
   - Has it been tested on a data set of adequate size to predict its scalability?
   - Has it been tested against likely countermeasures or changes in technologies, threats, and society?

5. Is there a clearly stated set of operational or business processes that comprehensively specify how the information-based program should operate in the organization?

6. Is the information-based program capable of being integrated in practice with related systems and tools?
   - Does the system interact effectively with the sources of information on which it relies?

- If it requires combining data, can it do so in practice to yield meaningful results and at the speed necessary?
- Can the end product of the system be acted on meaningfully by people or other systems?

7. Is the information-based program robust?
   - Can it easily be compromised by user errors?
   - Can it easily be circumvented by countermeasures?

8. Are there appropriate guarantees that the data on which the information-based program depends are appropriate and reliable?
   - Are there adequate guarantees of the information's validity, provenance, availability, and integrity?
   - Are the data easily compromised or manipulated so that the system can be defeated?

9. Does the information-based program provide for appropriate data stewardship?
   - Are the data protected from unlawful or unauthorized disclosure, manipulation, or destruction?
   - Are there technologies and/or procedures built into the system to ensure that privacy, security, and other data stewardship and governance policies are followed?

10. Are there adequate guarantees of objectivity in the testing and assessment of the information-based program?
    - Has there been peer review or its equivalent?
    - Has the program been evaluated by entities with no stake in its success?
    - Have test results been evaluated by independent experts?
    - Was testing blind—to both researchers and research subjects—whenever possible?

11. Is there ongoing assessment of the information-based program?
    - Are there mechanisms for detecting and reporting errors?
    - Are there monitoring tools and regular audits to assess system and operator performance?

12. Have the effectiveness of the information-based program and its compliance with these key requirements been documented?
    - Has the documentation been examined by an entity capable of evaluating the scientific evidence of effectiveness outside the agency promoting the new system?

### 2.5.2 For Evaluating Consistency with Laws and Values

**The Agency**

1. Does the agency have in place a policy-level privacy officer?
2. Does the agency report to Congress not less than annually, or more frequently as required by law, on the use of its information-based programs, their effectiveness, the nature and use of redress mechanisms, and the integrity of the programs and the data on which they rely? Is that report made public to the greatest extent possible?
3. Have all persons engaged in developing or using information-based programs been trained in their appropriate use and the laws and regulations applicable to their use?

**The Program**

4. Is the objective of the information-based program clearly stated? Is that objective lawful for the agency developing, deploying, or using the program to pursue?
5. Does the information-based program comply fully with applicable existing law?
6. Has the information-based program been demonstrated to be effective in achieving the intended objective? Is that demonstration based on scientifically valid criteria?
7. Has the information-based program been demonstrated to yield a rate of false positives that is acceptable in view of the purpose of the search, the severity of the effect of being identified, and the likelihood of further investigation?
8. Is there a process in place for identifying the frequency and effects of false positives and for dealing with them (e.g., reporting false positives to developers to improve the system, correcting incorrect information if possible, remedying the effects of false positives as quickly as practicable), as well as a specific locus of responsibility for carrying out this process?
9. Have the likely effects on individuals identified through the information-based program been defined clearly (e.g., they will be the subject of further investigation for which a warrant will be sought, they will be subject to additional scrutiny before being allowed to board an aircraft, and so on)?
10. Does the information-based program operate with the least personal data consistent with its objective? Does it access, disseminate, and retain only minimally necessary data? Have data by which specific individuals can be commonly identi-

fied (e.g., name, address, telephone number, Social Security number, unique title) been removed, encrypted, or otherwise obscured whenever possible?

11. Does the information-based program create a permanent, tamper-resistant record of when data have been accessed and by whom? Does it provide for continuous, automated analysis of audit records?

12. Is the information-based program developed, deployed, and operated with the greatest transparency possible, consistent with its objective?

13. Is the information-based program secured against accidental or deliberate unauthorized access, use, alteration, or destruction? Is access to the information-based program restricted to persons with a legitimate need and protected by appropriate access controls, taking into account the sensitivity of the data?

14. Has (or will) a senior agency official, preferably one subject to Senate confirmation, certified (or will certify) in writing that the information-based program complies with the requirements of this framework?

15. If the information-based program relies on sensitive personally identifiable information, personally identifiable information collected surreptitiously, personally identifiable information that has been obtained from a third party without individual consent, or personally identifiable information that is being used for a purpose that is incompatible with that for which it was originally collected, have its deployment and use been conditioned on authorization from a source external to that in which the information-based program will exist, and have they been approved by an external authority (e.g., an appropriate court or other authority)?

16. Is the information-based program audited not less than annually to ensure compliance with the provisions of the proposed framework and other applicable laws and regulations?

## The Data

17. Are personal data necessary to accomplish the objective of a given information-based program? Are the specific uses for personal data clearly identified? Are there equally effective means of achieving the same purpose that rely less on personal data (or on less personal data)?

18. Are the sources of personal data clearly identified? Is it lawful for the source to supply the data and for the agency to obtain the data?

19. Are the personal data appropriate for the intended use, taking into account the purpose(s) for which the data were collected, their age, and the conditions under which they have been stored and protected? Do the data relate solely to the exercise of rights protected by the First Amendment (i.e., freedom of expression, the press, assembly, religion, and petition)?

20. If an information-based program uses personal data from other government agencies or from private industry, are the following additional protections in place?

    - Have the purpose for which the data were collected, their age, and the conditions under which they have been stored and protected been taken into account when determining whether the proposed information-based program is appropriate?
    - If data are to be used for purposes that are inconsistent with those for which they were originally collected, has the agency specifically evaluated whether the inconsistent use is justified and whether the data are appropriate for such use?
    - Are the data being left in place whenever possible? If this is impossible, are they being returned or destroyed as soon as practicable?
    - Is the agency reasonably compensating private entities that provide data to the government on request or subject to judicial process for the costs they incur in complying with the government's request or order?

### 2.5.3 For Developing New Laws and Policies

1. Agency competency
    - Is the agency being authorized to operate or use the information-based program competent to do so?
    - Is the program consistent with the agency's mission?
    - Is the agency staffed appropriately?
    - Are its staff trained appropriately?
    - Does it have a policy-level chief privacy officer?
    - Does it have a culture of respecting the law and civil liberties?

2. Purpose
   - Does the information-based program have a clearly articulated purpose against which its effectiveness and impact on civil liberties can be assessed?
   - Are there appropriate protections to guard against mission creep or repurposing of the program without careful deliberation?
   - Will the program's purpose remain valid in the face of countermeasures or likely technological changes?
   - Are there procedures in place for reevaluating the program's purpose?

3. Effectiveness
   - Has the information-based program been demonstrated to be effective in achieving the intended objective?
   - Is that demonstration based on scientifically valid criteria?
   - Are there credible processes in place to measure effectiveness and to ensure continual assessment of effectiveness and efforts to improve effectiveness?
   - Are measures of effectiveness documented?

4. Authorization
   - Are there requirements in place for authorization by an identified, accountable official both before an information-based program is created, procured, or deployed and before such programs are applied to personal data about a specific individual?
   - Does the authorization for applying the program to a specific individual come from a court or other source external to the agency operating the program, especially if the data gathering or use is covert?

5. Data
   - Are personal data necessary to accomplish the objective of a given information-based program?
   - Are the specific uses for personal data clearly identified?
   - Are there equally effective means of achieving the same purpose that rely less on personal data (or on less personal data)?
   - Are there protections to ensure that only necessary personal data are used, retained no longer than necessary, and protected against accidental or deliberate misuse?

- Does the information-based program operate with the least personal data consistent with its objective?
- Does the program access, disseminate, and retain only necessary data?
- Have data by which specific individuals can be commonly identified (e.g., name, address, telephone number, Social Security number, unique title, and so on) been removed, encrypted, or otherwise obscured whenever possible?
- Are there reasonable guarantees that the personal data to be used by an information-based program are appropriate, sufficiently accurate for the stated purpose, and reliably available?
- Are the sources of those personal data clearly identified?
- Is access to the information-based program restricted to persons with a legitimate need and protected by appropriate access controls, taking into account the sensitivity of the data?
- Is it lawful for the source to supply the data and for the agency to obtain the data?
- Are the data and the manner in which they are obtained consistent with U.S. values?
- Does their use deter the exercise of constitutionally protected rights?
- If an information-based program uses personal data from other government agencies or from private industry, are the appropriate additional protections in place?

6. Redress
   - Is there a process in place for identifying the frequency and effects of false positives and for dealing with them (e.g., reporting false positives to developers to improve the system, correcting incorrect information if possible, remedying the effects of false positives as quickly as practicable, and so on)?
   - Have the likely effects on individuals identified through the information-based program been defined clearly (e.g., they will be the subject of further investigation for which a warrant will be sought, they will be subject to additional scrutiny before being allowed to board an aircraft)?
   - Has the information-based program been demonstrated to yield a rate of false positives that is acceptable in view of the purpose of the search, the severity of the effect of being identified, and the likelihood of further investigation?

- Are there robust systems in place to identify errors, such as false positives, use them systematically to improve information-based programs, and provide rapid, effective redress to affected individuals?

7. Assessment
   - Are there reliable tools for assessing the performance of information-based programs and their compliance with applicable laws and regulations, as well as for acting on those assessments?
   - Does the information-based program create a permanent, tamper-resistant record of when data have been accessed and by whom?
   - Does it provide for continuous, automated analysis of audit records?
   - Is the information-based program audited not less than annually to ensure compliance with the provisions of this framework and other applicable laws and regulations?
   - Are the results of ongoing assessment documented?

8. Oversight
   - Is the information-based program subject to meaningful oversight from both inside and outside the agency, including from Congress?
   - Are the program and its oversight mechanism transparent to the public and the press to the greatest extent possible?
   - If transparency is impossible, are there reliable means for heightened independent agency, judicial, and/or congressional oversight?

# 3

# Conclusions and Recommendations

## 3.1 BASIC PREMISES

The committee's work was informed by a number of basic premises. These premises framed the committee's perspective in developing this report, and they can be regarded as the assumptions underlying the committee's analysis and conclusions. The committee recognizes that others may have their own analyses with different premises, and so for analytical rigor, it is helpful to lay out explicitly the assumptions of the committee.

**Premise 1.** *The United States faces two real and serious threats from terrorists. The first is from terrorist acts themselves, which could cause mass casualties, severe economic loss, and social dislocation to U.S. society. The second is from the possibility of inappropriate or disproportionate responses to the terrorist threat that can do more damage to the fabric of society than terrorists would be likely to do.*

The events of September 11, 2001, provided vivid proof of the damage that a determined terrorist group can inflict on U.S. society. All evidence to date suggests that the United States continues to be a prime target for such terrorist groups as Al Qaeda, and future terrorist attacks could cause major casualties, severe economic loss, and social disruption.[1] The danger of future terrorist attacks on the United States is both real and serious.

At the same time, inappropriate or disproportionate responses to the terrorist threat also pose serious dangers to society. History demonstrates that measures taken in the name of improving national security, especially in response to new threats or crises, have often proven to be both ineffective and offensive to the nation's values and traditions of liberty and justice.[2] So the danger of unsuitable responses to the terrorist threat is also real and serious.

Given the existence of a real and serious terrorist threat, it is a reasonable public policy goal to focus on preventing attacks before they occur—a goal that requires detecting the planning for such attacks prior to their execution. Given the possibility of inappropriate or disproportionate responses, it is also necessary that programs intended to prevent terrorist attacks be developed and operated without undue compromises of privacy.

**Premise 2.** *The terrorist threat to the United States, serious and real though it is, does not justify government authorities conducting activities or operations that contravene existing law.*

The longevity of the United States as a stable political entity is rooted in large measure in the respect that government authorities have had for the rule of law. Regardless of the merits or inadequacies of any legal regime, government authorities are bound by its requirements until the legal regime is changed, and, in the long term, public confidence and trust in government depend heavily on a belief that the government is indeed adhering to the laws of the land. The premises above would not change even if the United States were facing exigent circumstances. If existing legal authorities (including any emergency action provisions, of which there are many) are inadequate or unclear to deal with a given situation

---

[1]For example, the National Intelligence Estimate of the terrorist threat to the U.S. homeland provides a judgment that "the U.S. Homeland will face a persistent and evolving terrorist threat over the next three years. The main threat comes from Islamic terrorist groups and cells, especially al-Qa'ida, driven by their undiminished intent to attack the Homeland and a continued effort by these terrorist groups to adapt and improve their capabilities." See *The Terrorist Threat to the U.S. Homeland*, National Intelligence Estimate, July 2007, available from Office of the Director of National Intelligence, Washington, D.C.

[2]Consider, for example, the 1942 internment of U.S. citizens of Japanese origin in the wake of the Pearl Harbor attack. The United States formally apologized to the Japanese American community for this act in 1988, and beginning in 1990 paid reparations to surviving internees.

or contingency, government authorities should seek to change the law rather than to circumvent or disobey it.

A willingness of U.S. government authorities to circumvent or disobey the law in times of emergency is not unprecedented. For example, recently declassified Central Intelligence Agency (CIA) documents indicate widespread violations of the agency's charter and applicable law in the 1960s and 1970s, during which time the CIA conducted surveillance operations on U.S. citizens under both Democratic and Republican presidents that were undertaken outside the agency's charter.[3]

The U.S. Congress has also changed laws that guaranteed confidentiality in order to gain access to individual information collected under guarantees. For example, Section 508 of the USA Patriot Act, passed in 2001, allows the U.S. Department of Justice (DOJ) to gain access to individual information originally collected by the National Center for Education Statistics under a pledge of confidentiality. In earlier times, the War Powers Act of 1942 retrospectively overrode the confidentiality provisions of the Census Bureau, and it is now known that bureau officials shared individually identifiable census information with other government agencies for the purposes of detaining foreign nationals.[4]

Today, many laws provide statutory protection for privacy. Conforming to such protections is not only obligatory, but it also builds necessary discipline into counterterrorism efforts that serves other laudable purposes. By making the government stop and justify its effort to a senior official, a congressional committee, or a federal judge, warrant requirements and other privacy protections often help bring focus and precision to law enforcement and national security efforts. In point of fact, courts rarely refuse requests for judicial authorization to conduct surveillance. As government officials often note, one reason for these high success rates is the quality of internal decision making that the requirement to obtain judicial authorization requires.

**Premise 3.** *Challenges to public safety and national security do not warrant fundamental changes in the level of privacy protection to which nonterrorists are entitled.*

The United States is a strong nation for many reasons, not the least of which is its commitment to the rule of law, civil liberties, and respect

---

[3]M. Mazzetti and T. Weiner, "Files on illegal spying show C.I.A. skeletons from Cold War," *New York Times*, June 27, 2007.

[4]W. Seltzer and M. Anderson, "Census confidentiality under the second War Powers Act (1942-1947)," paper prepared for the Annual Meeting of the Population Association of America, March 30, 2007, Population Association of America, New York, available at http://www.uwm.edu/~margo/govstat/Seltzer-AndersonPAA2007paper3-12-2007.doc.

for diversity. Especially in times of challenge, it is important that this commitment remain strong and unwavering. New technological circumstances may necessitate an update of existing privacy laws and policy, but privacy and surveillance law already includes means of dealing with national security matters as well as criminal law investigations. As new technologies become more commonly used, these means will inevitably require extension and updating, but greater government access to private information does not trump the commitment to the bedrock civil liberties of the nation.

Note that the term "privacy" has multiple meanings depending on context and interpretation. Appendix L ("The Science and Technology of Privacy Protection") explicates a technical definition of the term, and the term is often used in this report, as in everyday discourse, with a variety of informal meanings that are more or less consistent with the technical definition.

**Premise 4.** *Exploitation of new science and technologies is an important dimension of national counterterrorism efforts.*

Although the committee recognizes that other sciences and technologies are relevant as well, the terms of reference call for this report to focus on information technologies and behavioral surveillance techniques. The committee believes that when large amounts of information, personal and otherwise, are determined to be needed for the counterterrorist mission, the use of information technologies will be necessary and counterterrorist authorities will need to collect, manage, and analyze such information. Furthermore, it believes that behavioral surveillance techniques may have some potential for inferring intent from observed behavior if the underlying science proves sound—a capability that could be very useful in counterterrorist efforts "on the ground" if realized in the future.

**Premise 5.** *To the extent reasonable and feasible, counterterrorist programs should be formulated to provide secondary benefits to the nation in other domains.*

Counterterrorism programs are often expensive and controversial. In some cases, however, a small additional expenditure or programmatic adjustment may enable them to provide benefits that go beyond their role in preventing terrorism. Thus, they would be useful to the nation even if terror attacks do not occur. For example, hospital emergency reporting systems can improve medical care by prompt reporting of influenza, food poisoning, or other health problems, as well as alerting officials of bioterrorist and chemical attacks.

At the same time, policy makers must be aware of the phenomenon of "statutory mission creep"—in which the goals and missions of a program are expanded explicitly as the result of a specific policy action, such as congressional amendment of an existing law—and avoid its snares. In some instances, such as hospital emergency reporting systems, privacy interests may not be seriously compromised by their application to multiple missions. But in others, such as the use of systems designed for screening terrorists to identify ordinary criminals, privacy interests may be deeply implicated because of the vast and voluminous new data sets that must be brought to bear on the expanded mission. Mission creep may also go beyond the original understandings of policy makers regarding the scope and nature of a program that they initially approve, and thus effectively circumvent careful scrutiny. In some cases, a sufficient amount of mission creep may even result in a program whose operation is not strictly legal.

## 3.2  CONCLUSIONS REGARDING PRIVACY

The rich digital record that is made of people's lives today provides many benefits to most people in the course of everyday life. Such data may also have utility for counterterrorist and law enforcement efforts. However, the use of such data for these purposes also raises concerns about the protection of privacy and civil liberties. Improperly used, programs that do not explicitly protect the rights of innocent individuals are likely to create second-class citizens whose freedoms to travel, engage in commercial transactions, communicate, and practice certain trades will be curtailed—and under some circumstances, they could even be improperly jailed.

### 3.2.1  Protecting Privacy

**Conclusion 1.** *In the counterterrorism effort, some degree of privacy protection can be obtained through the use of a mix of technical and procedural mechanisms.*

The primary goal of the nation's counterterrorism effort is to prevent terrorist acts. In such an effort, identification of terrorists before they act becomes an important task, one that requires the accurate collection and analysis of their personal information. However, an imperfect understanding of which characteristics to search for, not to mention imperfect and inaccurate data, will necessarily draw unwarranted attention to many innocent individuals.

Thus, records containing personal information of terrorists cannot be examined without violating the privacy of others, and so absolute privacy protection—in the sense that the privacy of nonterrorists cannot be compromised—is not possible if terrorists are to be identified.

This technical reality does not preclude putting into place strong mechanisms that provide substantial privacy protection. In particular, restrictions on the use of personal information ensure that innocent individuals are strongly protected during the examination of their personal information, and strong and vigorous oversight and audit mechanisms can help to ensure that these restrictions are obeyed.

How much privacy protection is afforded by technical and procedural mechanisms depends on critical design features of both the technology and the organization that uses it. Two examples of relevant technical mechanisms are encryption of all data transports to protect against accidental loss or compromise and individually logged[5] audit records that retain details of all queries, including those made by fully authorized individuals to protect against unauthorized use.[6] But the mere presence of such mechanisms does not ensure that they will be used, and such mechanisms should be regarded as one enabler—one set of necessary but not sufficient tools—for the robust independent program oversight described in Recommendation 1c below.

Relevant procedural mechanisms include restrictions on data collection and restrictions on use. In general, such mechanisms govern important dimensions of information collection and use, including an explication of what data are collected, whether collection is done openly or covertly, how widely the data are disseminated, how long they are retained, the decisions for which they are used, whether the processing is

---

[5]"Individually logged" refers to audit records designed to monitor system usage by individual users and maintain individual accountability. For example, consider a personnel office in which users have access to those personnel records for which they are responsible. Individually logged audit trails can reveal that an individual is printing far more records than the average user, which could indicate the selling of personal data.

[6]Note that audit records documenting accesses to a database are conceptually distinct from the data contained within a database. An audit record typically identifies the party that took some specific action now captured in the audit record and the nature of the data involved in that action, but it does not specify the content of the data involved. (For example, a database of financial transactions is routinely updated to include all of the credit card purchases of John Smith for the last year. Since today is April 14, 2008, the database contains all of his purchases from April 14, 2007, to April 13, 2008. An audit record relevant to those records might include the fact that on January 17, 2004, Agent Mary Doe viewed John Smith's credit card purchases—that is, she looked at his purchases from January 17, 2003, to January 16, 2004.) One result of this distinction is that the data within a database may be purged within a short period of time in accordance with a specified data retention policy, but audit records describing accesses to that data may be kept for the entire life of the database.

performed by computer or human, and who has the right to grant permissions for subsequent uses.

Historically, privacy from government intrusion has been protected by limiting what information the government can collect: voice conversations collected through wiretapping, e-mail collected through access to stored data (authorized by the Electronic Communications Privacy Act, passed in 1986 and codified as 18 U.S.C. 2510), among others. However, in many cases today, the data in question have already been collected and access to them, under the third-party business records doctrine, will be readily granted with few strings attached. As a result, there is great potential for privacy intrusion arising from analysis of data that are accessible to government investigators with little or no restriction or oversight. In other words, powerful investigative techniques with significant privacy impact proceed in full compliance with existing law—but with significant unanswered privacy questions and associated concerns about data quality.

Analytical techniques that may be justified for the purpose of national security or counterterrorism investigations, even given their potential power for privacy intrusion, must come with assurances that the inferences drawn against an individual will not then be used for normal domestic criminal law enforcement purposes. Hence, what is called for, in addition to procedural safeguards for data quality, are usage limitations that provide for full exploitation on new investigative tools when needed (and justified) for national security purposes, but that prevent those same inferences from being used in criminal law enforcement activity.

An example—for illustration only—of the latter is the use of personal data for airline passenger screening. Privacy advocates have often expressed concerns that the government use of large-scale databases to identify passengers who pose a potential risk to the safety of an airplane could turn into far-reaching enforcement mechanisms for all manner of offenses, such as overdue tax bills or child support payments. One way of dealing with this privacy concern would be to apply a usage-limiting privacy rule that allows the use of databases for the purpose of counterterrorism but prohibits the use of these same databases and analysis for domestic law enforcement. Those suspicious of government intentions are likely to find a rule limiting usage rather less comforting than a rule limiting collection, out of concern that government authorities will find it easier to violate a rule limiting collection than a rule limiting collection. Nevertheless, well-designed and diligently enforced auditing and oversight processes may help over time to provide reassurance that the rule is being followed as well as to provide some actual protection for individuals.

Finally, in some situations, improving citizen privacy can have the result of improving their security and vice versa. For example, improvements in the quality of data (i.e., more complete, more accurate data) used in identifying potential terrorists are likely to increase security by enhancing the effectiveness of information-based programs to identify terrorists *and* to decrease the adverse consequences that may occur due to confidentiality violations for the vast majority of innocent individuals. In addition, strong audit controls that record the details of all accesses to sensitive personal information serve both to protect the privacy of individuals *and* to reduce barriers to information sharing between agencies or analysts. (Agencies or analysts are often reluctant to share information, even among themselves, because they feel a need to protect sources and methods, and audit controls that limit information access provide a greater degree of reassurance that sensitive information will not be improperly distributed.)

**Conclusion 2.** *Data quality is a major issue in the protection of the privacy of nonterrorists.*

As noted in Chapter 1, the issue of data quality arises internally as a result of measurement errors within databases and also as a consequence of efforts to link data or records across databases in the absence of clear, unique identifiers. Sharing personal information across agencies, even with "names" attached, offers no assurances that the linked data are sufficiently accurate for counterterrorism purposes; indeed, there are no metrics for accuracy that appear to be systematically used to assess such linking efforts.

Data of poor quality severely limit the value of data mining in a number of ways. First, the actual characteristics of individuals are often collected in error for a wide array of reasons, including definitional problems, identify theft, and misresponse on surveys.

These errors could obviously result in individuals being inaccurately represented by data mining algorithms as a threat when they are not (with the consequence that personal and private information about them might be inappropriately released for wider scrutiny). Second, poor data quality can be amplified during file matching, resulting in the erroneous merging of information for different individuals into a single file. Again, the results can be improper treatment of individuals as terrorist threats, but here the error is compounded, since entire clusters of information are now in error with respect to the individual who is linked to the information in the merged file.

Such problems are likely to be quite common and could greatly limit the utility of data mining methods used for counterterrorism. There are no obvious mechanisms for rectifying the current situation, other than collecting similar information from multiple sources and using the duplicative nature of the information to correct inaccuracies. However, given that today the existence of alternate sources is relatively infrequent, correcting individual errors will be extraordinarily difficult.

### 3.2.2 Distinctions Between Capability and Intent

**Conclusion 3.** *Inferences about intent and/or state of mind implicate privacy issues to a much greater degree than do assessments or determinations of capability.*

Although it is true that capability and intent are both needed to pose a real threat, determining intent on the basis of external indicators is inherently a much more subjective enterprise than determining capability. Determining intent or state of mind is inherently an inferential process, usually based on indicators such as whom one talks to, what organizations one belongs to or supports, or what one reads or searches for online. Assessing capability is based on such indicators as purchase or other acquisition of suspect items, training, and so on. Recognizing that the distinction between capability and intent is sometimes unclear, it is nevertheless true that placing people under suspicion because of their associations and intellectual explorations is a step toward abhorrent government behavior, such as guilt by association and thought crime. This does not mean that government authorities should be categorically proscribed from examining indicators of intent under all circumstances—only that special precautions should be taken when such examination is deemed necessary.

### 3.3 CONCLUSIONS REGARDING THE ASSESSMENT OF COUNTERTERRORISM PROGRAMS

**Conclusion 4.** *Program deployment and use must be based on criteria more demanding than "it's better than doing nothing."*

In the aftermath of a disaster or terrorist incident, policy makers come under intense political pressure to respond with measures intended to prevent the event from occurring again. The policy impulse to *do something* (by which is usually meant something new) under these circumstances is understandable, but it is simply not true that doing something new is always better than doing nothing. Indeed, policy makers may deploy new information-based programs hastily, without a full consideration of (a) the actual usefulness of the program in distinguishing people or characteristic patterns of interest for follow-up from those not of interest, (b) an assessment of the potential privacy impacts resulting from the use of the program, (c) the procedures and processes of the organization that will use the program, and (d) countermeasures that terrorists might use to foil the program.

The committee developed the framework presented in Chapter 2 to help decision makers determine the extent to which a program is effective in achieving its intended goals, compliant with the laws of the nation, and reflective of the values of society, especially with regard to the protection of data subjects' privacy. This framework is intended to be applied by taking into account the organizational and human contexts into which any given program will be embedded as well as the countermeasures that terrorists might take to foil the program.

The framework is discussed in greater detail in Chapter 2.

## 3.4  CONCLUSIONS REGARDING DATA MINING[7]

### 3.4.1  Policy and Law Regarding Data Mining

**Conclusion 5.** *The current policy regime does not adequately address violations of privacy that arise from information-based programs using advanced analytical techniques, such as state-of-the-art data mining and record linkage.*

The current privacy policy regime was established prior to today's world of broadband communications, networked computers, and enormous databases. In particular, it relies largely on limitations imposed on the collection and use of certain kinds of information, and it is essentially silent on the use of techniques that could be used to process and analyze already-collected information in ways that might compromise privacy.

For example, an activity for counterterrorist purposes, possibly a data mining activity, is likely to require the linking of data found in multiple databases. The literature on record linkage suggests that, even assuming the data found in any given database to be of high quality, the data derived from linkages (the "mosaic" consisting of the collection of linked data) are likely to be error-prone. Certainly, the better the quality of the individual lists, the fewer the errors that will be made in record linkage, but even with high-quality lists, the percentage of false matches and false nonmatches may still be uncomfortably high. In addition, it is also the case that certain data mining algorithms are less sensitive to record linkage errors as inputs, since they use redundant information in a way that can, at times, identify such errors and downweight or delete them. Again, even in the best circumstances, such problems are currently extremely

---

[7]Additional observations about data mining are contained in Appendix H.

difficult to overcome. Error-prone data are, of course, both a threat to privacy (as innocent individuals are mistakenly associated with terrorist activity) and a threat to effectiveness (as terrorists are overlooked because they have been hidden by errors in the data that would have suggested a terrorist connection).

The committee also notes that the use of analytical techniques such as data mining is not limited to government purposes; private parties, including corporations, criminals, divorce lawyers, and private investigators, also have access to such techniques. The large-scale availability of data and advanced analytical techniques to private parties carries clear potential for abuses of various kinds that might lead to adverse consequences for some individuals, but a deep substantive examination of this issue is outside the primary focus of this report on government policy.

### 3.4.2 The Promise and Limitations of Data Mining

Chapter 1 (in Section 1.6.1) notes that data mining covers a wide variety of analytical approaches for using large databases for counterterrorist purposes, and in particular it should be regarded as being much broader than the common notion of a technology underlying automated terrorist identification.

**Conclusion 6.** *Because data mining has proven to be valuable in private-sector applications, such as fraud detection, there is reason to explore its potential uses in countering terrorism. However, the problem of detecting and preempting a terrorist attack is vastly more difficult than problems addressed by such commercial applications.*

As illustrated in Appendix H ("Data Mining and Information Fusion"), data mining has proven valuable in a number of private-sector applications. But the data used by analysts to track sales, banks to assess loan applications, credit card companies to detect fraud, and telephone companies to detect fraud are fundamentally different from counterterrorism data. For example, private-sector applications generally have access to a substantial amount of relatively complete and structured data. In some cases, their data are more accurate than government data, and, in others, large volumes of relevant data sometimes enable statistical techniques to compensate[8] to some extent for data of lower quality—thus, either way, reducing the data-cleaning effort required. In addition, a few false positives and false negatives are acceptable in private-sector

---

[8] A fact that underlies the ability of Internet search engines to propose correct spellings of many incorrectly spelled words.

applications, because a few false positives can usually be cleared up by contact with clients without a significant draw on resources, and a few false negatives are tolerable. Ground truth—that is, knowledge of what is actually true that can be used to validate or verify a new measurement or technique—is available in many private-sector applications, a point that enables automated learning and refinement to take place. All of the relevant data are available—at once—in private-sector applications.

These attributes are very different in the counterterrorism domain. Ground truth is rarely available in tracking terrorists, in large part because terrorists and terrorist activity are rare. Data specifically associated with terrorists (targeted collection efforts) are sparse and mostly collected in unstructured form (free text, video, audio recordings). The availability of much of the relevant data depends on the specific nature of data collected earlier (e.g., information may be needed to obtain a search warrant that then leads to additional information). Data tracks of terrorists in commercial and government administrative databases (as contrasted with government intelligence databases) are co-mingled with enormously larger volumes of similar data associated with innocent individuals, and they are not in any way apparent or obvious from the fact of their collection—that is, it is generally unknown who is a terrorist in any such database. And links among records in databases of varying accuracy will tend to reflect accuracies characteristic of the most inaccurate of the databases involved.

Such differences are not described here to argue that data mining for counterterrorist applications is ipso facto unproductive or operationally useless. But the existence of these differences underscores the difficulty of productively applying data mining techniques in the counterterrorist domain.

**Conclusion 7.** *The utility of pattern-based data mining is found primarily if not exclusively in its role in helping humans make better decisions about how to deploy scarce investigative resources, and action (such as arrest, search, denial of rights) should never be taken solely on the basis of a data mining result. Automated terrorist identification through data mining (or any other known methodology) is neither feasible as an objective nor desirable as a goal of technology development efforts.*

As noted in Appendix H, subject-based data mining and pattern-based data mining have very different characteristics. The common example of pattern-based data mining is what might be called automated terrorist identification, by which is meant an automated process that examines large databases in search of any anomalous pattern that might indicate a terrorist plot in the making. Automated terrorist identification is not technically feasible because the notion of an anomalous pattern—in the absence of some well-defined ideas of what might constitute a threatening pattern—is likely to be associated with many more benign activities than terrorist activities. In this situation, the number of false leads is likely to exhaust any reasonable limit on investigative or analytical resources. For these reasons, the desirability of technology development efforts aimed at automated terrorist identification is highly questionable.

Other kinds of pattern-based data mining may be useful in helping analysts to search for known patterns of interest (i.e., when they have a basis for believing that such a pattern may signal terrorist intent). For example, analysts may determine that a pattern is suggestive of terrorist activity on the basis of historical experience. By searching for patterns known to be associated with (prior) terrorist incidents, it may well be possible to uncover tangible and useful evidence of similar terrorist plots in the making. The significance of uncovering such plots, even if they are similar to those that have occurred in the past, should not be underestimated. Terrorists learn from their past failures and successes, and to the extent that counterterrorist activities can force them to develop new—and unproven—approaches, they will be placed at a significant disadvantage.

Patterns of interest may also be identified by analysts thinking about sets of activities that are indicative of or associated with terrorist activity, even if there is no historical precedent for such associations. Under some circumstances, terrorists might well be limited in the options they might pursue in attacking a specific target. If so, it might be reasonable to search for patterns associated with the planning and execution of those options.

Still, patterns of interest identified using these techniques should be regarded as indicative rather than authoritative, and they should be used only to suggest when further investigation may be warranted rather than as definitive indications of terrorist activity. The committee believes that data mining routines should never be the sole arbiter prior to actions that have a substantial impact on people's lives. Data mining should be used to help humans make decisions when the combination of human judgment and automated data mining results in better decisions than human judgment alone. But even when this is the case, it does not negate the fact that data mining routines, on their own, can make obvious mistakes in deciding the rankings and that the use of human judgment can dramatically reduce the rate of errors.

**Conclusion 8.** *Although systems that support analysts in the identification of terrorists can be designed with features and functionality that enhance privacy* *protection without significant loss to their primary mission, privacy-preserving examination of individually identifiable records is fundamentally a contradiction in terms.*

Systems can often be designed in ways that enhance privacy without compromising their primary mission. For example, in searching for a weapon at a checkpoint, a scanner might generate anatomically correct images of a person's body in graphic detail. Since what is of interest is not those images but rather the presence or absence of weapons, a system could be designed to detect the presence or absence of a weapon in a particular scan and that fact (presence or absence) reported rather than the image itself. Procedural protections could also be put into place: for example, an individual might be given the choice of going through an imaging scanner or undergoing a pat-down search. (Note also that a different and broader set of privacy implications arises if images are stored for further use, as they may well be for system assessments.)

Nevertheless, in the absence of a near-perfect profile of a terrorist, it is not possible, even in principle, to somehow examine the records of an individual (who might or might not be a terrorist) but to expose those records only if he or she actually *is* a terrorist. (A profile of a terrorist is intended to enable the sorting of individuals into those who match the profile and those who do not. If the profile is perfect, and the data contained in individual records are entirely accurate, all of those who match can be regarded with certainty as terrorists and all of those who do not match can be regarded with certainty as nonterrorists. In practice, profiles are never perfect and data are not entirely accurate, and so the notion of degrees of match is much more relevant than the notion of simply match or nonmatch.)

As a result, any realistic system examining databases containing information about terrorists will bring a mix of terrorists and nonterrorists to the attention of analysts, who will decide whether these individuals warrant further investigation. "Further investigation" in this nonroutine context necessarily results in an examination of the private personal information for these individuals, and it may result in tangible inconvenience and loss of various freedoms.

**Conclusion 9.** *Research and development on data mining techniques using real population data are inherently invasive of privacy to some extent.*

Much of data mining is focused on looking for patterns of behavior, characteristics, or transactions that are a priori plausible (i.e., plausible on the basis of expert judgment and experience) as possible indicators of terrorist activity. But these expert judgments about patterns of interest must be empirically valid if they are to have significant operational utility, whereby validity is measured by a high true positive rate in identifying terrorist activity and a low false positive rate.

On one hand, a degree of empirical validity can be obtained through the use of synthetic and anonymized data or historical data. For example, large population databases can be seeded with data created to resemble data associated with real terrorist activity. Although such data are, by definition, based on assumptions about the nature and character of terrorist activities, the expert judgment of experienced counterterrorism analysts can provide such data with significant face validity.[9] By testing various algorithms in this environment, the simulated terrorist signatures provide a measure of ground truth against which various data mining approaches can be tested.

On the other hand and by definition, the use of synthetic data to simulate terrorist signatures does not provide real-world empirical validation. Only real data can be the basis for real-world empirical validation. Thus, another approach is to use historical data on terrorists. For example, a great deal is known today about the actual behavioral and activity signatures of the September 11, 2001, terrorists. Seeding large population databases with such data and requiring various algorithms to identify known terrorists provide a complementary approach to validation.

The use of historical data on terrorists is limited in one fundamental respect: it does not account for unprecedented events. But it is entirely reasonable to suggest that the successful application of proposed tools and techniques to known past events is a minimum and necessary (though not sufficient) metric of success.

Using real population databases—large databases filled with actual behavioral and activity data on actual individuals—presents a serious privacy issue. Almost all of these individuals will have no connection to terrorists, and the use of such data in this context means that their private personal information will indeed be compromised.

It is a policy decision as to whether the risks to privacy inherent in conducting research and development (R&D) on data mining techniques for counterterrorism using real population data are outweighed by the potential operational value of using those techniques. The committee

---

[9]Generally speaking, procedures that produce sensible outputs in response to given, often extreme inputs (often best-case and worst-case scenarios) are said to have gained face validity. For example, input data for fictitious individuals that are designed to provoke an investigation given current procedures, and which are ranked as being of high interest using a data mining algorithm, provide some degree of face validity for that procedure and vice versa. The same is true for fictitious inputs for cases that would be of no interest to counterterrorism analysts for further investigation.

recommends that such R&D should be conducted on synthetic data (see Section 3.7), but if the decision is made to use real population data, the committee urges that policy makers face, acknowledge, and report on this issue explicitly.

## 3.5  CONCLUSIONS REGARDING DECEPTION DETECTION AND BEHAVIORAL SURVEILLANCE

**Conclusion 10.** *Behavioral and physiological monitoring techniques might be able to play an important role in counterterrorism efforts when used to detect (a) anomalous states (individuals whose behavior and physiological states deviate from norms for a particular situation) and (b) patterns of activity with well-established links to underlying psychological states.*

Scientific support for linkages between behavioral and physiological markers and mental state is strongest for elementary states (simple emotions, attentional processes, states of arousal, and cognitive processes), weak for more complex states (deception), and nonexistent for highly complex states (terrorist intent and beliefs). The status of the scientific evidence, the risk of false positives, and vulnerability to countermeasures argue for behavioral observation and physiological monitoring to be used at most as a preliminary screening method for identifying individuals who merit additional follow-up investigation. Indeed, there is no consensus in the relevant scientific community nor on the committee regarding whether any behavioral surveillance or physiological monitoring techniques are ready for use at all in the counterterrorist context given the present state of the science.

**Conclusion 11.** *Further research is warranted for the laboratory development and refinement of methods for automated, remote, and rapid assessment of behavioral and physiological states that are anomalous for particular situations and for those that have well-established links to psychological states relevant to terrorist intent.*

A number of techniques have been proposed for the machine-assisted detection of certain behavioral and physiological states. For example, advances in magnetic resonance imaging (MRI), electroencephalography (EEG), and other modern techniques have enabled measures of changes in brain activity associated with thoughts, feelings, and behaviors.[10] Research in image analysis has yielded improvements in machine recog-

---

[10]P. Root Wolpe, K.R. Foster, and D.D. Langleben, "Emerging neurotechnologies for lie-detection: Promises and perils," *The American Journal of Bioethics* 5(2):39-49, March 2005.

nition of faces under a variety of circumstances (e.g., when a face is smiling or when it is frowning) and environments (e.g., in some nonlaboratory settings).

However, most of the work is still in the basic research stage, with much of the underlying science still to be validated or determined. If real-world utility of these techniques is to be realized, a number of issues—practical, technical, and fundamental—will have to be addressed, such as the limits to understanding, the largely unknown measurement validity of new technologies, the lack of standardization in the field, and the vulnerability to countermeasures. Public acceptability regarding the privacy implications of such techniques also remains to be demonstrated, especially if the resulting data are stored for unknown future uses or undefined lengths of time.

For example, the current state-of-the-art of functional MRI technology can identify changes in the hemodynamics in certain regions of the brain, thus signaling activity in those regions. But such results are not necessarily consistent across individuals (i.e., different areas in the brains of different individuals may be active under the same stimulus) or even in the same individual (i.e., a slightly different part of the brain may become active even in the same individual under the same stimulus). Certain regions of the brain may be active under a variety of different stimuli. In short, understanding of what these regions do is still primitive. Furthermore, even if simple associations can be made reliably in laboratory settings, this does not necessarily translate into usable technology in less controlled situations. Behavior of interest to detect, such as terrorist intent, occurs in an environment that is very different from the highly controlled behavioral science laboratory.

**Conclusion 12.** *Technologies and techniques for behavioral observation have enormous potential for violating the reasonable expectations of privacy of individuals.*

Because the inferential chain from behavioral observation to possible adverse judgment is both probabilistic and long, behavioral observation has enormous potential for violating the reasonable expectations of privacy of individuals. It would not be unreasonable to suppose that most individuals would be far less bothered and concerned by searches aimed at finding tangible objects that might be weapons or by queries aimed at authenticating their identity than by technologies and techniques whose use will inevitably force targeted individuals to explain and justify their mental and emotional states. Even if behavioral observation and physiological monitoring are used only as a preliminary screening methods for identifying individuals who merit additional follow-up investigation, these individuals will be subject to suspicion that would not fall on others not so identified.

## 3.6 CONCLUSIONS REGARDING STATISTICAL AGENCIES

**Conclusion 13.** *Census and survey data collected by the federal statistical agencies are not useful for terrorism prevention: such data have little or no content that would be useful for counterterrorism. The content and sampling fractions of household surveys as well as the lack of personal identifiers makes it highly unlikely that these data sets could be linked with any reasonable degree of precision to other databases of use in terrorism prevention.*

The content of the data collected by the federal statistical agencies under the auspices of survey and census programs is generally inconsistent with the needs of counterterrorist activities, which require individually identifiable data. Even ignoring issues of access, the value of the data collected on national household or business surveys for terrorism prevention is minimal.

The reasons are several:

- Censuses collect little information beyond name, address, and basic demographic data on age, sex, and race; such data are unlikely to be of much value for identifying terrorists or terrorist behavior.
- Because a substantial proportion of individuals move frequently, the 10-year cycle of censuses means that the census information is unlikely to be timely, even in supplying current addresses.
- The census long form, which has been collected on a sample basis (and its successor program, the American Community Survey, ACS) have more information but still very little that is directly relevant to predicting terrorist activity. Moreover, because these data are collected only for a sample, the probability that those of interest would be in the sample for a given year of the ACS is very slight, and, furthermore, the ability to match files without identifiers into other record systems would be limited. At best, these data might provide background information to provide a description of the socioeconomic make-up of a clustered group of blocks.
- Other household surveys also collect little of direct relevance to terrorism prevention, and because they typically draw on much less than 1 percent of the population, the chances of identifying new information on an individual of interest are rather low.

Regarding establishment surveys, for terrorism detection one might be interested in businesses that have increased activity with people in various parts of the world, but such information is not contained on federal statistical system business censuses and surveys.

A variety of surveys collect information relevant to crime prevention and public health. Data collections on criminal activity, such as the National Crime Victimization Survey and the Uniform Crime Reports, contain data on victims of crime, and they are most useful in identifying geographic areas in which such criminal activity seems to be prevalent. Health surveys, such as the National Health Information Survey, the National Health and Nutrition Examination Survey, and the National Ambulatory Medical Care Survey (largely collected by the National Center for Health Statistics) have value in broader public health programs, but they cannot provide timely information for purposes of biosurveillance or for addressing a bioterrorist attack.

In addition, statistical agencies often collect information under a promise of confidentiality, and the costs of altering or relaxing the rules for confidentiality protection are quite substantial. The quality of the data collected could be adversely affected as a consequence of respondents' decreased willingness to cooperate. Statistical agencies typically collect information under a promise of confidentiality, and reneging on such officially provided assurances could substantially reduce the quality of the data collected, resulting in much poorer data on the state of the nation.[11]

Aside from census and survey data, statistical agencies also hold considerable administrative data (which they have collected from other agencies); such data may be merged with data collected for statistical purposes and thus create the potential for data sets and databases that could at some point conceivably be useful for purposes of counterterrorism. While these derived data sets are currently protected by pledges of

---

[11]At times, even the use of nonpersonally identifiable information collected by the statistical agencies for counterterrorism purposes can lead to public protest and endanger the cooperation of the public in providing information to statistical agencies. For example, in August 2002 and December 2003, the U.S. Department of Homeland Security asked the U.S. Census Bureau to provide information on the number of Arab Americans living in the United States by small-area tabulations, and the Census Bureau complied with this request. Although this request violated no law, it caused a public furor and led the Census Bureau to rethink its dissemination policy, even though no personally identifiable information was involved. In addition, groups representing Arab Americans threatened to withhold their future cooperation in the collection of data by the Census Bureau. (See L. Clemetson, "Homeland security given data on Arab-Americans," *New York Times*, July 30, 2004; L. Clemetson, "Threats and responses: Privacy; Coalition seeks action on shared data on Arab-Americans," *New York Times*, August 13, 2004; E. Lipton, "Panel says census move on Arab-Americans recalls World War II internments," *New York Times*, November 10, 2004.) If such public concern emerges from data requests that are entirely consistent with the confidentiality guarantees provided under existing law, it is not difficult to imagine that actions to weaken these guarantees might lead to similar controversy.

confidentiality if any of the component data sets are so protected, some additional consideration needs to be given to such constructs and how to respond to requests for them from other government agencies.

## 3.7  RECOMMENDATIONS

In light of the conclusions presented above, the committee has two central recommendations. The first recommendation has subparts a-d.

### 3.7.1  Systematic Evaluation of Every Information- Based Counterterrorism Program

**Recommendation 1.  U.S. government agencies should be required to follow a systematic process (such as the one described in the framework proposed in Chapter 2) to evaluate the effectiveness, lawfulness, and consistency with U.S. values of every information-based program, whether classified or unclassified, for detecting and countering terrorists before it can be deployed, and periodically thereafter.**

Appendix J ("The Total/Terrorist Information Awareness Program") recounts the story of the Total Information Awareness (TIA) program of the Defense Advanced Research Projects Agency (DARPA) and the intense controversy it engendered—which was a motivation for launching this study. The committee notes that in December 2003, the Department of Defense (DOD) inspector general's (IG) audit of TIA concluded that the failure to consider privacy adequacy during the early development of TIA led DOD to "risk spending funds to develop systems that may not be either deployable or used to their fullest potential without costly revision."[12] The DOD-IG report noted that this was particularly true with regard to the potential deployment of TIA for law enforcement: "DARPA need[ed] to consider how TIA will be used in terms of law enforcement to ensure that privacy is built into the developmental process."[13] Greater consideration of how the technology might be used not only would have served privacy but also would probably have contributed to making TIA more useful.

The committee believes that a systematic approach to the development, procurement, and use of information-based counterterrorism programs is necessary if their full value is to be obtained. The framework

---

[12]U.S. Department of Defense, Office of the Inspector General, *Information Technology Management: Terrorism Information Awareness Program* (D-2004-033), Washington, D.C., 2003, p. 4.

[13]Id. at 7.

developed by the committee and provided in Chapter 2 is intended as a template for government decision makers to use in evaluating the effectiveness, appropriateness, and validity of every information-based counterterrorism program and system. The U.S. Department of Homeland Security (DHS)—and all agencies of the U.S. government with counterterrorism responsibilities—should adopt the framework described in Chapter 2, or one similar to it, as a central element in their decision making about new deployments and existing programs in use. Failure to adopt such a systematic approach is likely to result in reduced operational effectiveness, wasted resources, privacy violations, mission creep, and diminished political support, not only for those programs but also for similar and perhaps for not-so-similar programs across the board.

To facilitate accountability, such evaluations (and the data on which they are based) should be made available to the broadest audience possible. Broad availability implies that these evaluations should be unclassified to the maximum extent possible—but even if evaluations are classified, they should still be performed and should be made available to those with the requisite clearances.

Such evaluations should be independent and comprehensive, and in particular they should assess both program effectiveness and privacy together, involving independent experts with the necessary technical, legal, and policy expertise to understand each of these areas and how interactions among them might affect the evaluation. For example, the meaning of privacy is in part technical, and an assessment of privacy cannot be left exclusively to individuals lacking such technical understanding.

Chapter 2 noted that much of the committee's framework is not new and also that government decision makers have failed to implement many of the guidelines embedded in the framework even when they have been cognizant of them. It is the committee's hope that by presenting to policy makers a comprehensive framework independent of any particular program, the pressures and exigencies associated with specific crises can be removed from the consideration and adoption of such a framework for application to all programs.

The committee also calls attention to four subrecommendations that derive from Recommendation 1.

**Recommendation 1a. Periodically after a program has been operationally deployed, and in particular before a program enters a new phase in its life cycle, policy makers should apply a framework such as the one proposed in Chapter 2 to the program before allowing it to continue operations or to proceed to the next phase.**

A systematic approach such as the framework in Chapter 2 is not intended to be applied only once in the life cycle of any given program. As noted in Appendix D ("The Life Cycle of Technology, Systems, and Programs"), a program undergoes a number of different phases in its lifetime: identification of initial needs, research and technology development, systems development, and operational deployment and continual operational monitoring. Each of these phases provides a desirable opportunity for applying the framework to help decide whether and how the program should transition to the next phase. Each of the framework's questions should still be asked. But the answers to those questions as well as the interpretation of the answers will vary depending on the phase. Such a review may result in a significant modification or even a cancellation of a given program.

The committee calls special attention to the importance of operational monitoring, whose purpose is to ensure that the initial deployed capability remains both effective at contributing to the mission for which it was designed and acceptable from a privacy standpoint. Often after initial deployment, the operational environment changes. Improved base technologies or entirely new technologies become available. Existing threat actors change their tactics, or entirely new threats emerge. Executive branch policies change, or new administrations take office. Analysts gain experience, or new analysts arrive. Interpretations of existing law change through court decisions, or new legislation is passed. Data-based models may change simply because more data have become available that change the parameters and estimates on which the models are based. Error rates may change for similar reasons. Because every program is necessarily embedded in this milieu, the net result is that successful programs are almost always dynamic, in that they evolve in response to such changing circumstances.

An evolved program is, by definition, not the same as the original program—and it is a fair question to ask whether the judgments made about any program in its original form would be valid for an evolved program. For these reasons, a policy regime is necessary that provides for periodic reassessment and reevaluation of a program after initial deployment, at the same time promoting and fostering necessary changes—whether technological, procedural, legal, ethical, or other.

Recommendation 1a is important to programs currently in existence—that is, programs in existence today, and especially programs that are operationally deployed today should be evaluated against the framework. To the best of the committee's knowledge, no such evaluations have been performed for any data mining or deception detection programs in operation, although this is not to say that none have been done. If such evaluations have been performed, they should be made available to policy makers (senior officials in the executive branch or the U.S. Congress), and if possible, the public as well. If not, they should be undertaken with all due speed. And if they cannot be performed without access to classified information, an independent group of experts with the requisite clearances should be chartered to perform such assessments.

**Recommendation 1b. To protect the privacy of innocent people, the research and development of any information-based counterterrorism program should be conducted with synthetic population data. If and when a program meets the criteria for deployment in the committee's illustrative framework described in Chapter 2, it should be deployed only in a carefully phased manner, e.g., being field tested and evaluated at a modest number of sites before being scaled up for general use. At all stages of a phased deployment, data about individuals should be rigorously subjected to the full safeguards of the framework.**

Almost by definition, technology in the R&D stage is nascent and unproven. Nascent and unproven technologies are not sufficiently robust or reliable to warrant risking the privacy of individuals—that is, the very uncertain (perhaps nonexistent) benefit that would be derived from their use does not justify the very real cost to privacy that would inevitably accompany their widespread use in operational settings. Thus, the committee advocates R&D based on synthetic population data whose use poses very little risk of violating the privacy of innocent individuals. In addition, the successful use of synthetic data in many fields, such as epidemiology, medicine, and chemistry, for testing methods provides another reason to explore its potential uses in counterterrorism.

The committee believes that realistic synthetic population data could probably be created along the lines originally suggested in Rubin and in Little and further developed by Fienberg et al. and Reiter,[14] for the specific purpose of providing the background against which terrorist signatures are sought. Furthermore, because it is difficult to create from entirely synthetic data large databases that are useful for testing and (partially) validating data mining techniques and algorithms, a partial substitute for entirely synthetic data is data derived from real population data in such a way that the individual identities of nonterrorists are masked

---

[14]D.B. Rubin, "Discussion: Statistical disclosure limitation," *Journal of Official Statistics* 9(2):461-468, 1993; R.J.A. Little, "Statistical analysis of masked data," *Journal of Official Statistics* 9(2):407-426, 1993; S.E. Fienberg, U.E. Makov, and R.J. Steele (with discussion by P. Kooiman and a response), "Disclosure limitation using perturbation and related methods for categorical data," *Journal of Official Statistics* 14:485-511, 1998; J.P. Reiter, "Inference for partially synthetic, public use microdata sets," *Survey Methodology* 29:181-188, 2003.

while preserving some of the important large-scale statistical properties of those data.

Using synthetic population data as the background, a measure of the utility of various data mining approaches can be obtained in R&D. Such results must be evaluated in the most rigorous and independent manner possible in order to determine if the program should move into deployment.

If the results are determined to be sufficiently promising (e.g., with sufficiently low false positive and false negative rates) that they offer significant operational capability, it is reasonable to apply the new capabilities to real data in an operational context.[15] But the change from synthetic to real data must be accompanied by a full array of operational safeguards that protect individuals from harm to their privacy, as suggested by the committee's proposed framework. Put differently, if real data are to be used, they—and the individuals with whom they are associated—deserve the full benefit of the privacy protections associated with the program in question.

Transitioning to an operational context from R&D must also be done carefully and is best undertaken in small phases. The traditional approach to acquisition generally involves the deployment of operational capabilities in large blocks of capability (i.e., large functional components deployed on a wide scale). Experience indicates that this approach is often slow and cumbersome, and it increases technical, programmatic, and budgetary risks. The operational environment often changes significantly in the time between initial requirements specification and first deployment—thus, the capability may even be obsolete when it is first deployed. And deploying systems on a large scale before they are deployed on a small scale is almost always problematic, because small-scale operational trials are needed to shake out the inevitable bugs when R&D technologies meet the real world.

By contrast, phased deployment is based on a philosophy of "build-a-little, test-a-little, deploy-a-little." Phased deployment recognizes that kinks and problems in the deployment of any new capability are inevitable, positing that by making small changes, system developers will be able to more easily identify and correct these problems than when everything changes all at once. Small changes are easier to reverse, should that become necessary. It also becomes feasible to test new capabilities offered by small changes in parallel with the baseline version, so that ground

---

[15]Note, however, that to the best of the committee's knowledge, current data mining programs for counterterrorist purposes have not been evaluated for operational effectiveness in such a manner, either with synthetic data or with real data.

truth provided by the baseline version can be used to validate the new capabilities when their domain of operation is the same.

The committee recognizes that, under this approach, operational capabilities will not have been subject to real-world empirical validation before deployment, although they will have had as much validation as possible with synthetic population data. And the phased deployment of privacy-sensitive capabilities reduces the likelihood of inappropriate or improper compromises of privacy from what they would have been under a more traditional acquisition model.[16]

The approach recommended above (synthetic data before deployment, deployment only in measured phases) places a high premium on two actions. First, every effort must be made to create good synthetic data that are useful for testing the validity of machine-learning tools and are simultaneously very realistic. For synthetic terrorist data, both historical data and expert judgment play a role in developing signatures that might plausibly be associated with terrorist activity, and plausibility should be assessed through independent panels of judges without a vested interest in any given scenario. Such judges must also be trained in or experienced with evasion or obfuscation techniques. For synthetic population data, every use must be made of known techniques for confidentiality protection and statistical disclosure limitation[17] to reduce the likelihood that the privacy of individuals is compromised, and further research on the creation of better synthetic data to represent large-scale populations is certainly warranted.

Second, evaluation of R&D results must truly be independent and rigorous, with high standards of performance needed for a decision to deploy. As noted in Conclusion 4, the rule that "X is better than doing nothing" often drives deployment decisions, and, given the high potential costs to individual privacy of deployment, the benefits afforded by deployment must be more than marginal to warrant the potential cost.

**Recommendation 1c.** **Any information-based counterterrorism program of the U.S. government should be subjected to robust, independent oversight of the operations of that program, a part of which would**

---

[16]The qualifier of "privacy-sensitive" increments of capability is important, because it would be all too easy for a program manager to shortchange privacy considerations in attempts to "get something working" for demonstration purposes. That is, privacy functionality must be built into the system from the start, rather than being an add-on to be deployed at the end "when everything else works."

[17]Additional discussion of some of these techniques can be found in the National Research Council publications *Expanding Access to Research Data: Reconciling Risks and Opportunities* (The National Academies Press, Washington, D.C., 2005) and *Private Lives and Public Policies* (National Academy Press, Washington, D.C., 1993).

**entail a practice of using the same data mining technologies to "mine the miners and track the trackers."**

In practice, operational monitoring is generally the responsibility of the program managers and operational personnel. But as discussed in Appendix G ("The Jurisprudence of Privacy Law and the Need for Independent Oversight"), oversight is necessary to ensure that actual operations have been conducted in accordance with stated policies.

The reason is that, in many cases, decision makers formulate policies in order to balance competing imperatives. For example, the public demands both a high degree of effectiveness in countering terrorism and a high degree of privacy. Program administrators themselves face multiple challenges: motivating high performance, adhering to legal requirements, staying within budget, and so on. But if operational personnel adhere to some elements of a policy and not to others, the balance that decision makers intended to achieve will not be realized in practice.

The committee emphasizes that independent oversight is necessary to ensure that commitments to minimizing privacy intrusions embedded in policy statements are realized in practice. The reason is that losses of privacy are easy to discount under the pressure of daily operations, and those elements of policy intended to protect privacy are more likely to be ignored or compromised. Without effective oversight mechanisms in place, public trust is less likely to be forthcoming. In addition, oversight can support continuous improvement and guide administrators in making organizational change.

For example, program oversight is essential to ensure that those responsible for the program do not bypass procedures or technologies intended to protect privacy. Noncompliance with existing privacy-protecting laws, regulations, and best practices diminishes public support and creates an environment in which counterterrorism programs may be curtailed or eliminated. Indeed, even if shortcuts and bypasses increase effectiveness in a given case, in the long run scandals and public outcry about perceived abuses will reduce the political support for the programs or systems involved—and may deprive the nation of important tools useful in the counterterrorist mission. Even if a program is effective in the laboratory and expected to be so in the field, its deployment must be accompanied by strong technical and procedural safeguards to ensure that the privacy of individuals is not placed at undue risk.

Oversight is also needed to protect against abuse and mission creep. Experience and history indicate that in many programs that collect or use personal information, some individuals may violate safeguards intended to protect individual privacy. Hospital clerks have been known to examine the medical records of celebrities without having a legitimate reason for doing so, simply because they are curious. Police officers have been known to examine the records of individuals in motor vehicle information systems to learn about the personal lives of individuals with whom they interact in the course of daily business. And, of course, compromised insiders have been known to use the information systems of law enforcement and intelligence agencies to further nefarious ends.

The phenomenon of mission creep is illustrated by the Computer-Assisted Passenger Prescreening System II (CAPPS II) program, initially described in congressional testimony as an aviation security tool and *not* a law enforcement tool but which morphed in a few months to a system that would analyze information on persons "with [any] outstanding state or federal arrest warrants for crimes of violence."[18]

To guard against such practices, the committee advocates program oversight that mines the miners and tracks the trackers. That is, all operation and command histories and all accesses to data-based counterterrorism information systems should be logged on an individual basis, audited, and mined with the same technologies and the same zeal that are applied to combating terrorists. If, for example, such practices had been in place during Robert Hanssen's tenure at the Federal Bureau of Investigation (FBI), his use of its computer systems for unauthorized purposes might have been discovered sooner.

Finally, the committee recognizes the phenomenon of statutory mission creep, as defined above in the discussion of Premise 5. It occurs, for example, because in responding to a crisis, policy makers will naturally focus on adapting existing programs and capabilities rather than creating new ones. On one hand, if successful, adaptation often promises to be less expensive and faster than creating a new program or capabilities from scratch. On the other hand, because an existing program is likely to be highly customized for specific purposes, adapting that program to serve other purposes effectively may prove difficult—perhaps even more difficult than creating a program from scratch. As importantly, adapting an existing program to new purposes may well be contrary to agreements and understandings established in order to initiate the original program in the first place.

---

[18]An initial description of CAPPS II by Deputy Secretary of DHS Admiral James Loy, then administrator of the Transportation Security Administration, assured Congress that CAPPS II was intended to be an aviation security tool, not a law enforcement tool. Testimony of Admiral James Loy before House Government Reform Subcommittee on Technology, Information Policy, Intergovernmental Relations and the Census (May 6, 2003). Morphed system—Interim Final Privacy Act Notice, 68 Fed. Reg. 45265 (August 1, 2003).

The committee does not oppose expanding the goals and missions of a program under all circumstances. Nevertheless, it cautions that such expansion should not be undertaken hastily in response to crisis. In the committee's view, following diligently the framework presented in Chapter 2 is an important step in exercising such caution.

**Recommendation 1d. Counterterrorism programs should provide meaningful redress to any individuals inappropriately harmed by their operation.**

Programs that are designed to balance competing interests (in the case of counterterrorism, collective security and individual privacy and civil liberties) will naturally be biased in one direction or another if their incentive/penalty structure is not designed to reflect this balance. The availability of redress to the individual harmed thus acts to promote the goal of compliance with stated policy—as does the operational oversight mentioned in Recommendation 1c—and to provide incentives for the government to improve the policies, technologies, and data underlying the operation of the program.

Although the committee makes no specific recommendation concerning the form of redress that is appropriate for any given privacy harm suffered by innocent individuals as the result of a counterterrorism program, it notes that many forms of redress are possible in principle, ranging from apology to monetary compensation. The most appropriate form of redress is likely to depend on the nature and purpose of the specific counterterrorism program involved. However, the committee believes that, at a minimum, an innocent individual should always be provided with at least an explicit acknowledgment of the harm suffered and an action that reduces the likelihood that such an incident will ever be repeated, such as correcting erroneous data that might have led to the harm. Note that responsibilities for correction should apply to the holder of erroneous data, regardless of whether the holder is the government or a third party.

The availability of redress might, in principle, enable terrorists to manipulate the system in order to increase their chances of remaining undetected. However, as noted in Item 7 of the committee's framework on effectiveness, information-based programs should be robust and not easily circumvented by adversary countermeasures, and thus the possibility that terrorists might manipulate the system is not a sufficient argument against the idea of redress.

### 3.7.2 Periodic Review of U.S. Law, Policy, and Procedures for Protection of Privacy

**Recommendation 2. The U.S. government should periodically review the nation's laws, policies, and procedures that protect individuals' private information for relevance and effectiveness in light of changing technologies and circumstances. In particular, Congress should reexamine existing law to consider how privacy should be protected in the context of information-based programs (e.g., data mining) for counterterrorism.**

The technological environment in which policy is embedded is constantly changing. Although technological change is not new, the pace of technological change has dramatically increased in the digital age. As noted in *Engaging Privacy and Information Technology in a Digital Age*, advances in information technology make it easier and cheaper by orders of magnitude to gather, retain, and analyze information, and other trends have enabled access to new kinds of information that previously would have been next to impossible to gather about another individual.[19] Furthermore, new information technologies have eroded the privacy protection once provided through obscurity or the passage of time. Today, it is less expensive to store information electronically than to decide to get rid of it, and new and more powerful data mining techniques and technologies make it much easier to extract and identify personally identifiable patterns that were previously protected by the vast amounts of data "noise" around them.

The security environment is also constantly changing. New adversaries emerge, and counterterrorist efforts must account for the fact that new practices and procedures for organizing, training, planning, and acquiring resources may emerge as well. Most importantly, new attacks appear. The number of potential terrorist targets in the United States is large,[20] and

---

[19]National Research Council, *Engaging Privacy and Information Technology in a Digital Age*, The National Academies Press, Washington, D.C., 2007.

[20]Analysts and policy makers have debated the magnitude of this number. In one version of the Total Information Awareness program, the number of important and plausible terrorist targets was estimated at a few hundred, while other informed estimates place the number at a few thousand. Still other analysts argue that the number is virtually unlimited, since terrorists could, in principle, seek to strike anywhere in their attempts to sow terror. There is evidence on both sides of this point. Some point to Al Qaeda planning documents and other intelligence information to suggest that it continues to be very interested in large-scale strikes on targets that are media-worthy around the world, such as targets associated with air transportation. Others point out that the actual history of terrorist attacks around the world has for the most part involved attacks on relatively soft and undefended targets, of which there are very many.

although the different types of attack on these targets may be limited, attacks might be executed in myriad ways.

As an example of a concern ripe for examination and possible action, the committee found common ground in the proposition that policy makers should seriously consider restrictions on how personal information is used in addition to restrictions on how records are collected and accessed. Usage restrictions could be an important and useful supplement to access and collection limitation rules in an era in which much of the personal information that can be the basis for privacy intrusion is already either publicly available or easily accessible on request without prior judicial oversight. Privacy protection in the form of information usage restrictions can provide a helpful tool that balances the need to use powerful investigative tools, such as data mining, for counterterrorism purposes and the imperative to regulate privacy intrusions of such techniques through accountable adherence to clearly stated privacy rules. (Appendix G elaborates on this aspect of the recommendation.)

Such restrictions can serve an important function in helping to ensure that programs created to address a specific area stay focused on the problem that the programs were designed to address and in guarding against unauthorized or unconsidered expansion of government surveillance power. They also help to discourage mission creep, which often expands the set of purposes served by the program without explicit legislative authorization and into areas that are poorly matched by the original program's structure and operation. An example of undesirable mission creep would be the use of personal data collected from the population acquired for counterterrorist purposes to uncover tax evaders or parents who have failed to make child support payments. This is not to say that finding such individuals is not a worthy social goal, but rather that the mismatch between such a goal and the intrusiveness of data collection measures for counterterrorist purposes is substantial indeed. Without clear legal rules defining the boundaries for use between counterterrorism and inappropriate law enforcement uses, debates over mission creep are likely to continue without constructive resolution.

A second example of a concern that may be ripe for legislative action involves the current legal uncertainty supporting private-sector liability for cooperation with government data mining programs. Such uncertainty creates real risk in the private sector, as indicated by the present variety of private lawsuits against telecommunications service providers,[21] and private-sector responsibilities and rights must be clarified along

---

[21] For example, the Electronic Frontier Foundation filed a class-action lawsuit against AT&T on January 31, 2006, claiming that AT&T violated the law and the privacy of its customers by collaborating with the National Security Agency in the Terrorist Surveillance Program.

with government powers and privacy protections. What exists today is a mix of law, regulation, and informal influence in which the legal rights and responsibilities of private-sector entities are highly uncertain and not well understood.

A coherent, comprehensive legal regime regulating information-intensive surveillance such as government data mining, would do much to reduce such uncertainty. As one example, such a regime might address the issue of liability limitation for private-sector data sources (database providers, etc.) that provide privacy-intrusive information to the government.

Without spelling out the precise scope and coverage of the comprehensive regime, the committee believes that to the extent that the government legally compels a private party to provide data or a private party otherwise complies with an apparently legal requirement to disclose information, it should not be subject to liability simply for the act of complying with the government compulsion or legal requirement. Any such legal protection should not extend to the content of the information it supplies, and the committee also believes that the regime should allow incentives for data providers to invest reasonable effort in ensuring the quality of the data they provide. Furthermore, they should provide effective legal remedies for those individuals who suffer harm as a result of provider negligence. Furthermore, the regime would necessarily preserve the ability of individuals to challenge the constitutionality of the underlying data access statute.

Listed below are other examples of how the adequacy of privacy-related law might be called into question by a changing environment (Appendix F elaborates on these examples).

- *Conducting general searches.* On one hand, the Fourth Amendment forbids general searches—that is, searches that are not limited as to the location of the search or the type of evidence the government is seeking—by requiring that all searches and seizures must be reasonable and that all warrants must state with particularity the item to be seized and the place to be searched. On the other hand, machine-aided searching of enormous digital transaction records is in some ways analogous to a general search. Such a search can be a dragnet that sweeps through millions or billions of records, often containing highly sensitive information. Much like a general search in colonial times was not limited to a particular person or place, a machine-aided search through digital databases can be very broad. How, if at all, should database searches be regulated by the Fourth Amendment or by statute?

A related issue is that the historical difficulty of physical access to ostensibly public information has provided a degree of privacy protection for that information—what might be known as privacy through obscurity. But a search-enabled digital world erodes some of these previously inherent protections against invasions of privacy, changing the technological milieu that surrounds privacy jurisprudence.

• *Increased access to data; searches and surveillance of U.S. persons outside the United States.* The Supreme Court has not yet addressed whether the Fourth Amendment applies to searches and surveillance for national security and intelligence purposes that involve U.S. persons[22] who are connected to a foreign power or that are conducted wholly outside the United States.[23] Lower courts, however, have found that there is an exception to the Fourth Amendment's warrant requirement for searches conducted for intelligence purposes within the United States that involve only non-U.S. persons or agents of foreign powers.[24] The Supreme Court has yet to rule on this important issue, and Congress has not supplied any statutory language to fill the gap.

• *Third-party records.* Two Supreme Court cases (*United States v. Miller*, 1976, and *Smith v. Maryland*, 1979)[25] have established the precedent that there is no constitutionally based reasonable expectation of privacy for information held by a third party, and thus the government today has access unrestricted by the Fourth Amendment to private-sector records on every detail of how people live their lives. Today, these third-party transactional records are available to the government subject to a very low threshold—through subpoenas that can be written by almost any government agency without prior judicial oversight—and are one of the primary data feeds for a variety of counterterrorist data mining activities. Thus, the public policy response to privacy erosion as a result of data mining used with these records will have to address some combination of the scope of use for the data mining results, the legal standards for access to and use of transactional information, or both.[26] (See also Appendix G for

---

[22]A U.S. person is defined by law and Executive Order 12333 to mean "a United States citizen, an alien known by the intelligence agency concerned to be a permanent resident alien, an unincorporated association substantially composed of United States citizens or permanent resident aliens, or a corporation incorporated in the United States, except for a corporation directed and controlled by a foreign government or governments."

[23]J.H. Smith and E.L. Howe, "Federal legal constraints on electronic surveillance," pp. 133-148 in *Protecting America's Freedom in the Information Age*, Markle Foundation Task Force on National Security in the Information Age, Markle Foundation, New York, N.Y., 2002.

[24]See *United States v. Bin Laden*, 126 F. Supp. 2d 264, 271-72 (S.D.N.Y. 2000).

[25]*United States v. Miller*, 425 U.S. 435 (1976); *Smith v. Maryland*, 442 U.S. 735 (1979).

[26]Transactional information is the data collected on individuals from their interactions (transactions) with outside entities, such as businesses (e.g., travel and sales records), public facilities and organizations (e.g., library loans), and Web sites (e.g., Internet usage). Aggregate information, in contrast, is information in summary form (e.g., total visitors and sales) that does not contain data that would permit the identification of a specific individual.

discussion of how usage limitations can fill gaps in current regulation of the confidentiality of third-party records.)

• *Electronic surveillance law.* Today's law regarding electronic surveillance is complex. Some of the complexity is due to the fact that the situations and circumstances in which electronic surveillance may be involved are highly varied, and policy makers have decided that different situations call for different regulations. But it is an open question as to whether these differences, noted and established in one particular set of circumstances, can be effectively maintained over time. Although there is broad agreement that today's legal regime is not optimally aligned with the technological and circumstantial realities of the present, there is profound disagreement about whether the basic principles underlying today's regime continue to be sound as well as in what directions changes to today's regime ought to occur.

In making Recommendation 2, the committee intends the government's reexamination of privacy law to cover the issues described above but not be limited to them. In short, Congress and the president should work together to ensure that the law is clear, appropriate, up to date, and responsive to real needs.

Greater clarity and coherence in the legal regime governing information-based programs would have many benefits, both for privacy protection and for the counterterrorist mission. It is perhaps obvious that greater clarity helps to protect privacy by eliminating what might be seen as loopholes in the law—ambiguities that can be exploited by well-meaning national security authorities, thereby overturning or circumventing the intent of previously established policy that balanced competing interests. But the benefits of greater clarity from the standpoint of improving the ability of the U.S. government to prosecute its counterterrorism responsibilities are less obvious and thus deserve some elaboration.

First and most importantly from this perspective, greater legal clarity would help to reduce public controversy over potentially important tools that might be used for counterterrorist purposes. Although many policy makers might wish that they had a free hand in pursuing the counterterrorist mission and that public debate and controversy would just go away, the reality is that public controversy does result when the government is seen as exploiting ambiguities and loopholes.

As discussed in Appendix I ("Illustrative Government Data Mining Programs and Activity"), a variety of government programs have been shut down, scaled back, delayed, or otherwise restricted over privacy considerations: TIA, CAPPS II for screening airline passengers, MATRIX (Multistate Anti-Terrorism Information Exchange) for linking law enforcement records across states with other government and private-sector databases, and a number of data-sharing experiments between the U.S. government and various airlines. Public controversy about these efforts may have prematurely compromised counterterrorism tools that might have been useful. In addition, they have also made the government more wary of national security programs that involve data matching and made the private sector more reluctant to share personal information with the government in the future.

In this regard, this first rationale for greater clarity is consistent with the conclusion of the Technology and Privacy Advisory Committee: "[privacy] protections are essential *so that* the government can engage in appropriate data mining when necessary to fight terrorism and defend our nation. And we believe that those protections are needed to provide clear guidance to DOD personnel engaged in anti-terrorism activities."[27]

Second, greater legal clarity and coherence can enhance the effectiveness of certain information-based programs. For example, the Privacy Act of 1974 requires that personal data used by federal agencies be accurate, relevant, timely, and complete. On one hand, these requirements increase the likelihood that high-quality data are stored, thus enhancing the effectiveness of systems that use data subject to those requirements. On the other hand, both the FBI's National Crime Information Center and the passenger screening database of the Transportation Security Administration have exemptions from some of these requirements;[28] to the extent that these exemptions result in lower-quality data, these systems are likely to perform less well.

Third, the absence of a clear legal framework is likely to have a profound effect on the innovation and research that are necessary to improve the accuracy and effectiveness of information-based programs. Such clarity is necessary to support the investment of financial, institutional, and human resources in often risky research that may not pay dividends for

---

[27]Technology and Privacy Advisory Committee, *Safeguarding Privacy in the Fight Against Terrorism*, U.S. Department of Defense, Washington, D.C., March 2004, p. 48, available at http://www.cdt.org/security/usapatriot/20040300tapac.pdf.

[28]The Department of Justice and the Transportation Security Administration have published notices on these programs in the *Federal Register*, exempting them from certain provisions of the Privacy Act that are allowed under the act. In March 2003, the DOJ exempted the FBI's National Crime Information Center from the Privacy Act's requirements that data be "accurate, relevant, timely and complete," *Privacy Act of 1974*; Implementation, 68 Federal Register 14140 (2003) (DOJ, final rule). In August 2003, the Department of Homeland Security exempted the TSA's passenger screening database from the Privacy Act's requirements that government records include only "relevant and necessary" personal information, *Privacy Act of 1974*: Implementation of Exemption, 68 Federal Register 49410 (2003) (DHS, final rule). Outside these exceptions, the Privacy Act otherwise applies to these programs. (Under the act, exemptions have to be published to be effective, and so the committee assumes that there are no "secret" exemptions.)

decades. But that type of research is essential to counterterrorism efforts and to finding better ways of protecting privacy.

Finally, a clear and coherent legal framework will almost certainly be necessary to realize the potential of new technologies to fight terrorism. Because such technologies will operate in the political context of an American public concerned about privacy, the public—and congressional decision makers—will have to take measures that protect privacy when new technologies are deployed. All technological solutions will require a legal framework within which to operate, and there will always be gaps left by technological protections, which law will be essential to fill. Consequently, a lack of clarity in that framework may not only slow their development and deployment, as described above, but also make technological solutions entirely unworkable.

# Appendixes

# A

# Acronyms

| | |
|---|---|
| 9-11 | reference to the terrorist attacks on the U.S. that occurred September 11, 2001 |
| ACE | Automated Commercial Environment |
| ACS | Automated Commercial System |
| ADVISE | Analysis, Dissemination, Visualization, Insight, and Semantic Enhancement |
| All-WME | all weapons of mass effect |
| ANFO | ammonium nitrate/fuel oil |
| ANNM | ammonium nitrate nitromethane |
| ANS | autonomic nervous system |
| APS | Advance Passenger Information System |
| AQ | Al Qaeda |
| ARCOS | Automation of Reports of Consolidated Orders System |
| ARPA | Advanced Research Projects Agency |
| ATF | Bureau of Alcohol, Tobacco, and Firearms |
| ATM | automated teller machine |
| ATS | Automated Targeting System |
| BATS | Bomb Arson Tracking System |
| BKC | Biodefense Knowledge Center |
| BSA | Bank Secrecy Act |

| | |
|---|---|
| CAAIOPEE | (KDD application) |
| CAGR | Compound Annual Growth Rate |
| CALEA | Communications Assistance for Law Enforcement Act of 1994 |
| CAPPS-II | Computer-Assisted Passenger Prescreening System II |
| CART | Computer Analysis and Response Team |
| CBP | Customs and Border Protection |
| CCTV | closed circuit television |
| CDC | Centers for Disease Control and Prevention |
| CDR | call data record |
| CDT | Center for Democracy and Technology |
| CIA | Central Intelligence Agency |
| CIPSEA | Confidential Information Protection and Statistical Efficiency Act |
| CMIR | International Transportation of Currency or Monetary Instruments Report |
| CMS | Centers for Medicare and Medicaid Services |
| COBIT | Control Objectives for Information and Related Technologies |
| CONUS | continental United States |
| COTS | commercial off-the-shelf |
| CPNI | Customer Proprietary Network Information |
| CRS | Congressional Research Service |
| CTR | Currency Transaction Report |
| CTRC | Currency Transaction Report by Casinos |
| CVS | Crew Vetting System |
| | |
| DARPA | Defense Advanced Research Projects Agency |
| DARTTS | Data Analysis and Research for Trade Transparency System |
| DEA | Drug Enforcement Administration |
| DEP | Designation of Exempt Person |
| DHS | Department of Homeland Security |
| DME | durable medical equipment |
| DMV | Department of Motor Vehicles |
| DNA | deoxyribonucleic acid |
| DOD | Department of Defense |
| DOD-IG | Department of Defense Inspector General |
| DOJ | Department of Justice |
| DOT | Department of Transportation |
| DT&E | developmental test and evaluation |
| DTL | Drug Theft Loss |

| DTO | Disruptive Technology Office |
|-----|------------------------------|
| ECPA | Electronic Communications Privacy Act |
| ED | emergency department |
| EDW | Enterprise Data Warehouse |
| EEG | electroencephlalograph |
| EFF | Electronic Frontier Foundation |
| EMG | electromyelogram |
| ETA | Euskadi Ta Askatasuna (terrorist organization) |
| FAA | Federal Aviation Administration |
| FACTS | Factual Analysis Criminal Threat Solution |
| FBAR | Foreign Bank and Financial Accounts Report |
| FBI | Federal Bureau of Investigation |
| FCC | Federal Communications Commission |
| FDA | Food and Drug Administration |
| FDLE | Florida Department of Law Enforcement |
| FDNS-DS | Fraud Detection and National Security Data System |
| FinCEN | Financial Crimes Enforcement Network |
| FISA | Foreign Intelligence Surveillance Act |
| fMRI | functional magnetic resonance imaging |
| FOIA | Freedom of Information Act |
| FTC | Federal Trade Commission |
| FTTTF | Foreign Terrorist Tracking Task Force |
| GAO | Government Accountability Office |
| HHS | Department of Health and Human Services |
| HIPPA | Health Insurance Portability and Accountability Act |
| HTF | High Terrorist Factor |
| I2F | Intelligence and Information Fusion |
| IAO | Information Awareness Office |
| IC3 | Internet Crime Complaint Center |
| ICE | Immigration and Customs Enforcement |
| ICEPIC | ICE Pattern Analysis and Information Collection System |
| ICHAST | Interagency Center for Applied Homeland Security |
| ICU | Intensive Care Unit |
| ID | identification |
| IDW | Investigative Data Warehouse |
| IIR | Institute for Intergovernmental Research |
| IP | Internet Protocol |

| | |
|---|---|
| IRA | Irish Republican Army |
| IRS | Internal Revenue Service |
| IRSS | Institute for Research in Social Science |
| ISO | International Organization for Standards |
| ISP | Internet service provider |
| IT | information technology |
| ITIL | IT Infrastructure Library |
| IV&V | independent verification and validation |
| | |
| KDD | knowledge, discovery in databases |
| | |
| LEA | law enforcement agencies |
| LI | lawful intercepts |
| LLNL | Lawrence Livermore National Laboratory |
| LTTE | Liberation Tigers of Tamil Eelam (terrorist organization) |
| | |
| MATRIX | Multi-State Anti-Terrorism Information Exchange |
| MEG | magneto-encephalography |
| MERGE-PURGE | (KDD application) |
| MSB | Money Service Business |
| | |
| NASA | National Aeronautics and Space Administration |
| NCTC | National Counterterrorism Center |
| NETLEADS | Law Enforcement Analytic Data System |
| NIISO | National Immigration Information Sharing Office |
| NIMD | Novel Intelligence from Massive Data |
| NIPS | Numerical Integrated Processing System |
| NIST | National Institute of Standards and Technology |
| NORA | Non-Obvious Relationships Awareness |
| NRC | National Research Council |
| NSA | National Security Administration |
| NW3C | National White Collar Crime Center |
| NYC | New York City |
| NYCDOH | New York City Department of Health |
| | |
| OCDETF | Organized Crime and Drug Enforcement Task Force |
| OECD | Organization for Economic Cooperation and Development |
| OIP | Online Investigative Project |
| OMB | Office of Management and Budget |
| OTA | Office of Technology Assessment |
| OTC | over-the-counter |

| OT&E | operational test and evaluation |
|---|---|
| PARC | Palo Alto Research Center |
| PDA | personal digital assistant |
| PET | Positron Emission Tomography |
| PII | personally identifiable information |
| PKK | Kurdistan Workers Party (terrorist organization) |
| PNR | Passenger Name Record |
| PUMS | Public Use Microdata Sample |
| QID | Questioned Identification Documents |
| R&D | research and development |
| RAF | Red Army Faction (terrorist organization) |
| RDD | random-digit-dialed |
| RFID | radio-frequency identification |
| RR3 | response rate (category 3)[1] |
| RR4 | response rate (category 4)[2] |
| RTAS | Remote Threat Alerting System |
| S&T | science and technology |
| SAR | Suspicious Activity Report |
| SEC | Securities and Exchange Commission |
| SKYCAT | (KDD application) |
| SQL | Structured Query Language |
| SSN | Social Security number |
| STAR | System-To-Assess Risk |
| SWIFT | Society for Worldwide Interbank Financial Telecommunication |
| TAPAC | DOD Technology and Privacy Advisory Committee |
| TASA | Telecommunications Alarm-Sequence Analyzer (KDD application) |
| TB | terabyte |
| TECS | Treasury Enforcement Communications System |
| TIA | Total/Terrorist Information Awareness program |
| TISS | Tactical Information Sharing System |
| TSA | Transportation Security Administration |
| TVIS | Threat Vulnerability Integration System |

---

[1]See more information at http://www.pol.niu.edu/response.html.
[2]See more information at http://www.pol.niu.edu/response.html.

| | |
|---|---|
| URL | uniform resource locator |
| USA PATRIOT | Uniting and Strengthening America by Providing Appropriate Tools Required to Intercept and Obstruct Terrorism |
| USSOCOM | United States Special Operations Command |
| VoIP | Voice over Internet Protocol |
| WMD | weapons of mass destruction |
| WME | weapons of mass effect |

# B

# Terrorism and Terrorists

## B.1 THE NATURE OF TERRORISM

Terrorism is the deliberate targeting of noncombatants for a political purpose. It is the means used, and not the ends pursued, that determine whether or not a group is a terrorist group. Terrorism is a weapon of the weak. Because terrorist groups are both outmanned and outgunned by their opponents, they use violence against civilians, not in the expectation of defeating their adversary but rather to communicate a political message.[1] The choice of symbolic and particularly vulnerable targets enhances the psychological impact of their actions and thereby compensates for their relative weakness. Put differently, terrorism is often the strategy of choice for parties without the capability to achieve their

---

NOTE: This appendix provides some essentials about terrorism, but the reader is urged to consult more authoritative references, including D. Benjamin and S. Simon, *The Age of Sacred Terror, Radical Islam's War Against America*, Random House, New York, 2003; M. Crenshaw, ed., *Terrorism in Context*, Pennsylvania State University Press, University Park, Pa., 1995; R. Gunaratna, *Inside Al Qaeda's Global Network of Terror*, Columbia University Press, New York, 2002; B. Hoffman, *Inside Terrorism*, 2nd edition, Columbia University Press, New York, 2006; W. Reich, ed., *Origins of Terrorism: Psychologies, Ideologies, Theologies, States of Mind*, Woodrow Wilson Center Press, Washington D.C., 1998; L. Richardson, *What Terrorists Want, Understanding the Enemy, Containing the Threat*, Random House, New York, 2006; Marc Sageman, *Understanding Terror Networks*, University of Pennsylvania Press, Philadelphia, Pa., 2004.

[1]Taking this point seriously means that not all acts that create terror in the population are terrorist acts. For example, although the Washington, D.C., sniper incident was widely reported as a terrorist incident, the D.C. snipers did not act with any known political motivation or purpose in mind, and so the D.C. sniper incident does not meet the definition of a terrorist act.

objectives otherwise. It is an "asymmetric" response in the face of greater power of more conventional forms. Box B.1 draws the contrast between conventional war and a war against terrorists.

Unlike the case for perpetrators of other forms of political violence, for terrorists the victims of their violence and the audience they seek to influence are not the same. Victims are chosen either at random or as

---

### BOX B.1
### The Contrast Between Conventional War
### and Counterterrorist Efforts

The struggle against terrorism differs from historical norms for providing for our security. In the past, we have raised armies to defend against state-organized military forces and to enforce our security and interests outside our borders. These "enemy" forces were most often easily identifiable as enemies and we created a set of rules for monitoring their activities, for defending against them, and for attacking them. These rules regularly call for the violation of the laws of other countries.

We have also provided for our security against those who break our laws through the application of law enforcement techniques by federal, state, and local governments. Those who are found guilty of breaking laws are regarded as "criminals." Until they are found guilty, they are afforded all the rights of the innocent and can be found guilty only by a rigorous process of evidence and judicial process.

The rules for armies dealing with "enemies" in battle conditions have evolved to be quite different from those that apply to law-enforcement agencies dealing with "potential criminals" within the domestic borders of the United States. Armies permit their elements and members to destroy "enemies" upon "recognition." They do so quickly and without "due process" by any separate jurisdictional structure.

One of the most demanding new attributes of our current struggle with terrorists is that some of the "enemy" is imbedded in our day-to-day midst. The "enemy" has advertised its intent to destroy our society as a necessary part of defending its own and has demonstrated the potential to be a serious threat to our way of life. The "criminal" is seeking to satisfy selfish interest at the expense of others but is not attempting to destroy society. Although the difference between these motives is profound, our processes for dealing with "enemies within" and "criminals" is not much different.

A criminal act that produces localized terror is different from the serious national threat posed by a terrorist group from a foreign organization. We cannot realistically prevent all manner of tragedies that are the result of criminal behavior, stupidity, or random acts—U.S. highway fatalities, for example, total something like 30,000 per year. However, those conditions create a quality of anxiety very different from the citizen's sense of insecurity that results from his/her government being unable to provide safety in the face of the threat of foreign organizations that wish to do them and their country harm. One involves the routine and understood risks of daily life that can be addressed by each individual in his/her own way. The other is a frighteningly ever-present risk of unknown harm from an uncertain deliverer.

representatives or symbols of a larger group in order to influence the behavior of a third party, usually a government. What sets terrorism apart from other forms of political violence, even the most proximate forms like guerrilla warfare, is the deliberate targeting of civilians, not as an unintended consequence of warfare, but as deliberate strategy.

## B.2 SOME TACTICS OF TERRORISM

The operational code of the current generation of transnational terrorists is to use the strengths of Western democracies against us. For example, they exploit a free press to amplify their actions and spread the fear their operations inspire. And they exploit the openness of society to operate covertly. Although basic training and recruitment may occur in the open (much as Al Qaeda operated in Afghanistan), operational planning and training are often undertaken in an undercover manner. That is, the individuals, the organizations, and their leadership attempt to keep their identities, communications, plans, and locations from being known to the targeted nation even when the terrorists are within the borders of the nation being targeted.

In addition, terrorists are prepared often to give up their lives, take the lives of innocent bystanders, and disavow other conventional forms of value in pursuit of their goals. This ferocity of commitment makes deterrence more complicated than it might be for nation-to-nation confrontation. In stark terms, what might serve to deter a suicide bomber? Whatever the answer is to this question, death is not it.

Lastly, for most practical purposes, terrorists do not appear to place many limits on the violence that they are willing to perpetrate,[2] and so the specter of terrorists with weapons of mass destruction looms large in counterterrorist efforts. Likewise, the highly interdependent nature of modern society leaves the United States (and other developed nations)

---

[2]A memo found on an Al Qaeda computer in late 2001 appeared to indicate that even Al Qaeda recognized *some* limits on the extent of violence they were willing to perpetrate. The memo said that:

> Because of Saddam and the Baath Party, America punished a whole population. Thus its bombs and its embargo killed millions of Iraqi Muslims. And because of Osama bin Laden, America surrounded Afghans and bombed them, causing the death of tens of thousands of Muslims . . . God said to assault whoever assaults you, in a like manner . . . In killing Americans who are ordinarily off limits, Muslims should not exceed four million non-combatants, or render more than ten million of them homeless. We should avoid this, to make sure the penalty [that we are inflicting] is no more than reciprocal. God knows what is best.

Cited in A. Cullison, "Inside Al Qaeda's Hard Drive," *The Atlantic Monthly*, pp. 55-70, September 2004.

more vulnerable than many other societies. Our access to power, communications, information, transportation and ultimately food and water are very vulnerable to attack. This asymmetric mismatch between modern dependence on attackable infrastructure and the relatively lower dependence of terrorist adversaries on such infrastructure lessens the ability to deter through conventional forms of retaliation.

## B.3  A HISTORICAL PERSPECTIVE ON TERRORISM

Terrorism is not a new phenomenon. Documented groups like the Zealots and the Sicarii date as far back as the first century in the Common Era. Like their contemporary successors they had a mix of religious and political motives and sought to ignite a general revolt among the masses against the established authorities. The Medieval Assassins, who operated from the 11th to the 13th century, provide an early example of state-sponsored terrorism as well as an early example of a culture of martyrdom among terrorists. Generally speaking, terrorist groups prior to the French Revolution tended to mix religious and political motives, whereas in the 19th and 20th centuries terrorists groups, reflecting the broader secularization of society, tended to focus on political objectives. This changed in the 1970s with the impact of the Iranian revolution and the popularization of the ideas of fundamentalist Islamic writers like Maulana Mawdudi and Sayyid Qutb that again fused religion and politics and inspired the founders of contemporary Islamic terrorist groups.

Although the primary terrorist threat to the United States emanates from Islamic extremists, terrorists have belonged to most religious traditions and to none. There have been Christian terrorists, like the ETA in Spain and the IRA in Ireland. There have been Jewish terrorists, like the Zealots, the Sicarii, and the Stern Gang in Palestine. There have been Hindu terrorists, like the Thugi in India. There have been atheist terrorists, like the RAF in Germany, Action Directe in France, and the Red Brigades in Italy, and there have been secular terrorists like the Shining Path in Peru, the PKK in Turkey, and the LTTE in Sri Lanka.

## B.4  EXPLAINING TERRORISM

Terrorism is a tactic employed by many different groups in many parts of the world in pursuit of many different objectives. There are no simple or uniform explanations of its causes. The fact that terrorism has been so widespread, used by Peruvian peasants, German professors, Saudi imams, Egyptian intellectuals, Tamil teenagers, and young cricket players from Britain, suggests that no single cause can explain the actions of such a diverse group. Yet, the actual practitioners of terrorist tactics are very few. Meta-explanations like poverty, inequality, or alienation thus cannot adequately explain the behavior of small groups, when these conditions are so widespread. Factors like inequality and alienation are better understood as risk factors that increase the likelihood of terrorism and increase the likelihood that once a terrorist group forms it will gain adherents, rather than as causes of terrorism per se. The essential requirements for terrorism are a disaffected individual, a complicit community, and a legitimizing ideology.

It is helpful to think in terms of terrorists as having both primary and secondary motives, or underlying and immediate motives. The primary motives differ with the type of group: ethno-nationalist groups, for example, want autonomy or secession; social revolutionary groups want to overthrow capitalism; religious groups want to bring about the apocalypse or to replace secular law with religious law. Terrorist groups have been singularly unsuccessful in achieving these underlying or primary objectives. However, terrorists have been quite successful in achieving their secondary or more immediate objectives: revenge, renown, and reaction.

The single most powerful motive of the terrorist is the desire for revenge. This holds true no matter what the precise political objective is or where in the world the terrorist is operating. Sometimes this is revenge for a perceived wrong inflicted on the individual or his family; more often it is a wrong inflicted on a group with which the terrorist identifies. Second, terrorists seek renown. This implies publicity, but much more than that, it implies glory in an effort to redress the perceived humiliation a person, or the group with which he identifies, has suffered. Finally, terrorists seek to provoke their adversaries into a reaction, preferable an overreaction. Terrorists do not have territory or even armies; all that they have is their action, and it is how they communicate with the world. By reacting, their adversary demonstrates their importance. By provoking the war on terror, terrorists have succeeded in exacting revenge, in attaining renown, and in eliciting a reaction.

## B.5  AL QAEDA AND THE TERRORIST THREAT TO THE UNITED STATES

Ever since the terrorist attacks of September 11, 2001, the threat of further attacks has become the primary national security concern of the United States and many of its allies. The scale, ferocity, and nature of the attack were unprecedented in the lengthy annals of terrorism. The fact that the attack took place on American soil, targeted American civilians, and inflicted casualties greater than the attack on Pearl Harbor in 1941 has led to a serious re-evaluation of U.S. national security strategy.

For the first time in U.S. history nonstate actors have both demonstrated a capacity to inflict serious harm on the United States and have articulated a desire to do so. Previously U.S. foreign and defense policy has been based on the assumption that our adversaries were other states or alliances of other states, but now we face a threat from transnational substate actors. The evolving nature of this threat requires that the United States develop new strategies in response.

Operation Enduring Freedom was launched in fall 2001 in response to the attacks of 9/11. This campaign succeeded in toppling the Taliban regime in Afghanistan, which had harbored Al Qaeda, the group responsible for the attacks. The military campaign also succeeded in destroying the central command-and-control structure of the group. The group and the ideology to which it adheres, however, survive.

Today, the most salient and serious terrorist threat to the United States is Al Qaeda, even though throughout history there have been many other types of terrorist movements in many parts of the world.[3] Its senior leadership has explicitly argued that the continued success of the Western way of life, as exemplified by the United States, will inevitably lead to the erosion of traditional Islamic values and their way of life. In the face of this proposition, Al Qaeda has resorted to terrorism as a means to achieve its ends because that choice does not confront U.S. economic and military strength and it leverages the safeguards of U.S. domestic freedoms to their advantage. Other movements or organizations may emerge in the future to challenge the United States in this way, but for the moment, Al Qaeda is the primary terrorist adversary of the United States.

The basis of Al Qaeda's strength is twofold. They have a motivating ideology that a great many people find appealing, and they have demonstrated extraordinary organizational agility. Al Qaeda's ideology is an eclectic and inconsistently articulated mix of Islamic fundamentalism; animus toward the West and secular Muslim regimes; objections to specific U.S. policies in the Middle East, in particular U.S. support for Israel; and grandiose aspirations for a return to a mythical caliphate stretching from Spain to Indonesia. The breadth of these criticisms of the West means that disaffected Muslims all over the world can identify with some part of the ideology while the religious basis provides a legitimacy and coherence to the appeal. Unlike earlier terrorist groups that tended to start with local grievances and then build from there, part of the success of Al Qaeda has been the ease with which the ideology has infused local conditions, thereby gaining adherents for the transnational cause. The religious nature of the ideology has facilitated the elevation of the conflict with the West into cosmic terms, thereby eliminating previous constraints on the

---

[3]Even the United States has been the target of non-Al Qaeda terrorism, as occurred on April 19, 1995, when the Alfred P. Murrah Federal Building was bombed in Oklahoma City, Okla.

behavior of individuals and legitimizing and rationalizing the infliction of mass casualties.

The organizational agility has been demonstrated by the ease with which Al Qaeda has adapted to the destruction of its central command structure and training bases and re-emerged in an entirely new organizational form: a diffuse network of like-minded individuals bent on destruction of the West. The new incarnation of Al Qaeda is made possible by the existence of new technologies, the very attributes of the globalization they are so quick to decry but are so adept at exploiting. The current organization of Al Qaeda, therefore, is unprecedented as it is entirely dependent on a set of new technologies that were unavailable to its predecessors.

It is impossible to know how many terrorists there are who wish to harm the United States. Ten of thousands of Mujahadeen fought against the Soviet Union in Afghanistan. After the war many returned to their country of origin, radicalized local militant groups, and fought to overthrow local secular regimes. Others turned their attention to the fight against the West, which was blamed for propping up corrupt regional leaders.

After the first Gulf War the deployment of U.S. troops in Saudi Arabia, designed to serve as a trip wire in the event of another Iraqi invasion, served as a rallying cry for extremists who perceived the deployment as humiliating and were convinced that the United States was determined to take over the Muslim world. Al Qaeda re-established training camps in Afghanistan under the Taliban and recruited young Muslims from the Middle East and the Muslim diaspora in Europe to come and be trained in the militant arts of jihad.

The war in Iraq has served to swell the ranks of those who wish to attack the West. Young men from North Africa and the Gulf states are flocking to Iraq to take part in the war against the United States while the unpopularity of the war is radicalizing young Muslims resident in Europe to attack their compatriots, as occurred in Madrid on March 11, 2004, and in London on July 7, 2005. The numbers, therefore, appear to be growing.

The numbers, however, tell only part of the story. One of the more disturbing trends is the fact that weapons of greater and greater lethality can now fall into the hands of smaller and smaller groups. The problem for the security services is that the smaller the group, the more difficult it will be to detect.

Up until now, terrorist groups, with the singular exception of the Aum Shinriku cult in Japan that released sarin gas on the Tokyo subway in March 1995, have evinced little interest in using weapons of mass destruction, specifically nuclear, radiological, chemical, or biological weapons. But earlier terrorist groups have not attempted to inflict mass casualties. The perpetrators of 9/11, however, clearly wished to kill as many people as possible. The actions of the Al Qaeda leadership and their statements both suggest that if they could acquire these weapons, they would deploy them. The difficulties of acquiring, transporting, and successfully deploying these types of weapons are such that this is a very-low-probability event. But the consequences of a successful deployment of even the easiest of them, a radiological device, would be so catastrophic that there is no responsible option but to defend against them.

We cannot accurately predict how terrorists will next choose to attack us. Historically terrorists have been very conservative in their use of tactics, preferring simple tried technologies, given the conditions of uncertainty in which they operate. Hence, bombs and bullets have been tools of choice. However, the psychological payback of a successful deployment of even a crude chemical or radiological device is such that some terrorists are likely to try to acquire these weapons.

Still, the probability is higher that terrorists will attempt an attack with conventional explosives that can be acquired very easily but when strategically deployed can inflict significant casualties and even great psychological damage. The diffuse nature of the threat, and the fact that many militants operate largely independently of any central command, suggest the need to be prepared for all types of attack as well as the fact that different types of attack could be planned simultaneously.

Particular branches of fundamentalist Islam have proven to be very successful in attracting a range of different individuals to the jihadi cause and in so doing making it impossible to single out those to track. These range from poor uneducated young men from the Middle East and North Africa, to middle class, computer-literate, and Western-educated young men from the region, as well as first- and second-generation Muslims living in the West. They have also won converts to the cause via the Web, prisons, and personal networks. There is no simple profile of the terrorist; rather, the background of those participating in violence is constantly expanding and increasingly including formerly excluded categories of individuals. The anonymity of the web, for example, permits the participation of women.

## B.6 TERRORISTS AND THEIR SUPPORTING TECHNOLOGIES

The most important of these new technologies is information technology, and in particular the Internet—Al Qaeda could not function without it. Today, Islamic fundamentalist terrorist groups rely on the Internet to communicate with their members, their supporters, and one another across the globe. They use the Internet to recruit members by hosting Web sites detailing the iniquities of their adversaries, the successes of their attacks against the West, and the path to action. In this way they have successfully won adherents to the cause in counties as diverse as Algeria, the United Kingdom, Saudi Arabia, the United States, Pakistan, and Spain. Once recruited, they use the Internet to train their followers by providing online education manuals as well as directions to training camps. They produce propaganda videos both to sustain the converted and to intimidate Western publics as well as to win recruits and to raise funds. They use the Internet to create a virtual community of support for the militant wherever he or she may be and to sustain their commitment to the cause. They also use the Internet to plan and carry out their attacks, as was effectively demonstrated on September 11th.

Just as important, terrorists—by their very modes of operation—intermingle with the society they target. Not only do they use the indigenous information technology infrastructure and the Internet to interact with each other, they must also interact with society at large. Thus, they use cell phones, pay with credit cards, travel commercially, rent vehicles and apartments, and otherwise engage in conventional commercial activities—all of which are activities that leave a digital footprints that may subsequently be tracked.

Lastly, terrorists have more or less lost the territorial bases they used to house their own institutional infrastructure. Given the anonymity, affordability, and ease of access to the Internet, they have created a command-and-control structure in cyberspace. Given the determination of Western governments to deny terrorist groups safe physical havens within which to operate and to train with impunity, it seems certain that their reliance on new information technologies will only increase.

## B.7  LOOKING TO THE FUTURE

We can reasonably expect the threat from terrorists groups to continue for the foreseeable future. There are no signs of the abatement of the threat. Terrorism, like other tactics, will continue to be deployed as long as it proves effective. Terrorists have been unsuccessful in achieving the fundamental political change they seek, but they have been particularly successful in achieving their more immediate goals: exacting revenge for real or perceived grievances, achieving renown for themselves and their cause, and provoking a reaction from the authorities. As long as terrorists continue to be successful in achieving their objectives of revenge, renown, and reaction, they are likely to continue to use terrorist tactics.

# C

# Information and Information Technology

## C.1 THE INFORMATION LIFE CYCLE

As Chapter 1 points out, digital information in use typically goes through a seven-step life cycle. These steps include collection, correction and cleaning, storage, use or analysis, publication or sharing, monitoring and evaluation, and retention or deletion.

### C.1.1 Information Collection

The information collected for a program must be appropriate to its purpose. Data minimization requires that only information critical to that purpose be collected, though minimization often conflicts with the temptation to gather more information "just in case" it might be useful later in easing the relevant analytical tasks or even for other possibly relevant purposes. Legislation, regulation, or other governance rules may require that internal or external authorization to collect the information be obtained, including from relevant third parties. The information source(s) and the information itself must be verified as reliable, objective, and compliant with relevant laws.

The government collects information for counterterrorism from many other sources, primarily as extracts from information systems. The government mandates or requests information from many industries: Customs and Border Protection obtains manifests for trucks entering the United States from trucking firms; the Department of Homeland Security (DHS), including the Transportation Security Administration, and the National Aeronautics and Space Administration obtain passenger names and records from airlines; the Justice Department obtains Web search terms, URLs, and other records from the information technology (IT) and telecommunications industries; the National Security Agency obtains phone call records from communications providers; and the Treasury Department obtains suspicious activity reports from the financial community.

In addition, employers, retailers, banks, and travel and telecommunications companies collect data directly from customers as well as from many other government and private sources. The largest databases in the world are click-streams collected from Web interactions, second only to retail and scientific databases. For example, it is conventional practice for companies to collect extensive information on prospective employees from financial and educational institutions, law enforcement, former employers, and so forth. Information collection is a significant and growing sector of the information economy.

Finally, the government obtains a great deal of data from private data brokers, who aggregate data on individuals from all legally available sources. Because the data are collected by private parties, much of the data are not subject to existing restrictions on government collection efforts.

### C.1.2 Information Correction and Cleaning

A significant practical and research challenge is to ensure that the information is correct, accurate, and reliable. This is aided by ensuring reliable information provenance and the use of automated and human data validation techniques. For example, automated techniques could be used easily to recognize as anomalous an indicator of pregnancy in the medical records of a male.

Moreover, in certain instances, laws govern the rights of an individual to correct information errors in commercial applications, for example in one's credit report. If the individual finds what he or she believes to be an error, documentation of that error can be provided and the error corrected. If the party providing the data does not agree that it made an error, the individual has the right to insert into the record a statement of limited length providing his side of the story.

To the best of the committee's knowledge, individuals negatively affected by counterterrorism programs as the result of data errors have no comparable ability. Indeed, for national security reasons, individuals are not permitted to review the data on which adverse decisions are based, even though they may experience the negative consequences (e.g., by being denied boarding a plane).

### C.1.3  Information Storage

To be used subsequent to collection, information must be stored in some information repository, often an electronic database. The storage mechanism must maintain the data quality, reliability, and accuracy while ensuring operational characteristics such as robustness to failure and scalability to accommodate both data and processing volumes. In addition, since information systems have vulnerabilities and are subject to threats, appropriate data stewardship must be enforced.

Whereas banks and telecommunications companies rate highest in information protection, many industries and the government in particular rate considerably lower. Increasingly, laws or regulations govern the storage and management of information both at rest (i.e., on a storage device) and in motion (i.e., as it traverses communications networks), thus mandating improvements in data stewardship. For example, regulations requiring the encryption of information on a detachable storage medium or transmitted through a communications channel can be used to protect information in transit and at rest.

### C.1.4  Information Analysis and Use

The step of information analysis and use involves the use of the program during its operational lifetime to deliver the services defined in the purpose and the rational basis and tested in the experimental basis. As with information storage, information processing must meet operational requirements such as robustness and scalability. As stated in the committee's proposed framework (see Chapter 2) and others, a program must be used solely as defined in the approved purpose and rational basis (i.e., requirements).

Additional uses must be reviewed and approved as an extension to the approved purpose. For example, if a law enforcement program were applied to counterterrorism, that new use should be reviewed under the relevant laws and regulations. Unfortunately, unless protected by a privacy policy, commercial information systems are often used for purposes unanticipated by customers, e.g., customers receiving marketing and promotional material unrelated to the ticket that they purchased from an airline. In approving additional uses of information, one need not specify the precise method of analysis, since that is often difficult to anticipate—only the general purpose to which the information will be directed needs to be specified.

### C.1.5  Information Sharing

A major counterterrorism theme that has emerged since September 11 (9/11) is the notion of information sharing—that U.S. counterterrorist efforts will be more effective when the relevant agencies can easily and effectively cooperate and share information.[1] The National Counter-terrorism Center (NCTC) was established to serve as a multiagency center analyzing and integrating all intelligence pertaining to terrorism, including threats to U.S. interests at home and abroad. NCTC also is responsible for developing, implementing, and assessing the effectiveness of strategic operational planning efforts to achieve counterterrorism objectives.

Compared to the relevant policy and practices, the technology for sharing information is relatively well developed. Today, modern information systems live in an ecosystem of other information systems and services, accessible enterprise-wide over an intranet or worldwide over the Internet, and it is increasingly common for both raw information and analytical results to be published electronically.

A modern information system obtains information and services from many other information systems, in some cases thousands of information systems, and reciprocally provides information and services. Such ecosystems developed originally to increase automation by eliminating paper or electronic reports that were exchanged with humans or other systems by largely human means. Currently such ecosystems permit organizations to modify and enhance their businesses with great speed and agility. Customers have the convenience of reserving a trip with a travel agent and having all of the relevant hotels, car rental agencies, airlines, credit card companies, and banks handled transparently. While information systems' interoperation and information sharing are a convenience for a customer, they are a business-critical requirement in almost every business.

Clear civil liberties concerns arise when information is shared and repurposed without restriction. Hence, the committee's framework lists the criteria and best practices that are required to protect civil liberties, including appropriateness, agency and external authorization, defined purpose, and assessment, as discussed below.

### C.1.6 Information Monitoring

An information program must be continuously monitored and assessed to ensure that it is effective in achieving its purpose and that

---

[1]See for example, National Security Council, *National Strategy for Combating Terrorism*, National Security Council, Washington, D.C., September 2006, available at http://www.whitehouse.gov/nsc/nsct/2006/; National Commission on Terrorist Attacks upon the United States, *9/11 Commission Report*, U.S. Government Printing Office, Washington, D.C., July 2004; and three reports of the Markle Foundation Task Force on National Security in the Information Age, Markle Foundation, New York, N.Y., available at http://www.markletaskforce.org/: *Protecting America's Freedom in the Information Age* (2002), *Creating a Trusted Network for Homeland Security* (2003), and *Mobilizing Information to Prevent Terrorism: Accelerating Development of a Trusted Information Sharing Environment* (2006).

it complies with all relevant laws, regulations, and governance. The committee's framework lists several relevant criteria for which there are best practices, including audit trails, auditing for compliance with existing laws, ensuring reporting and redress of false positives and related impacts on individuals, and having in place a privacy officer, training, agency authorization, and external authorization.

One of the most challenging aspects of information-intensive systems is evaluating their efficacy or their effectiveness relative to their purpose. The growth in data, transactions, and analytical volumes is a direct measure of the value and the efficacy of data and information processing. The continued growing investment in these programs is a direct measure of their effectiveness in promoting economic competitiveness in the marketplace.[2] More specifically, each industry and application domain, such as telecommunications billing, has well-defined measures of efficacy or business effectiveness. For example, two of the many telecommunications billing metrics include time and cost to produce. An extreme example involves Wall Street arbitrageurs who search the entire history of stock market trades and simultaneous trades as they occur in all U.S. trading floors and find, on a regular basis, investment opportunities in 100ths of seconds. Typically there are best practices and defined standards for assessing effectiveness, as called for in the committee's framework. Following information system best practices, counterterrorism programs should have efficacy metrics defined for them against which they can be assessed.

### C.1.7  Information Retention

The final step of the information life cycle involves the retention or deletion of information based on a defined retention period, data quality, data minimization, or other criteria.[3] Data retention refers to the period of time during which an organization can or must retain data in its automated and manual records. A data retention requirement may be that data

---

[2]In 2005, the information technology products sector accounted for $640 billion or 2.8 percent of the U.S. Gross Domestic Output, while the communications sector accounted for $514 billion or 2.25 percent. The IT sector has experienced a 2.7 percent compound annual growth rate (CAGR) since 1998, and the communications sector a 6.5 percent CAGR (U.S. Department of Commerce, Bureau of Economic Analysis, "Gross Domestic Product: Fourth Quarter 2006 (Advance)," available at http://www.bea.gov/newsreleases/national/gdp/2007/gdp406a.htm; Andrew Bartels, *U.S. IT Spending Summary: Q3 2006*, Forrester Research, Inc., Cambridge, Mass., November 29, 2006).

[3]Data Privacy and Integrity Advisory Committee, *Framework for Privacy Analysis of Programs, Technologies, and Applications*, Report No. 2006-01, U.S. Department of Homeland Security, Washington, D.C., adopted March 7, 2006.

can be kept no longer than the defined period or that it must be kept at least until the defined period is over. When a data item is to be deleted, all copies of the item must be found and deleted from all automated and manual records. In the context of this report, data retention is a privacy and civil liberties issue when applied to personally identifiable information (PII) such as name plus Social Security number.

The increased digitization of individuals' personal and professional lives has led to dramatic increases in the amount of PII that is stored in automated and manual records. While this information provides significant value and convenience, it also exposes people to risks such as identity theft, one of the most frequent crimes in the United States, and to other digital crimes and loss of privacy. One report indicates that over 168 million data records have been compromised due to security breaches in the United States from January 2005 to October 2007.[4] To protect the public from such crimes, state and federal governments have passed many laws and regulations[5] and are continuing to draft new laws and regulations in response to the increased risks related to the growth of retained PII and the power of current technologies. These laws and regulations define data retention periods for specific types of data.

Information retention poses complex and unresolved business, legal, and technical issues. In the normal course of business, data must be retained relative to the relevant business cycle, e.g., to monthly, quarterly, or annual billing cycles, and to the much longer, e.g., 10 years, statute of limitations periods during which legal disputes could arise and be prosecuted. At the same time organizations may want to delete data to reduce their exposure to compliance irregularities or potential legal discovery by data forensic techniques, data such as e-mail trials in the Enron case and voice mails in a case involving Hewlett Packard. Businesses must meet the requirements of relevant regulations; Sarbanes-Oxley is one of hundreds that are applicable to specific data types in specific business contexts.

Legal issues include evolving and conflicting laws, regulations, and government requests. Within the United States, there are more than 45 different state data security and privacy laws and several evolving federal laws. Government agencies make conflicting requests. The Department of Justice (DOJ) and DHS requested lengthy retention periods to fight child pornography, e.g., 20 years, and terrorism, e.g., forever, respectively. At

---

[4]Privacy Rights Clearing House, "A Chronology of Data Breaches," posted April 20, 2005, available at http://www.privacyrights.org/ar/ChronDataBreaches.htm#CP.

[5]See, for example, U.S. Congressional Research Service, *Data Security: Protecting the Privacy of Phone Records*, RL33287, Congressional Research Service, Library of Congress, Washington, D.C., updated May 17, 2006.

the same time, the Federal Communications Commission (FCC) and the Federal Trade Commission (FTC) requested shortened retention periods, e.g., 90 days, to protect privacy and other civil liberties.

Technical issues involve keeping up with evolving data retention requirements, mediating between conflicting requirements, and simply implementing data retention policies covering unimaginable volumes of data. Information sharing causes information to be copied and distributed to other systems within an organization or via the Internet across the world. One form of information distribution is to publish it on paper or digital media, as reports, or for technical purposes such as backup and disaster recovery. Implementing a data retention policy requires that all copies be traced or identified so that they can be deleted compliant with the relevant policy. As the requirements change, so must technical solutions for managing the data retention policy as it applies to all copies. Entirely new content and record management technologies are being developed to automate data retention policies. Positive impacts of data retention laws and regulations include data minimizatoin—eliminating all data that are not essential to the relevant business purpose—and raising the previously low priority of data protection and security in all organizations.

### C.1.8  Issues Related to Data Linkage

Additional issues arise when information is assembled or collected from a variety of sources for presentation to an application. Assembling such a collection generally entails linking records based on data fields such as unique identifiers (if present and available) or less perfect identifiers (such as combinations of name, address, and date of birth). In practice, it is often the case that data may be linked with little or no control for accuracy or ability to correct errors in these fields, with the likely outcome that many records will be linked improperly and/or that many other records that should be linked are not linked. Without checks on the accuracy of such linkages, there is no way of understanding how errors resulting from linkage may affect the quality of the subsequent analysis. (For more on issues related to data linkage, see Appendix H.)

### C.1.9  Connecting the Information Life Cycle to the Framework

The framework defined in Chapter 2 of this report provides guidance on information practices to achieve efficacy of counterterrorism programs while ensuring adequate civil liberties protections. All information practices related to information-based programs can be considered in the context of the typical information life cycle. Each step of the life cycle is governed by prevailing laws, regulations, and governance rules intended to protect confidentiality, intellectual property, and, for example, in the intelligence community, classified information.

Efficacy and civil liberties issues arise in each step of the information life cycle. Hence, the effective and appropriate use of information programs involves the use of relevant best practices in each step.[6] The term "best practice" refers to a practice or solution that was known to have worked well according to the requirements. The name "best practice" is misleading, since a best practice is seldom proven to be best nor to work in all circumstances. Even if best practices were effective, they are used in less than 30 percent of applicable cases. These issues and practices also arise in and pose challenges for information-intensive programs in the private sector.

For example, most commercial enterprises publish a privacy policy that defines how they treat customer information in each step of the information life cycle. Privacy policies generally define what information is collected, indicate customer rights to correct the information, state that the information is stored and used by the enterprise (typically at their discretion), describe what information will be shared under specific circumstances, pledge to monitor its appropriate use, and finally, say how long the information will be retained. Hence, the committee's framework calls for a privacy officer to oversee these issues for each counterterrorism program.

The main criterion on which a program is evaluated is its purpose or objective. All other evaluation criteria are based on the program's stated purpose or objective. Due to the investment in resources and the impact programs can have, programs require a sound rational and experimental basis. In information systems terminology, the rational basis is expressed in terms of systems requirements that define precisely what the information system is to do and how it is to operate. The purpose and rational basis must be evaluated relative to the relevant real-world requirements and the prevailing laws and regulations. Once approved, this acts as the approved basis for the program. It is the nature of programs that their requirements evolve constantly. When they do, they must be evaluated and approved, as were the original requirements. The experimental basis is proven, objectively, during various testing and user acceptances tests in which the information system is tested in all possible environments against the outcomes defined in the systems requirements. The purpose,

---

[6]D. Aron and A. Rowsell-Jones, *Success with Standards*, Gartner EXP, Stamford, Conn., May 2006; IT Governance Institute (ITGI), *IT Governance Global Status Report—2006*, ITGI, Rolling Meadows, Ill., 2006.

along with the rational and experimental bases, must cover all steps of the information life cycle and be fully documented.

## C.2 THE UNDERLYING COMMUNICATIONS AND INFORMATION TECHNOLOGY

### C.2.1 Communications Technology

Twenty-first century communications technology is in a continuing phase of rapid growth, evolution, and transformation. Today, there are more than 5,600 telecommunications providers in the United States. Whereas in the past providers were distinguished by the technology of the communications medium involved, more recently deregulation and advances in technology have led to a convergence of technologies and companies, and today any company can become a telecommunications provider, thus expanding both the number of service providers and the types of communications services. For example, the Shell Oil Company is treated for certain purposes as a communications service provider because it provides its customers Internet-based services with which to check or modify heating or other electrical appliances in their home.

The scale of communications network usage is almost beyond imagination and growing rapidly. In the United States, the average annual growth rate in wireless calls, VoIP calls, and e-mail has been around 50 percent. In addition to these conventional forms of communication there is a wide range of new services such as instant messaging, small messaging service, video messaging, and a plethora of new business services communicated over the Internet. These communications are also enormous in data volume. A 2003 rough estimate[7] of annual data volumes claimed over 9 exabytes of wireline calls and over 2 exabytes of wireless calls, with over 1.5 petabytes of Internet traffic. A rough approximation of an exabyte is 100,000 times the data volume that corresponds to the more than 19 million books in the Library of Congress.

The data associated with telecommunications fall into three categories:

- *The actual communication or content of the communication.* In general but depending on the nature of the service, communications providers are generally precluded from examining content except for technical reasons such as improving quality of service.

---

[7]P. Lyman and H.R. Varian, *How Much Information, 2003*, retrieved from http://www.sims.berkeley.edu/how-much-info-2003 on May 13, 2008.

- *The information required to manage and process the call*, e.g., the source number, the destination number, the start time, and the end time, called call data records (CDR). (Such information is generally known as customer proprietary network information (CPNI).) Communications providers retain the management data for billing and other technical and business purposes, such as detection and prevention of telecommunications fraud, and thus maintain vast data repositories of CDRs (in the petabyte range). For example, in 2001 AT&T reported generating more than 300 million CDRs per day for 100 million long-distance accounts.

- *Subscriber information*, such as address, credit and billing information, and descriptions of services provided. As services become more sophisticated, the need for additional subscriber information grows to further define services and increase ease of use. For example, customer profiles kept by service providers on the Internet often include detailed preferences so that the automated service can meet customer needs without having to request that information on each use.

Telecommunications companies collect data in all three categories. Access to CPNI is strictly governed by federal and other legislation and by telecommunications regulations with severe penalties for each violation. Due to the significant growth in the types of communications services and a continuing large growth in communications volumes, as well as significant advances in technology, the nature, management, and governance of CPNI must be constantly updated, and laws, regulations, and practices must be revised to reflect new and emerging opportunities and threats, including those related to counterterrorism and civil liberties. One illustration of the need for rebalancing is an ongoing tension between the FCC, the FTC, and civil liberties interests (who have argued for reducing the time that service providers retain CPNI) and DHS and DOJ (which have argued to increase retention time in case it is required for terrorist, legal, or other security purposes).

Access to data in the other categories provides a more highly revealing portrait of personal behavior and is covered by law (although not telecommunications law).

## C.2.2 Information Technology

For most citizens in daily life, the world is increasingly digital. Citizens apply electronically for government services, such as passports and licenses. In an increasingly cashless society, consumers engage in numerous financial transactions that are precisely recorded, often including the location and time. Whether for entertainment, personal, or professional purposes, clicks on the Internet are recorded for future use. Every trip is recorded, from the airline, hotel, and car rental reservations to the actual events of the trip. Increasingly people and organizations publish detailed aspects of themselves, including electronic calendars, photographs, videos, music, and aspects of their personal lives. Increasingly activities in public places, stores, and enterprises are recorded and stored by surveillance systems. Educational institutions, e.g., flight schools, record their members' activities. Employers record and retain extensive information on employees. With the increasing use of technologies such as RFID (radio frequency identification) tags, objects that people own and use provide personal information that can be read at a distance; for example, automobile and appliance parts, articles of clothing, retail products, and electronic devices such as telephones, personal data assistants, and computers can communicate information such as location, status, and temperatures.

Moreover, the very types of personal information that can be collected are proliferating. For most of the 20th century, digital information referred to structured information such as name, address, telephone number, purchase order number, and the like. In the 21st century, digital information has expanded to include anything that can be represented digitally such as graphics, music, and video. There is a dramatic growth in unstructured information, captured, for example, by the 4.2 million closed-circuit television (CCTV) cameras in Britain—about one for every 14 people and other surveillance cameras in the United States, much of it stored for future processing.

The scale of information processing undertaken in the United States is unimaginably large. Fortune 500 companies and large federal agencies are likely to have more than 5,000 information systems each with one or more databases. It would be rare to find any business of any size in the United States that did not have a significant investment in information systems and databases. The largest databases in the world, according to the 2005 bi-ennual Winter Corporation survey,[8] exceeded 23 terabytes (TB) for transactional databases and more than 100 TB with 3 trillion entries for data warehouses, which is equivalent in data volume to 10 times the contents of the Library of Congress. Growth rates over 2 years for these databases were between a factor of 2 for transactional databases and a factor of 3 for the largest data warehouse. Over the past 4 years the average database size rose 243 percent, while the maximum size rose 578 percent. The use of these databases, or workloads, is equally staggering. The largest transactional workload was 1 billion SQL statements (e.g., a database query) per hour, with an average of 35 million and 30 million for the largest data warehouse (query only) workload, at an average of 2 million

---

[8]K. Auerbach, *2005 TopTen Program Summary: Select Findings from the TopTen Program*, Winter Corporation, Waltham, Mass., May 2006.

per hour. (SQL is a computer language for accessing and querying databases.) Winter estimated in 2005 that by 2008 transactional workloads would have grown 174 percent while data warehouse workloads would have quadrupled. While individual databases and their use are growing dramatically, so is the total number of databases.

### C.2.3 Managing Information Technology Systems and Programs

There are many formally defined private-sector[9] and government[10] IT assessment frameworks, i.e., guidelines and best practices, for improving IT governance, transparency, and performance management, as well as improving specific areas, such as security,[11] privacy,[12] and information fairness.[13] These frameworks are intended to quantify difficult-to-evaluate information systems objectives such as information systems effectiveness, quality, availability, agility, reliability, accuracy, completeness, efficiency, compliance with applicable regulations, and confidentiality. Although these criteria are difficult to define and evaluate, they are common requirements that the IT industry must evaluate for all critical systems on a regular basis. While there is never a simple or discrete answer, the IT industry must make its best approximation.

Three of the 30 most widely followed frameworks are Control Objectives for Information and Related Technologies (COBIT), IT Infrastructure Library (ITIL), and International Organization for Standardization (ISO)

---

[9]D. Aron and A. Rowsell-Jones, *Success with Standards*, Gartner EXP, Stamford, Conn., May 2006; The IT Governance Institute (ITGI), *IT Governance Global Status Report—2006*, ITGI, Rolling Meadows, Ill., 2006.

[10]U.S. General Accounting Office (GAO), *Information Technology Investment Management: A Framework for Assessing and Improving Process Maturity*, GAO-04-394G, Version 1.1, GAO, Washington, D.C., March 2004.

[11]U.S. Office of Management and Budget, "Security of Federal Automated Information Resources," OMB Circular A-130, Appendix III, available at http://www.whitehouse.gov/omb/circulars/a130/a130appendix_iii.html, revises procedures formerly contained in Appendix III to OMB Circular No. A-130 (50 FR 52730; December 24, 1985) and incorporates requirements of the Computer Security Act of 1987 (P.L. 100-235) and responsibilities assigned in applicable national security directives; W.H. Ware, ed., *Security Controls for Computer Systems: Report of Defense Science Board Task Force on Computer Security*, AD # A076617/0, Rand Corporation, Santa Monica, Calif., February 1970, reissued October 1979; Federal Information Security Management Act of 2002 (FISMA, 44 U.S.C. § 3541, et seq.).

[12]Data Privacy and Integrity Advisory Committee, *Framework for Privacy Analysis of Programs, Technologies, and Applications*, Report No. 2006-01, U.S. Department of Homeland Security, Washington, D.C., adopted March 7, 2006.

[13]U.S. Department of Health, Education, and Welfare, Secretary's Advisory Committee on Automated Personal Data Systems, Records, Computers, and the Rights of Citizens, *Code of Fair Information Practices*, July 1973, available at http://aspe.hhs.gov/datacncl/1973privacy/tocprefacemembers.htm.

17799.[14] In comparison with COBIT, which has 34 high-level objectives that cover 215 control objectives, the committee's framework has two high-level objectives (i.e., effectiveness, and consistency with U.S. laws and values) that cover 30 control objectives. Although no one framework has the same high-level and control objectives as the committee's framework, they nevertheless provide guidance for achieving all of the committee's information and communications technologies criteria. Analysts advise that organizations judiciously select specific frameworks or criteria based on their relevance to well-defined objectives and the readiness of the organization to apply them.[15] This method applies also to implementing the committee's framework.

Most IT organizations surveyed worldwide[16] and in the United States[17] have adopted a framework. While many have developed their own, there is increasing adoption of formal frameworks based on reports of their efficacy, such as a 30 percent increase in productivity over 2 years through a consistent application of formal frameworks.[18] Failures with framework implementation are often related to inappropriate selection of criteria, as well as to formulaic implementations that emphasize process and checklists by those who do not understand the objectives or how to evaluate whether they have been achieved.

[14]The IT Governance Institute (ITGI), *IT Governance Global Status Report—2006*, ITGI, Rolling Meadows, Ill., 2006.

[15]D. Aron and A. Rowsell-Jones, *Success with Standards*, Gartner EXP, Stamford, Conn., May 2006.

[16]The IT Governance Institute (ITGI), *IT Governance Global Status Report—2006*, ITGI, Rolling Meadows, Ill., 2006.

[17]C. Symons, *IT Governance Survey Results: More Work to Be Done*, Forrester Research, Cambridge, Mass., April 14, 2005.

[18]D. Aron and A. Rowsell-Jones, *Success with Standards*, Gartner EXP, Stamford, Conn., May 2006.

# D

# The Life Cycle of Technology, Systems, and Programs

As noted in Chapter 2, the framework articulated for evaluating and deploying information-based technologies, programs, and systems acknowledges that the proposed inquiries are unlikely to yield definitive answers (i.e., "yes" or "no") at a given point in time and also that the answers may well change with time due to changes in the operational environment. This reality suggests that the policy regime—that is, what to make of and do with the answers to the questions provided by the framework—must be linked to the program life cycle. (In principle, the complete program life cycle begins with research, goes through development and deployment and then into operations, maintenance, and upgrade, and ends with program retirement.)

Mature models exist in other application areas that provide some guidance for how to proceed in this domain. For example, before new pharmaceuticals are approved by the Food and Drug Administration they must pass through multiple stages of testing designed to assess drug efficacy (therapeutic benefit) as well as safety (acceptable risks) in clinical trials. After approval is obtained for deployment, ongoing monitoring evaluates effectiveness and risks in the real-world environment; drugs may be recalled if they fall below acceptable standards. Similarly, product development programs typically rely on increasingly constrained testing regimes that, prior to deployment of a new system, mimic the real-world operating environment as nearly as is possible. Even product recall is not uncommon if, after deployment, a product is deemed to be ineffective or otherwise unacceptable (e.g., for safety reasons).

Similar processes exist to guide software development programs. For example, the National Aeronautics and Space Administration has for many years relied on independent verification and validation (IV&V) for safety-critical software applications. And the Software Engineering Institute and others have defined guidelines for verification and validation of large software applications. But these processes do not effectively address the complexities inherent in this class of information-based programs.

Multiple versions of what constitutes a program life cycle can be found in literature; here the committee describes a generic model with the following phases:

- *Identification of needs.* Analyze the current environment and solutions or processes currently in use; identify capability gaps or unmet needs.
- *Research and technology development.* Develop potential solutions to meet the identified needs.
- *Systems development and demonstration.* Develop and demonstrate the integrated system.
- *Operational deployment.* Complete production and full deployment of the program.
- *Operational monitoring.* Provide for ongoing monitoring to ensure that the deployed capability remains both effective and acceptable.
- *Systems evolution.* Institute upgrades to improve or enhance system functionality.

An effective policy regime should address each of the above phases in turn as indicated below:

- *Identification of needs.* During this phase, questions 1 and 2 from the summary of framework critera for evaluating effectiveness in Section 2.5.1 of Chapter 2 should be addressed—that is, the research should proceed only if a clear purpose and a rational basis are established and documented. Measures of effectiveness (benefit) and measures of performance (risk) should be drafted during this phase.
- *Research and technology development.* During this phase, testing should occur in a controlled laboratory setting—the equivalent of animal testing in the drug development process or developmental test and evaluation (DT&E) in traditional technology development programs. A key issue in testing information-based programs is access to data sets that adequately simulate real-world data such that algorithm efficacy can be evaluated. Ideally, standardized (and anonymized) data sets should be generated and maintained to support this phase of testing; the data sets maintained by the National Institute of Standards and Technology for

use in evaluating fingerprint and other biometric detection algorithms may serve as a useful model.[1] The program should proceed beyond this phase only after demonstration that a sound experimental basis exists in a laboratory setting; measures of effectiveness and measures of performance will likely require refinement during this phase of program development.

• *System development and demonstration.* During this phase, the program should be field-tested—subjected to the equivalent of human subject trials in the drug development process or operational test and evaluation (OT&E) in traditional technology development programs. The test environment must mimic real-world conditions as nearly as possible, and so both the simulation environment and requisite data sets must be designed and implemented with appropriate oversight. If it is necessary, for example, to use real-world data, then the test regime must provide appropriate protections to guard against inappropriate use of either the data or the results. During this phase of testing, the various elements of question 3 of the effectiveness criteria summary in Section 2.5.1 should be addressed (field-tested? tested to take into account real-world conditions? successful in predicting historical events? experimental successes replicated?), as should questions 4, 6, and 7 (scalability, capability for integration with relevant systems and tools, robustness in the field and against countermeasures). In addition, the development team should respond to questions 8 and 9 (guarantees regarding appropriateness and reliability of data, provision of appropriate data stewardship).

Also, given the class of programs under consideration in this report, a requirement for IV&V is needed at this phase of the life cycle. The IV&V process should review results from prior phases of testing and address the inquiries in question 10 (objectivity). Measures of effectiveness and measures of performance should be finalized for use in ongoing monitoring of the program if it is subsequently operationally deployed.

• *Operational deployment.* The final gate prior to operational deployment is an agency-level review of all items delineated in the summary of criteria for evaluating consistency with laws and values in Section 2.5.2, assurance that an ongoing monitoring process is in place, and definition of the conditions for operational deployment (e.g., threshold values for key measures). This review process should ensure that compliance is documented and reviewed in accordance with question 12 of the effectiveness criteria summary in Section 2.5.1.

---

[1]See http://www.itl.nist.gov/iad/894.03/databases/defs/dbases.html#finglist for more information.

• *Operational monitoring.* Once deployed, ongoing monitoring against established measures is vital to ensure that the system remains both effective and acceptable. If results in real-world operations suggest that, due either to changes in the external environment or to a lack of fidelity in the OT&E environment, system performance does not meet the established thresholds, an immediate agency-level review should be conducted to determine whether operational authorization should be revoked.

• *Systems evolution.* In general, information systems evolve or become obsolete. They evolve for many reasons. For example, new technologies may become available whose adoption can make the system more usable, or new applications may be required in a new operating environment, or new capabilities previously planned but not yet incorporated may be deployed. Because system evolution results in new functionality, reapplication of the framework described in Chapter 2 is usually warranted.

# E

# Hypothetical and Illustrative Applications of the Framework to Various Scenarios

This appendix illustrates how elements of the framework described in Chapter 2 might be applied to various hypothetical scenarios. Each scenario posits a particular kind of terrorist threat, a possible technological approach to addressing the threat, and some of the possible impacts on privacy entailed by that scenario. The scenarios are intended to illustrate how application of the framework draws out important questions to consider and answer when deciding on the deployment of a program. They are by no means exhaustive in their application of the framework, and they do not exemplify all the technologies considered in this report.

NOTE: The committee emphasizes that the descriptions of technological approaches in this appendix are NOT an endorsement of or a recommendation for their use.

## E.1 AIRPORT SECURITY

### E.1.1 The Threat

Terrorists continue to target air travel as an important objective. For the foreseeable future, aviation authorities will have to guard against the threat of an armed hijacking or the destruction of one or more fully loaded passenger planes.

### E.1.2 A Possible Technological Approach to Addressing the Threat

Checkpoint screening of airport passengers and their baggage to prevent the transport of weapons (e.g., firearms, explosives) will continue. However, with advancing technologies, future security checkpoints could be different from today's checkpoints in several ingenious respects:

- *Use of new sensors.* New imaging sensors could be introduced to reveal whether weapons are being hidden under clothing, although these sensors might also reveal anatomical features of the body. Retinal scans and other biometrics could be introduced to help validate passenger identity. Sensors for thermal imaging of the body or portions of the body could be introduced to detect signs of nervousness or excitement, and additional video cameras could be introduced with new software for face recognition and for analyzing body motion to search for signs of nervousness and other suspicious activity. Some of these sensors could be positioned so that passengers are aware they are being sensed, while others might be positioned so that passengers have no specific, explicit warning that they are being sensed.

- *Use of real-time networking to share data instantaneously across multiple airport security checkpoints (both within the same airport and at different airports), and to integrate data with information in other databases.* This approach would enable real-time sharing and fusion of information such as the detection that a nonstandard homemade briefcase containing unacceptable materials was found in airport A, and another similar event occurred in airport B, resulting in immediate transmission of information about the briefcase that would enable detecting other copies of it at other airports.

- *Use of data mining methods to draw inferences from a large shared data set, and to provide guidance to the human checkpoint operators.* For example, computer-based screening profiles for luggage and passengers might be improved continuously based on experience with millions of passengers across many airports. As one example, consider that today a human operator decides to hand inspect a certain fraction of luggage after it has passed through the x-ray scanner, perhaps because a suspicious-looking object is seen in the x-ray scan. Each time this occurs, the result of the hand inspection could be provided as a training example to a data mining program so that it could learn, from hundreds of thousands of such experiences, which x-ray images correspond to truly dangerous objects as opposed to false alarms. Computer-based machine learning algorithms could use such training data, collected from many security checkpoints at many airports, to formulate a potentially more accurate profile that could automatically estimate a risk level for each object seen in an x-ray scan and to assist the human screener with the goal of reducing the number of false alarms leading to invasive manual searches.

This imaginary future security checkpoint allows grounding many of the generic issues faced when deciding whether and how to introduce new information collection, fusion, and analysis systems and how to manage their potential impacts on civil liberties both in their specific implementation and in terms of general policies and legal frameworks.

### E.1.3 Possible Privacy Impacts

The privacy impact of detection technologies can vary significantly depending on choices made during deployment. The committee suggests that future regulations should differentiate systems and deployments based on features that can significantly affect perceived privacy impact, including:

- *Which data features are collected.* For example, when capturing images of baggage contents, the images might or might not be associated with the name or image of the passenger. Anonymous images of baggage, even if stored for future data mining, might be perceived as less invasive than baggage images associated with the owner.
- *Covertness of collection.* Images of passengers might be collected covertly, without the awareness of the individual, throughout the airport, or alternatively with the passenger's awareness and implicit consent at the security checkpoint. Many will consider the former to be more invasive of their privacy.
- *Data dissemination.* Data might be collected and used only for local processing, or disseminated more widely. For example, images of bags and passengers might be used only locally, or disseminated widely in a nationwide data store accessible to many agencies.
- *Retention.* Data might be required by regulations to be destroyed within a specified time interval, or kept forever.
- *Use.* Data might be restricted to a particular use (e.g., anatomically revealing images of airport passengers might be available for the sole purpose of checking for hidden objects), or unrestricted for arbitrary future use. The perceived impact on privacy can be very different in the two cases.
- *Use by computer versus human.* The data might be used (processed) by a computer, or alternatively by a human. For example, anatomically revealing images might be accessible only to a computer program that determines whether there is evidence of a hidden object under clothing. Alternatively, these images might be examined by the human security screener to manually search for hidden objects. The former case may be judged as less invasive by many passengers. Note that if a computer examination identifies a suspicious case, then a manual examination can be the next step. If the computer examination is sufficiently accurate, such

a two-stage computer-then-human process might significantly reduce the perceived privacy impact.

• *Control of permissions.* If data are retained for future uses, regulations might be placed on who can grant permission for subsequent dissemination and use (e.g., the collector of the data, a court, or the subject of the data). If the subject of the data is given a hand in granting permission, then the perceived privacy impact may be lessened.

### E.1.4  Applying the Framework

To illustrate the use of the framework proposed in Chapter 2 for evaluating the potential deployment of new systems, consider how it might be used to evaluate the possible deployment of one of the technologies suggested above. In particular, consider that company X comes to the U.S. government with a proposal to deploy a system that would (1) create a network to share images of baggage that are currently collected at all U.S. airport checkpoints, as well as the outcome of any manual searches of those bags by security screeners, and (2) use this nationwide database for two purposes: first, to perform data mining to identify homemade versus mass-produced luggage bags (based on their relative frequency of appearance at airports), and second, to use the results of the thousands of manual searches performed nationwide to automatically train more accurate software to spot suspicious items in x-ray images of baggage. How would the proposed framework apply to evaluating such a proposal?

**Effectiveness**

First, the framework asks for a clearly articulated purpose for the new system, an evaluation of why it may out perform current methods, and a thorough experimental evaluation of the system before full deployment. Note one might experimentally evaluate whether the data mining software of company X is capable of distinguishing home-made versus mass-produced luggage without going to the step of a full network deployment, by testing its use in one or two individual trial airports first. However, in many data mining applications, including this one, proving the value of collecting the full data set by testing on small sets is difficult, because performance sometimes improves as the size of the data set grows.

The framework would also raise issues regarding the rational basis for the program (Is it of significant value to spot custom-made luggage or custom-modified mass-produced luggage?, Does data mining the results of manual luggage inspections actually lead to more accurate automated luggage inspections and if so does this lead in turn to safer or less invasive screening?). It would raise issues about scalability (Can the computer system and human infrastructure handle the large volume of data from all U.S. airports in real time?), and data stewardship (Who will be responsible for the data collection?, and How will it be administered?).

### Compliance with Laws and Values

The framework asks whether an information-based program is consistent with U.S. law and values. The criteria for such consideration have been divided into three categories: data, programs, administration, and oversight. Does the proposed system operate with the least personal data consistent with goals of the system? Note that this question raises the issue of whether the owner of the luggage should be identified with each luggage image, and of evaluating the impacts of this on both system utility and on personal privacy. Does the system produce a tamper-resistant audit trail of who accesses which data? Is it secured against illegal tampering? What process is in place to assure monitoring the performance of the deployed system in terms of false positives and in terms of likely impacts on individuals? The framework asks questions about the agency collecting and deploying the system, perhaps the Transportation Security Administration in this case. Does this agency have a policy-level privacy officer, are its employees and others who might access the data trained appropriately, and are all of the uses of this nationwide luggage image dataset clearly articulated and in compliance with existing laws?

## E.2 SYNDROMIC SURVEILLANCE

### E.2.1 The Threat

A major issue for those concerned with ensuring public health is the early detection of an outbreak or attack capable of causing widespread disease, injury, or death. The presumption behind most early detection systems is that early warning would aid the rapid deployment of emergency resources and the initiation of public health and medical responses that would help to limit the spread of disease or any ill effects.

### E.2.2 A Possible Technological Approach to Addressing the Threat

In the past, officials have relied on hospitals and doctors to signal outbreaks by reporting disturbing or unusual trends and troubling cases or indicators. Today, however, with the increasing sophistication of technology, including data mining, sensors, and communications capabilities, many officials are investigating better ways of getting earlier warning of outbreaks or attacks. For example, could it be useful to monitor pharmacy sales of over-the-counter (OTC) drugs to get early warning for, say, something like an influenza epidemic? Perhaps it could be helpful to monitor school or work absentee rates for indications of widespread illness or biological attack. These forms of so-called "syndromic surveillance" are geared toward achieving the earliest possible detection of public health emergencies.[1]

Syndromic surveillance requires access to many different kinds of data. For example, in a large city, the data streams into a syndromic surveillance system might include digital records of common, OTC sales of medicines from pharmacies in the city, absentee records from city schools and some select businesses, counts of 911 calls to the city categorized into more than 50 call types (e.g., "influenza like illness," "breathing problems," and so on), and records of chief complaints from hospital emergency departments. In addition, these data streams could contain temporal and spatial information.

Such data streams would be monitored periodically (say, ever 24 hours) and compared automatically to archived data collected over the past. Changes from expected values would be automatically analyzed for statistical significance. The geographical data in the streams would also enable the system to identify the location of "hot spots" that might indicate possible outbreak points in the city.

Box E.1 describes how a syndromic surveillance system might be used in practice.

### E.2.3 Possible Privacy Impacts

From a privacy perspective, personal health data are among the most sensitive pieces of information. However, to generate initial indicators, only anonymized data are needed. Follow-up may be needed, as might be the case if interviews with patients or providers are necessary, and undertaking follow-up is impossible if anonymity rules. (In many pub-

---

[1]For more information on syndromic surveillance generally, as well as more information about previous efforts, see http://www.cdc.gov/mmwr/PDF/wk/mm53SU01.pdf. Overviews of syndromic surveillance can be found at http://iier.isciii.es/mmwr/preview/mmwrhtml/su5301a3.htm; K.D. Mandl, J.M. Overhage, M.M. Wagner, W.B. Lober, P. Sebastiani, F. Mostashari, J.A. Pavlin, P.H. Gesteland, T. Treadwell, E. Koski, L. Hutwagner, D.L. Buckeridge, R.D. Aller, and S. Grannis, "Implementing syndromic surveillance: A practical guide informed by the early experience," *Journal of the American Medical Informatics Association* 11(2):141-150, 2004; and J.W. Buehler, R.L. Berkelman, D.M. Hartley, and C.J. Peters, "Syndromic surveillance and bioterrorism-related epidemics," *Emerging Infectious Diseases* 9(10):1197-1204, 2003.

**BOX E.1**
**An Illustrative Operational Scenario for the**
**Use of Syndromic Surveillance**

On a winter afternoon, a GoodCity public health official conducting routine daily data analysis notes a spike in the number of hospital emergency department (ED) visits and pharmacy sales detected by GoodCity's syyndromic surveillance system, which is designed to detect early, indirect indicators of a possible bioterror attack. None of the other data streams indicate unusual patterns.

The health official, who has been specially trained to operate the statistical data mining software involved, analyzes the temporal and spatial distribution of ED visits using scan statistics and finds that two hospitals in the same zip code, and located within blocks of each other, accounted for most of the excess visits. A third hospital in the same area of the city experienced a normal volume of ED visits during the previous 24 hours. Further examination of available data reveals that respiratory illness was the chief complaint of a majority of the patients seen in the two EDs of interest. Further analysis shows that in the past 24 hours, both hospitals experienced higher rates of ED visits for "respiratory illness" than expected based on comparisons with hospital-specific rates gathered in previous years.

Meanwhile, the health officer's examination shows that over-the-counter (OTC) medicine sales, in particular medicines to treat cough and fever, are much increased compared to the previous week and compared to the same week of the previous year. The system tracks sales by store and zip code, but no pattern is evident. Past analyses have shown that increased purchases of OTC medications do not consistently presage a higher volume of ED visits.

Concerned that the increased incidence of respiratory complaints in a geographically discrete neighborhood of the city, combined with city-wide increases in the purchase of cough and fever medicines, might indicate the leading edge of an aerosolized anthrax attack or some other disease outbreak of public health significance, the health official assigns a public health nurse to conduct a telephonic descriptive review of the ED cases seen in the affected hospitals. The nurse will also query staff from a sample of hospitals that are not part of the surveillance system, looking for unusual presentations or higher-than-usual volume.

After several hours of phone calls, the public health nurse discovers that many of the excess ED visits were indeed for cough and respiratory complaints, but most patients were not deemed seriously ill and were sent home with a diagnosis of "viral illness." Early in her calls, the nurse heard of two young adult patients who had been extremely ill with apparent "pneumonia" and admitted to the intensive care unit. Since it is unusual for healthy young adults to require hospitalization for pneumonia, the nurse tracked down and interviewed the admitting physicians for both patients. In both cases, the patients involved had an underlying illness that explained their condition.

The hospital staff consulted reported that ED volume throughout the day was not abnormally high; today's syndromic surveillance data documenting ED visits city-wide would not be available for another 12 hours. Public health officials decided on the basis of these investigations to do nothing more, but to continue to closely monitor hospital ED visits and OTC sales over the coming days.

lished articles on syndromic surveillance, the emergency department (ED) data is the most important and useful data stream for both detecting and ruling out disease outbreaks.)

The efficacy of a surveillance system could be significantly enhanced through the potential inferential power of multivariate information about specific individuals arriving through different data streams. For example, two data streams might be purchases of OTC medications for coughs and school attendance records. Rather than simply analyzing these data streams separately and noting temporal correlations in them, considerably more inferential power would be available if it were possible to associate a specific child absent from school on Tuesday with the purchase of cough syrup on Tuesday by his father.

However, linking attendance records to drug store purchasing records in such a manner would require personal identifiers in each stream to enable such a match. Privacy interests would therefore be implicated as well. For example, while the Health Information Portability and Accountability Act allows the use of medical information for public health purposes, it is unclear how to interpret the privacy restrictions in the context of regular surveillance systems. Further, different laws govern access to or restrictions on data associated with educational systems and organizations, and grocery chains restrict access to proprietary information on customer purchases.

### E.2.4 Applying the Framework

Since a number of syndromic surveillance systems are in operation, the committee has been able to draw on public information and research in its reflection on the application of the framework presented in Chapter 2 and hence report some of that information to the reader. However, the illustration here does not constitute an endorsement or disapproval of such systems by this committee. The implementation of syndromic surveillance systems was prompted largely by the federal government when the U.S. Department of Health and Human Services (HHS) made bioterrorism preparedness monies available to state health agencies in 2002. Many such systems, of varying type and scope, were created—some by city health departments, some by state health agencies in collaboration with universities, and others by private contractors who not only designed but also operated the systems and then reported analyzed results to government officials.

#### Effectiveness

The framework asks for a clearly stated purpose. The purpose of most syndromic surveillance systems is to detect a covert bioterrorist attack before large numbers of victims seek medical care, in order to improve response and save lives. HHS did not specify any operational standards or explicit goals for the systems it helped fund, although some standards have been evolving.[2]

When considering a rational basis, the concept that lives might be saved if a bioattack were recognized earlier rather than later, thus lengthening the time available to get countermeasures (medicines and vaccines) to those infected, to conduct investigations into where the attack occurred and who is at risk, and so on gives merit to syndromic surveillance systems. However, the systems' integration into current practices in the field should be taken into account. There is good evidence that syndromic surveillance systems can detect large disease outbreaks, but it is less clear how and if such detection improves public health response. Health officials confronted with a spike in syndromic signals typically seek more definitive evidence of a true rise in illnesses among city residents before taking action.[3] This is in part because there is a lot of noise in the systems—illness rates, OTC medicine purchases, 911 reports—that varies widely even within a given season and location. Also, syndromic surveillance generates many false positives (discussed below), and the "signal" is not specific enough in most instances to guide action. Syndromic signals spur health officials to look harder but do not usually trigger a public health response.[4] Whether syndromic surveillance would actually improve the rapidity of the response to a bioattack compared to clinical case finding is unproven and probably not testable.[5] Recently, Buckeridge and colleagues attempted to compare clinical case finding and syndromic

---

[2]See the U.S. Department of Health and Human Services (HHS)-sponsored published reviews of specific systems' operating characteristics and evaluation challenges (M. Sosin and J. DeThomasis, "Evaluation challenges for syndromic surveillance—Making incremental progress," *Morbidity and Mortality Weekly Report* 53(Suppl):125-129, 2004) as well as a "decision-making framework" for implementing syndromic surveillance suggested by a Centers for Disease Control and Prevention Working Group (J.W. Buehler, R.S. Hopkins, J.M. Overhage, D.M. Sosin, and V. Tong, "Framework for evaluating public health surveillance systems for early detection of outbreaks," *Morbidity and Mortality Weekly Report* 53(RR-5):1-11, May 7, 2004).

[3]R. Heffernan, F. Mostashari, D. Das, M. Besculides, C. Rodriguez, J. Greenko, L. Steiner-Sichel, S. Balter, A. Karpati, P. Thomas, M. Phillips, J. Ackelsberg, E. Lee, J. Leng, J. Hartman, K. Metzger, R. Rosselli, and D. Weiss, "New York City syndromic surveillance systems," *Morbidity and Mortality Weekly Report* 53(Suppl):25-27, September 24, 2004.

[4]As has been evidenced from responses to signals from biosensors placed at the Salt Lake City Olympics (2002) and with Biowatch (an environmental sensor system deployed by the U.S. Department of Homeland Security).

[5]In a bioattack, it is likely that many people will become ill and appear in the health care system at the same time, making it apparent to clinicians that an unusual event is unfolding. The time between syndromic signal detection plus confirmation activities may in practice offer little if any advantage over clinical case finding.

surveillance for detection of inhalational anthrax due to a bioterror attack using a simulation study.[6] These investigators found that syndromic systems could be designed such that detection of an anthrax attack would be improved by one day, but when systems were sensitive enough to detect a substantial portion of outbreaks before clinical case finding, frequent false positives were also produced, which could impose a considerable burden on public health resources.

There are limits to the experimental basis for syndromic surveillance systems, and any gains should be weighed against the costs of developing and operating such systems. Observable behaviors that might precede patients seeking medical care for an illness are not precisely known. Although in 1993 a run on OTC medicines in Milwaukee famously preceded public health detection of a large, waterborne cryptosporidiosis outbreak, the purchase of nonprescription, OTC medicines does not reliably precede outbreaks of illness in populations.[7] Moreover, a retrospective analysis of 3 years of syndromic surveillance data gathered by the New York City Health Department concluded that "syndromic surveillance signals [for gastrointestinal disease outbreaks] occur frequently, [and] are difficult to investigate satisfactorily. . . ."[8]

The New York City Department of Health operates one of the country's most sophisticated syndromic surveillance systems, which has been in use since the late 1990s and has been continually upgraded. This system has been documented as detecting seasonal influenza a week before culture-positive samples of flu were found in New York City and has detected large sales of OTC antidiarrheal medicines which subsequent investigations associated with gastrointestinal illness and eating spoiled food after a city-wide blackout. The system failed, however, to detect either the unprecedented outbreak of West Nile Virus in 1999 or the anthrax cases of 2001.[9]

Syndromic surveillance systems should be developed from technical specifications, data flows, and types of signals that have been rigorously shown to be most reliable and productive. However, such development

---

[6]D.L. Buckeridge, D.K. Owens, P. Switzer, J. Frank, and M.A. Musen, "Evaluating detection of an inhalational anthrax outbreak," *Emerging Infectious Diseases* 12(12), 2006.

[7]R. Armstrong, P. Coomber, S. Prior, and A. Dincher, *Looking for Trouble: A Policymaker's Guide to Biosensing*, Center for Technology and National Security Policy, National Defense University, Washington, D.C., June 2004.

[8]S. Balter, D. Weiss, H. Hanson, V. Reddy, D. Das, and R. Heffernan, "Three years of emergency department gastrointestinal syndrome surveillance in New York City: What have we found?," *Morbidity and Mortality Weekly Report* 54(Suppl):175-180, August 26, 2005.

[9]"Syndromic surveillance for bioterrorism following the attacks on the World Trade Center—New York City, 2001" *Morbidity and Mortality Weekly Report* 51(Special Issue):13-15, September 11, 2002.

has many challenges. Because bioterrorist attacks are rare events, most of the "positive" signals syndromic surveillance produces will be false positives. Setting the system to be very sensitive (i.e., increasing the types and size of data streams) will generate more false positives, which can, over time, erode confidence in the system. The complex and larger data streams are also likely to increase the complexity of the investigations that follow the detection of syndromic "signals," which could further delay any response action.[10]

There are also great difficulties in doing real-time record linkage on multiple data streams. With static record linkage, all of the databases in question are available for analysis, which means that it is possible to perform cross-validation, error assessment, and careful blocking to reduce comparisons. With real-time linkage, only a limited data sample is applicable (i.e., those that relate to present cases), which means that the data available to revise parameter estimates and error rates are limited.

Finally, a key challenge in assessing the utility and efficacy of a syndromic surveillance system is to differentiate between the power of the particular algorithmic approach used in analyzing the data (which may be inadequate regardless of the quality of the data) and the quality of the data used in that particular approach (which may be too poor regardless of the power of the algorithm).

Assessing the scalability of such systems is also challenging. Consideration must be placed on whether this approach is viable for all localities of any size as well as whether some data streams are more important than others or must be of certain minimal scope. The trade-offs between the size of the signal (number and size of different data streams) and the sensitivity and specificity of the signal (i.e., number of false positives and negatives) must be taken into account.

A syndromic surveillance system should be designed to allow the enforcement of business processes; business processes define the ways in which the system is used, who the agents are, who are authorized to use it, and the steps taken in each individual task. Business processes can be different for different syndromic surveillance systems. For example, an agency in one city will allow anyone above a certain pay grade to execute a report but with the concurrence of the chief epidemiologist, whereas the comparable agency in another city will only allow the chief epidemiologist and two other delegated individuals to do so. Business processes will help determine how syndromic surveillance systems can be integrated into routine public health practice and what additional resources are

---

[10]A. Reingold, "If syndromic surveillance is the answer, what is the question?," *Biosecurity and Bioterrorism* 1(2):77-81, 2003; M. Stoto, "Syndromic surveillance," *Issues in Science and Technology*, Spring 2005.

required. When private contractors or university partners operate the syndromic surveillance systems, processes will have to be defined to ensure that health officials receive data and analyses in a timely manner with no uncertainties about the validity of analyses.

Syndromic surveillance systems have the potential to contain large amounts of data. Those operating such systems will have to consider how to guarantee appropriate and reliable data as well as appropriate data stewardship. Some questions to consider include: Is the system collecting only the data necessary to detect a threat? Can syndromic data be forwarded to health departments in a manner that protects patient privacy in routine uses but allows identification to subsequently interview particular patients, in keeping with routine public health practice, in crisis? Can the utility of the system be preserved if geographic aggregation or some other form of protection is done to protect individual privacy? What is known about the accuracy of data submitted from different sources? How long do data streams need to be retained? Can records of illness patterns be retained without individual data streams? If such data are retained for long periods, will clinical data about specific patients and their commercial records (e.g., drug purchases) be available in these systems? Who will have access to the data? What policies need to be established to protect from unlawful or unauthorized disclosure, manipulation, or destruction?

## Lawfulness

The framework asks to consider whether an information-based program, such as syndromic surveillance, is consistent with U.S. law and values. The criteria for such consideration have been divided into three categories: data, programs, and administration and oversight. For effective syndromic surveillance systems, the need for personal medical data from emergency rooms is clear, and in most (not all) current syndromic systems the data are anonymized before being sent to public health agencies. In many published articles on syndromic surveillance, the emergency room data constitute the most important and useful data stream for both detecting and ruling out disease outbreaks. Data from OTC purchases and attendance records seem useful to this system. However, they, as personal data, should be considered only if they are reasonably shown to prove the effectiveness of system. Within currently operating systems, data on OTC medications are used but are more easily associated with particular stores and less easily associated with individuals.[11] Linking

---

[11]For example, individuals may purchase over-the-counter medications with credit cards or store affinity cards. Though these individually identifiable purchase records are not rou-

such information to school absences and clinical information raises major privacy issues and has not been attempted as part of any biosurveillance program, to the committee's knowledge.

Public health agencies do have legal authority to release personal medical data if such information is pertinent to public health. Frequency of false positives is a major concern with these systems, as the scenario in Box E.1 demonstrates. In large public health agencies where resources exist to maintain and staff syndromic surveillance systems appropriately and where digitized data streams are available, such systems may be cost-effective. A bioattack alarm may lead to revelations of the names and medical conditions of specific patients seen in emergency rooms associated with syndromic reporting. In such "emergencies" the violation of an individual's privacy might be deemed acceptable given the public's right to know what is going on. However, agencies should have procedures in place for dealing with consequences of false positives. They should also assess and identify the impact on individuals in non-alarm routine operations. The system itself should produce a tamper-resistant audit trail, and all personnel authorized to use the system and its outputs should receive training in appropriate use and the laws and policies applicable to its use. The agency should employ a privacy officer to ensure compliance with laws, policies, and procedures designed to protect individual privacy. These are but a few considerations toward assessing whether syndromic surveillance systems are consistent with U.S. laws and values.

---

tinely made available to public health authorities, they do exist with the data that drugstores routinely collect and could be made available under some circumstances. Various authors have analyzed these data bases to illustrate the potential of early detection of bioterrorist attacks—e.g., A. Goldenberg, G. Shmueli, R.A. Caruana, and S.E. Fienberg, "Early statistical detection of anthrax outbreaks by tracking over-the-counter medication sales," *Proceedings of the National Academy of Sciences* 99(8):5237-5240, April 2002. Linking such information to school absences and clinical information raises major privacy issues and has not been attempted as part of any biosurveillance program to the committee's knowledge.

# F

# Privacy-Related Law and Regulation: The State of the Law and Outstanding Issues

The law intended to guide intelligence operations is complex and has failed to keep up with the significant changes in terrorist threats, surveillance technologies, and the volume, variety, and accessibility of digital data about individuals. The absence of a coherent and up-to-date legal framework has contributed to undermining trust in intelligence activities. A brief description of that law along with an explanation of its inadequacies will help illustrate why.

## F.1 THE FOURTH AMENDMENT

### F.1.1 Basic Concepts

The government has very broad power to obtain personal information. Historically, the primary constitutional limit on that power is the Fourth Amendment, which reflects the Framers' hostility to general searches. A general search is a search that is not based on specific evidence that allows the search to be targeted as to the location of the search or the type of evidence the government is seeking. The purpose of the Fourth Amendment was to forbid general searches by requiring that all search and seizures must be reasonable and that all warrants must state with particularity the item to be seized and the place to be searched.

The Fourth Amendment requires that warrants be issued only "upon probable cause, supported by oath or affirmation, and particularly describing the place to be searched, and the persons or things to be seized." Federal law defines "probable cause" to mean "a belief that an individual is committing, has committed, or is about to commit a particular offense" and that the information sought is germane to that crime.[1] The Supreme Court generally requires that the government provide the subject of a search with contemporaneous notice of the search.[2]

Collecting information from a person constitutes a search if it violates that individual's reasonable expectation of privacy. The Supreme Court has held that a person has a reasonable expectation of privacy in their homes, sealed letters, and the contents of their telephone calls. On the other hand, the Court has determined, for example, that warrants are not required to search or seize items in the "plain view" of a law enforcement officer,[3] for searches that are conducted incidental to valid arrests,[4] or to obtain records held by a third party, even if those records are held under a promise of confidentiality.[5] The Court has interpreted this last exception broadly to find that the Fourth Amendment is inapplicable to telecommunications "attributes" (e.g., the number dialed, the time the call was placed, the duration of the call, etc.), because that information is necessarily conveyed to, or observable by, third parties involved in connecting the call.[6]

Moreover, the Fourth Amendment poses no limits on how the government may use information, provided that it has been obtained legally, and some limits on the use of data obtained illegally. Consequently, personal data seized by the government in compliance with the Fourth Amendment may later be used in a context for which the data could not have been obtained lawfully. The rest of this section addresses two important examples of areas in which the evolution of technology and new circumstances suggest that current Fourth Amendment law and practice may be outdated or inadequate.

### F.1.2 Machine-Aided Searches

In some ways, machine-aided searching of enormous volumes of digital transaction records is analogous to a general search, especially if those records contain highly sensitive information. Much like a general search in colonial times was not based on specific evidence or limited to a particular person or place, a machine-aided search through digital databases can be very broad.

---

[1] 18 U.S.C. § 2518(3)(a).
[2] *Richards v. Wisconsin*, 520 U.S. 385 (1997).
[3] *Coolidge v. New Hampshire*, 403 U.S. 443 (1971).
[4] *United States v. Edwards*, 415 U.S. 800 (1974).
[5] *United States v. Miller*, 425 U.S. 435 (1976).
[6] *Smith v. Maryland*, 442 U.S. 735 (1979).

Existing Fourth Amendment law speaks to such searches only in limited contexts, however. The Fourth Amendment requires the government to obtain a search warrant when looking through a person's hard drive or private e-mail, for example. It also requires that the warrant specify the type of evidence the government is seeking. It may also require a warrant or a subpoena to collect information that is inside a database. However, if the government collects data in compliance with the Fourth Amendment, and then it aggregates the data into a database, the process of searching through the database is not itself regulated by the Fourth Amendment. Even if the government violates the Fourth Amendment when collecting the data, the data may be stored, aggregated, and used for any purpose other than that for which the data were wrongfully accessed. So, for example, the Court has allowed records illegally seized by criminal investigators to be used by tax investigators on the basis that restricting the subsequent use would not deter the original unconstitutional conduct.[7]

Broad machine-aided searches and the government's reuse of lawfully or unlawfully obtained data raise very important questions of public policy. What standards should govern access to or use of data that has already been collected? Should use of databases or specific analytical techniques such as data mining be regulated at all? If querying a database or running a data mining program on a database constitutes a search, when is such a search "reasonable"? Must the police have a specific individual in mind before searching a database for information on him or her? In the absence of clear standards or guidelines to govern their conduct or even to help them make reasonable judgments, the police cannot do their work. Moreover, what level of legal authorization should guide database queries? If a legal standard is used, is relevance the right standard? Or is something more like reasonable suspicion or probable cause the proper standard to use?

### F.1.3 Searches and Surveillance for National Security and Intelligence Purposes That Involve U.S. Persons Connected to a Foreign Power or That Are Conducted Wholly Outside the United States

The Fourth Amendment applies to searches and surveillance conducted for domestic law enforcement purposes within the United States, and those conducted outside of the United States if they involve U.S. citizens (although not necessarily permanent resident aliens). In a 1972 case commonly referred to as the *Keith* decision, the Supreme Court held that the Fourth Amendment also applies to searches and surveillance conducted for national security and intelligence purposes within the United

---

[7]*United States v. Janis*, 428 U.S. 433, 455 (1975).

States if they involve U.S. persons who do not have a connection to a foreign power.[8] The Court, however, recognized that "different policy and practical considerations" might apply in the national security context than in traditional law enforcement investigations, and specifically invited Congress "to consider protective standards for . . . [domestic security] which differ from those already prescribed for specified crimes in Title III."[9] The Court left open the question of whether the Fourth Amendment applies to searches and surveillance for national security and intelligence purposes that involve U.S. persons who are connected to a foreign power or are conducted wholly outside of the United States,[10] and the Congress has not supplied any statutory language to fill the gap.

### F.1.4 The Miller-Smith Exclusion of Third-Party Records

As noted in Chapter 1, some legal analysts believe that there is no better example of the impact of technological change on the law than the exemption from the Fourth Amendment created by the Supreme Court for records held by third parties. According to this perspective, such an exemption significantly reduces constitutional protections for personal privacy—not as the result of a conscious legal decision, but through the proliferation of digital technologies that make larger quantities of more detailed information available for inspection than ever before.

Other analysts suggest that as a general point, the protection of privacy is better founded as a matter of statute and regulation (that is, of policy choices) rather than as a matter of Constitutional right.[11] In this view, legislatures have many advantages that enable the legislative privacy rules regulating new technologies to be more balanced, comprehensive, and effective than judicially created rules. These advantages include the ability to act more quickly in the face of technological change than courts are able to do and to appreciate existing technology and the impact of different legal rules. In addition, and specifically relevant to the third party exemption for the privacy of records held by third par-

---

[8]*United States v. U.S. District Court for the Eastern District of Michigan*, 407 U.S. 297 (1972).

[9]Id. at 322.

[10]J.H. Smith and E.L. Howe, "Federal legal constraints on electronic surveillance," p. 133 in *Protecting America's Freedom in the Information Age* (Markle Foundation Task Force on National Security in the Information Age), Markle Foundation, New York, N.Y., 2002. Lower courts have found, however, that there is an exception to the Fourth Amendment's warrant requirement for searches conducted for intelligence purposes within the United States that involve only non-U.S. persons or agents of foreign powers. See *United States v. Bin Laden*, 126 F. Supp. 2d 264, 271-72 (S.D.N.Y. 2000).

[11]O.S. Kerr, "The Fourth Amendment and new technologies: Constitutional myths and the case for caution," *Michigan Law Review* 102:801-888, 2004.

ties, some analysts argue that without some ability for law enforcement officials to obtain some transactional data without a warrant, criminals and terrorists operating in cyberspace would be largely able to prevent law enforcement from obtaining probable cause to obtain indictments or to investigate more deeply.

## F.2 THE ELECTRONIC COMMUNICATIONS PRIVACY ACT

The Fourth Amendment is not the only restraint on the government's power to collect and use information through surveillance. The Electronic Communications Privacy Act (ECPA) is a collection of three different statutes that also regulates government collection of evidence in the context of telecommunications networks. The Wiretap Act is amended in Title I of ECPA, and as amended deals with the interception of telephone and Internet communications in transmission.[12] It applies to "wire communications," although not to video unaccompanied by sound. To intercept communications in transit requires a "'super' search warrant,"[13] unless an exception to the warrant requirement applies such as consent. A warrant can only be sought by designated federal officials and requires probable cause, details about the communication to be intercepted, minimization of any non-relevant communications inadvertently intercepted, and termination immediately upon completion. Information obtained in violation of these requirements can subject the responsible agent to minimum damages of $10,000 per violation and is subject to the exclusionary rule (except for e-mail) so that it cannot be used in a subsequent criminal prosecution.

Title II—the Stored Communications Act—which was adopted in 1986 deals with communications in electronic storage, such as e-mail and voice mail.[14] It contains rules that govern compelled disclosure of information from service providers as well as when providers can disclose information voluntarily. Traditional warrants are required to obtain access to communications stored 180 days or less. To obtain material stored for more than 180 days, the government need only provide an administrative subpoena, a grand jury subpoena, a trial subpoena, or a court order, all of which are easier to obtain than a traditional warrant. Non-content information, such as information about a customer's account maintained by a communications provider, can be obtained by the government either

---

[12]Wiretap Act, Public Law 90-351, 82 Stat. 197 (1968) (codified as amended at 18 U.S.C. §§ 2510-2522).

[13]O.S. Kerr, "Internet surveillance law after the USA Patriot Act: The big brother that isn't," *Northwestern University Law Review* 97(2):607-673, 2003.

[14]Stored Communications Act, Public Law 99-508, Title II, § 201, 100 Stat. 1848 (1986) (codified as amended at 18 U.S.C. §§ 2701-2711).

with a subpoena or by providing "specific and articulable facts showing that there are reasonable grounds to believe that . . . the records or other information sought are relevant and material to an ongoing criminal investigation."[15] Violations carry a minimum fine of $1,000; no exclusionary rule applies.

Title III—the Pen Register Act—which was also adopted in 1986, applies to "pen registers" (to record outgoing call information) and "trap and trace" devices (to record incoming call information).[16] To obtain information akin to what is contained in a phone bill or revealed by "Caller ID," e-mail header information (the "To," "From," "Re," and "Date" lines in an e-mail), or the IP address of a site visited on the Web, the government need only obtain a court order. The court must provide the order—there is no room for judicial discretion—if the government certified that "the information likely to be obtained by such installation and use is relevant to an ongoing investigation."[17] The exclusionary rule does not apply to violations of the act.

### F.3 THE FOREIGN INTELLIGENCE SURVEILLANCE ACT

While the ECPA regulates surveillance for law enforcement purposes, successive presidents insisted that it did not limit their power to engage in surveillance for national security purposes. In the aftermath of Watergate, the Senate created the Select Committee to Study Government Operations with Respect to Intelligence Activities, chaired by Senator Frank Church (D-Idaho). The Church Committee's final report, published in 1976, cataloged a wide array of domestic intelligence surveillance abuses committed under the protection of the president's national security authority.[18] While some must have been plainly understood at the time by their perpetrators to have involved wrong-doing, such as spying on political opponents, many involved what today would be called "mission creep."[19]

That report, the unresolved nature of the president's power to con-

---

[15]18 U.S.C. § 2703(d).

[16]Pen Register Act, Public Law 99-508, Title III, § 301(a), 100 Stat. 1868 (1986) (codified as amended at 18 U.S.C. §§ 3121-3127).

[17]18 U.S.C. § 3123(a).

[18]Senate Select Committee to Study Government Operations with Respect to Intelligence Activities, 94th Congress, *Final Report on Intelligence Activities and the Rights of Americans*, Book II, April 26, 1976; see also M.H. Halperin, J.J. Berman, R.L. Borosage, and C.M. Marwick, *The Lawless State: The Crimes of the U.S. Intelligence Agencies*, Penguin Publishing Company Ltd., London, U.K., 1976.

[19]Senate Select Committee to Study Government Operations with Respect to Intelligence Activities, 94th Congress, *Final Report on Intelligence Activities and the Rights of Americans*, Book II, April 26, 1976.

duct domestic surveillance, and the Supreme Court's 1972 invitation to Congress in the *Keith* decision to "consider protective standards" in this area all coalesced in enactment of the Foreign Intelligence Surveillance Act (FISA) of 1978.[20] The act creates a statutory regime governing the collection of "foreign intelligence" from a "foreign power" or "agent of a foreign power" within the borders of the United States.

The act created a special court—the Foreign Intelligence Surveillance Court—of seven (now eleven) federal district court judges. The court meets in secret and hears applications from the Department of Justice (DOJ) for ex parte orders authorizing surveillance or physical searches. All that the government must show is that there is "probable cause to believe that the target of the electronic surveillance is a foreign power or agent of a foreign power"[21] and that gathering foreign intelligence is "the purpose" of the requested order.[22] In 2001, the USA Patriot Act changed this standard to "a significant purpose."[23] This change and a decision from the three-judge FISA review court created by the statute to hear appeals brought by the government have resulted in making information obtained from FISA surveillance freely available in criminal prosecutions.[24] In 2003, for the first time, the federal government sought more surveillance orders under FISA than under ECPA.[25]

As this report is being written (November 2007), changes to the FISA act are being contemplated by the U.S. Congress. The final disposition of these changes remains to be seen.

### F.4  THE PRIVACY ACT

The Privacy Act of 1974 provides safeguards against an invasion of privacy through the misuse of records by federal agencies and establishes a broad regulatory framework for the federal government's use of personal information.[26] The Act requires federal agencies to store only relevant and necessary personal information and only for purposes required to be accomplished by statute or executive order; to collect information

---

[20]Public Law 95-511, 92 Stat. 1783 (1978) (codified at 50 U.S.C. § 1801-1811).

[21]50 U.S.C. § 1805(a)(3)(A).

[22]Id. § 1804(7) (prior to being amended in 2001).

[23]Uniting and Strengthening America by Providing Appropriate Tools Required to Intercept and Obstruct Terrorism Act of 2001, Public Law 107-56, § 204, 115 Stat. 272 (codified at 50 U.S.C. § 1804(a)(7)(B)).

[24]In re Sealed Case, 310 F.3d 717 (FISA Review Court 2002).

[25]P.P. Swire, "The system of foreign intelligence surveillance law," *George Washington Law Review* 72(6):1306-1308, 2004. This article provides analysis of the history and details of FISA generally.

[26]5 U.S.C. § 552a.

to the extent possible from the data subject; to maintain records that are accurate, complete, timely, and relevant; and to establish administrative, physical, and technical safeguards to protect the security of records.[27] The Privacy Act also prohibits disclosure, even to other government agencies, of personally identifiable information in any record contained in a "system of records," except pursuant to a written request by or with the written consent of the data subject, or pursuant to a specific exception.[28] Agencies must log disclosures of records and, in some cases, inform the subjects of such disclosures when they occur. Under the Act, data subjects must be able to access and copy their records, each agency must establish a procedure for amendment of records, and refusals by agencies to amend their records are subject to judicial review. Agencies must publish a notice of the existence, character, and accessibility of their record systems.[29] Finally, individuals may seek legal redress if an agency denies them access to their records.

The Privacy Act is far less protective of privacy than may first appear, because of numerous broad exceptions.[30] Twelve of these are expressly provided for in the Act itself. For example, information contained in an agency's records can be disclosed for "civil or criminal law enforcement activity if the activity is authorized by law."[31] An agency can disclose its records to officers and employees within the agency itself, the Census Bureau, the National Archives, Congress, the Comptroller General, and consumer reporting agencies.[32] Information subject to disclosure under the Freedom of Information Act is exempted from the Privacy Act.[33] And under the "routine use" exemption,[34] federal agencies are permitted to disclose personal information so long as the nature and scope of the routine use was previously published in the Federal Register and the disclosure of data was "for a purpose which is compatible with the purpose for which it was collected." According to the Office of Management

---

[27]Id.

[28]Id. § 552a(b).

[29]Id. § 552a(e)(4).

[30]S. Fogarty and D.R. Ortiz, "Limitations upon interagency information sharing: The Privacy Act of 1974," pp. 127-128 in *Protecting America's Freedom in the Information Age* (Markle Foundation Task Force on National Security in the Information Age), Markle Foundation, New York, N.Y., 2002.

[31]5 U.S.C. § 552a (b)(7).

[32]Id. § 552a(b).

[33]Id. § 552a(b)(2).

[34]Id. § 552a(b)(3).

and Budget, "compatibility" covers uses that are either (1) functionally equivalent or (2) necessary and proper.[35]

Moreover, the Privacy Act applies only to information maintained in a "system of records."[36] The Act defines "system of records" as a "group of any records under the control of any agency from which information is retrieved by the name of the individual or by some identifying number, symbol, or other identifying particular assigned to the individual."[37] The U.S. Court of Appeals for the District of Columbia Circuit held that "retrieval capability is not sufficient to create a system of records. . . . 'To be in a system of records, a record must . . . in practice [be] retrieved by an individual's name or other personal identifier.'"[38] This is unlikely to be the case with new antiterrorism databases, in which information may not be sufficiently structured to constitute a "system of records" in the meaning of the Privacy Act.

The Privacy Act has also been subject to judicial interpretations which have created new exceptions. For example, courts have found that the following entities do not constitute an "agency": a federally chartered production credit association, an individual government employee,[39] state and local government agencies,[40] the White House Office and those components of the Executive Office of the President whose sole function is to advise and assist the President,[41] grand juries,[42] and national banks.[43]

As a result, the Privacy Act plays little role in providing guidance for government intelligence activities or limiting the government's power to collect personal data from third parties. Moreover, the Privacy Act only

---

[35]Privacy Act of 1974, 5 U.S.C. § 552a; "Guidance on the Privacy Act Implications of 'Call Detail' Programs to Manage Employees' Use of the Government's Telecommunications Systems," 52 Fed. Reg. 12900, 12993 (1987) (OMB) (publication of guidance in final form); see generally S. Fogarty and D.R. Ortiz, "Limitations upon interagency information sharing: The Privacy Act of 1974," pp. 127-128 in *Protecting America's Freedom in the Information Age* (Markle Foundation Task Force on National Security in the Information Age), Markle Foundation, New York, N.Y., 2002.

[36]5 U.S.C. § 552a(b).

[37]Id. § 552a(a)(5).

[38]*Henke v. United States Department of Commerce*, 83 F.3d 1453, 1461 (D.C. Cir. 1996) (quoting *Bartel v. FAA*, 725 F.2d 1403, 1408 n.10 (D.C. Cir. 1984)).

[39]*Petrus v. Bowen*, 833 F.2d 581 (5th Cir. 1987).

[40]*Perez-Santos v. Malave*, 23 Fed. App. 11 (1st Cir. 2001); *Ortez v. Washington County*, 88 F.3d 804 (9th Cir. 1996).

[41]*Flowers v. Executive Office of the President*, 142 F. Supp. 2d 38 (D.D.C. 2001).

[42]*Standley v. Department of Justice*, 835 F.2d 216 (9th Cir. 1987).

[43]*United States v. Miller*, 643 F.2d 713 (10th Cir. 1981). See generally S. Fogarty and D.R. Ortiz, "Limitations upon interagency information sharing: The Privacy Act of 1974," pp. 127-128 in *Protecting America's Freedom in the Information Age* (Markle Foundation Task Force on National Security in the Information Age), Markle Foundation, New York, N.Y., 2002, supra at 128.

applies to federal agencies—it does not generally regulate the collection of personal information by private-sector entities. In short, the Privacy Act provides limited protection when government-collected data are involved, and very little when private-sector data are involved.

## F.5 EXECUTIVE ORDER 12333 (U.S. INTELLIGENCE ACTIVITIES)

Promulgated on December 4, 1981, Executive Order (EO) 12333 regulates the conduct of U.S. intelligence activities.[44] Section 2.2 of EO 12333 sets forth "certain general principles that, in addition to and consistent with applicable laws, are intended to achieve the proper balance between the acquisition of essential information and protection of individual interests." Using a definition of United States person specified in Section 3.4(i) of this order (a United States person is "a United States citizen, an alien known by the intelligence agency concerned to be a permanent resident alien, an unincorporated association substantially composed of United States citizens or permanent resident aliens, or a corporation incorporated in the United States, except for a corporation directed and controlled by a foreign government or governments"), Section 2.3 of EO 12333 establishes constraints on procedures for agencies within the intelligence community (IC) to collect, retain or disseminate information concerning United States persons.

Under EO 12333, only certain types of information may be collected, retained, or disseminated by IC agencies. These types of information include "information that is publicly available or collected with the consent of the person concerned; information constituting foreign intelligence or counterintelligence, including such information concerning corporations or other commercial organizations; information obtained in the course of a lawful foreign intelligence, counterintelligence, international narcotics or international terrorism investigation; information needed to protect the safety of any persons or organizations, including those who are targets, victims or hostages of international terrorist organizations; information needed to protect foreign intelligence or counterintelligence sources or methods from unauthorized disclosure; information concerning persons who are reasonably believed to be potential sources or contacts for the purpose of determining their suitability or credibility; information arising out of a lawful personnel, physical or communications security investigation; information acquired by overhead reconnaissance not directed at specific United States persons; incidentally obtained information that may indicate involvement in activities that may violate

---

[44]The full text of EO 12333 can be found at http://www.tscm.com/EO12333.html.

federal, state, local or foreign laws; and information necessary for administrative purposes."

Under Section 2.4 of EO 12333, IC agencies are required to use the least intrusive collection techniques feasible within the United States or directed against United States persons abroad. In addition, this section places certain limitations on various agencies. For example, the Central Intelligence Agency is forbidden to engage in electronic surveillance within the United States except for the purpose of training, testing, or conducting countermeasures to hostile electronic surveillance. In addition, no IC agency is allowed to conduct "physical surveillance of a United States person abroad to collect foreign intelligence, except to obtain significant information that cannot reasonably be acquired by other means." (See the full text of the EO for additional restrictions.)

## F.6 THE ADEQUACY OF TODAY'S ELECTRONIC SURVEILLANCE LAW

The law applicable to surveillance and intelligence gathering and the attention to limitations in the law suggests that the law suffers from what Professor Daniel Solove has described as "profound complexity."[45] Professor Orin Kerr has written that "the law of electronic surveillance is famously complex, if not entirely impenetrable."[46] Courts agree with these assessments and have "described surveillance law as caught up in a 'fog,' 'convoluted,' 'fraught with trip wires,' and 'confusing and uncertain.'"[47]

Why is today's law regarding electronic surveillance complex? Some of the complexity is certainly due to the fact that the situations and circumstances in which electronic surveillance may be involved are highly varied, and policy makers have decided that different situations and situations call for different regulations. That is, different treatment of electronic surveillance in different situations is a consequence of legislative and executive branch policy choices to treat these situations differently.

But it is another issue as to whether such differences, noted and established in a one particular set of circumstances, can be effectively maintained over time. First, circumstances evolve. For example, today's law includes major distinctions based on the location of the surveillance, the purposes for which the intercepted information is sought, and whether

---

[45]D.J. Solove, "Reconstructing electronic surveillance law," *George Washington Law Review* 72, 2004. The article provides a description and analysis of electronic surveillance law in the United States.

[46]O.S. Kerr, "Lifting the 'fog' of internet surveillance: How a suppression remedy would change computer crime law," *Hastings Law Journal* 54:805-820, 2003.

[47]D.J. Solove, op. cit., p. 1293.

the target is a "U.S. person" or a "non-U.S. person." Yet these distinctions are difficult to apply in a world of digital communications and networks that do not easily recognize national borders, terrorist threats of foreign origin that are planned or executed within the borders of the United States, and the growing integration of foreign intelligence, domestic intelligence, and law enforcement.

Another important distinction is the historical separation between criminal and national security investigations. Since September 11, 2001, some of the barriers separating criminal and national security investigations have been lowered (for example, the government is now freer to share information gathered by law enforcement in criminal investigations with national security authorities, and vice versa). However, the ECPA and the FISA are based on the existence of clear distinctions between criminal and national security investigations, as reflected in their disparate treatment of information that is collected and stored under each regime.

Second, evolving technologies also complicate the application of laws and precedents created in an earlier technological era, and at times existing law seems outpaced by technological change. In 2004, the Department of Defense Technology and Privacy Advisory Committee (TAPAC) wrote in its final report:

> Laws regulating the collection and use of information about U.S. persons are often not merely disjointed, but outdated. Many date from the 1970s, and therefore fail to address extraordinary developments in digital technologies, including the Internet. . . . Dramatic advances in information technology, however, have greatly increased the government's ability to access data from diverse sources, including commercial and transactional databases. . . .
>
> . . . Current laws are often inadequate to address the new and difficult challenges presented by dramatic developments in information technologies. And that inadequacy will only become more acute as the store of digital data and the ability to search it continue to expand dramatically in the future.[48]

As an example, the ECPA draws a sharp distinction regarding whether a message is "in transit" or "in storage." When ECPA was adopted in 1986, users downloaded e-mail from their service provider onto their local computer. Messages therefore were not stored centrally after being read. Today, many e-mail systems are accessed through Web interfaces, so e-mail is by default stored on servers belonging to third parties. Thus, according to an analysis by the Center for Democracy and Technology, "As a result of ECPA's complex rules, the same email mes-

---

[48]U.S. Department of Defense, Technology and Privacy Advisory Committee, *Safeguarding Privacy in the Fight Against Terrorism*, March 2004, p. 6.

sage will be subject to many different rules during its life span. These complex rules likely do not match the expectations of email users."[49]

The government exploits such distinctions. The Federal Bureau of Investigation's Key Logger System, which records individuals' keystrokes on their computers, was designed to collect data only when the users' machines are not connected to the Internet. When a user logs on, the keystroke recording stops, so that the agency argues that the device is not capturing communications "in transit," but merely "in storage," and therefore is not required to comply with Title I of the ECPA.[50]

A second example is that when the statutory authorization was adopted for the National Security Agency (NSA) to carry out electronic surveillance outside of the United States, it was highly unusual for ordinary persons in the U.S. to make international phone calls, and e-mail did not yet exist.[51] Today, the proliferation of information technology into the population at large means that many ordinary people in the U.S. make international phone calls and use e-mail, with the result that many more communications of ordinary people are potentially subject to NSA surveillance.[52] To be sure, a variety of regulations exist to prevent just such occurrences from intruding on the privacy of ordinary Americans, but it is undeniable that more communications involving Americans will fall within the ambit of electronic surveillance directed outside U.S. borders as global communications increase.

Third, the law today embeds in some significant inconsistencies. For example, the very high protection for communications under Title I of ECPA does not extend to video surveillance if sounds are not captured at the same time. Meanwhile, the much weaker protection of FISA does apply. "Foreign agents therefore receive protection against silent video surveillance whereas United States citizens do not."[53] Similarly, protection for stored communications hinges on whether the message has been stored for more than 180 days. Why? Telephone calls and e-mail receive significantly different protection from government surveillance without any apparent reason.

Fourth, key intelligence questions remain without clear answers. For example, do any of these laws apply to "data mining" or searches for keywords or relationships conducted by computer? Is it possible to show

---

[49]Center for Democracy and Technology (CDT), *Digital Search & Seizure: Updating Privacy Protections to Keep Pace with Technology*, CDT, Washington, D.C., 2006, p. 11.

[50]See *United States v. Scarfo*, 180 F. Supp. 2d 572 (D.N.J. 2001); see generally D.J. Solove, op. cit., pp. 1281-1282.

[51]Center for Democracy and Technology (CDT), *Digital Search and Seizure: Updating Privacy Protections to Keep Pace with Technology*, CDT, Washington, D.C., 2006.

[52]Ibid.

[53]D.J. Solove, op. cit., p. 1293.

probable cause, under either the high standard of Title I of ECPA or the weaker standard of FISA, for searches that target a pattern of behavior rather than an identified person? How should opened e-mail and voice mail messages be treated? DOJ argues that they are merely remotely stored files and therefore do not fall within the protection of Title II of ECPA.[54] Why aren't they simply stored communications that are directly covered by Title II (the Stored Communications Act)?[55]

Finally, the slow pace at which law has evolved in the face of changing technologies may have done more to undermine rather than enhance trust in information sharing. The Supreme Court initially refused to apply the Fourth Amendment to wiretapping at all,[56] and it took the Court 39 years to reverse that decision.[57] Conversely, in 1934 Congress prohibited wiretapping in any form and for any purpose.[58] It took 34 years before Congress recognized the potential of electronic surveillance, properly regulated, to aid law enforcement,[59] and another twelve before it statutorily authorized its use to advance national security.[60] Congress also receives only limited information about surveillance conducted under ECPA and FISA, and even less about the Administration's surveillance conducted outside of this statutory framework. There is no federal reporting requirement about electronic surveillance by states, which account for the majority of wiretaps, and only half of the states in fact report statistics about their wiretap orders.[61]

---

[54]Computer Crime and Intellectual Property Section, U.S. Department of Justice, *Manual on Searching and Seizing Computers and Obtaining Electronic Evidence in Criminal Investigations* III.B, 2001.

[55]For more detailed analyses of gaps and inconsistencies in statutory and Fourth Amendment protections, see P.L. Bellia, "Surveillance law through cyberlaw's lens," *George Washington Law Review* 72:1375, 2004; D.K. Mulligan, "Reasonable expectations in electronic communications: A Critical perspective on the Electronic Communications Privacy Act," *George Washington Law Review* 72:1557, 2004; D.J. Solove, "Reconstructing electronic surveillance law," *George Washington Law Review* 72:1264, 2004; P.P. Swire, "The system of foreign intelligence surveillance law," *George Washington Law Review* 72:1306, 2004; O.S. Kerr, "Internet surveillance law after the USA Patriot Act: The big brother that isn't," *Northwestern University Law Review* 97(2):607-673, 2003; O.S. Kerr, "Lifting the 'fog' of internet surveillance: How a suppression remedy would change computer crime law," *Hastings Law Journal* 54:805-820, 2003.

[56]*Olmstead v. United States*, 277 U.S. 438 (1928).

[57]*United States v. Katz*, 389 U.S. 347 (1967).

[58]Communications Act of 1934, ch. 652, § 605, 48 Stat. 1064 (codified as amended at 47 U.S.C. § 605).

[59]Omnibus Crime Control and Safe Streets Act of 1968, Public Law 90-351, § 802, 82 Stat. 212 (codified as amended at 18 U.S.C. § 2510-2520).

[60]Foreign Intelligence Surveillance Act of 1978, Public Law 95-511, 92 Stat. 1783 (codified at 50 U.S.C. § 1801-1811).

[61]D.J. Solove, op. cit., p. 1296.

What does the analysis above imply for changing today's law regarding electronic surveillance? There is broad agreement that today's legal regime is not optimally aligned with the technological and circumstantial realities of the present. But there is profound disagreement both about whether the basic principles underlying today's regime continue to be sound and about the directions in which changes to today's regime ought to occur. Some analysts believe that the privacy has suffered as the result of an increasing gap between technology/circumstances and the more slowly changing law, while others believe that technological change is upsetting the traditional balance away from the legitimate needs of law enforcement and national security.

### F.7  FURTHER REFLECTIONS FROM THE TECHNOLOGY AND PRIVACY ADVISORY COMMITTEE REPORT

Many of the issues discussed above were also flagged in the report issued by the TAPAC, a bipartisan panel of independent legal experts and former government officials appointed by Secretary of Defense Donald Rumsfeld in the wake of the TIA [Total/Terrorist Information Awareness program; see Appendix J] debacle. For example, the report noted that the risks to informational privacy of government data mining efforts were exacerbated by disjointedness in the laws applicable to data mining. Thus, programs that appear to pose similar privacy risks are subject to a variety of often inconsistent legal requirements. Such inconsistencies, the report argued, reflected "the historical divide in the United States between laws applicable to law enforcement and those applicable to foreign intelligence and national security activities, as well as the different departments, contexts, and times in which those programs were developed."

It also noted that depending on which department developed the tools, the use of data mining to protect the homeland was either required or prohibited and that today's laws regulating the collection and use of information about U.S. persons were created in the 1970s, and thus do not take into account recent developments in digital technologies, including the Internet. Pointing out that "the ubiquity of information networks and digital data has created new opportunities for tracking terrorists and preventing attacks," the report argued that "new technologies [also] allow the government to engage in data mining with a far greater volume and variety of data concerning U.S. persons, about whom the government has no suspicions, in the quest for information about potential terrorists or other criminals" and that then-current laws were "often inadequate to address the new and difficult challenges presented by dramatic developments in information technologies."

The TAPAC report concludes that "[t]hese developments highlight the need for new regulatory boundaries to help protect civil liberties and national security, and to help empower those responsible for defending our nation to use advanced information technologies—including data mining appropriately and effectively. It is time to update the law to respond to new challenges."[62]

---

[62]U.S. Department of Defense, Technology and Privacy Advisory Committee, *Safeguarding Privacy in the Fight Against Terrorism*, March 2004, p. ix.

# G

# The Jurisprudence of Privacy Law and the Need for Independent Oversight

Privacy protection rules regulating law enforcement and national security use of personal information can be usefully understood in two distinct categories: first, substantive rules that limit access to and usage of private information and, second, procedural rules that provide safeguards to encourage compliance and ensure accountability for compliance failures.

Neither the Constitution nor any statute can anticipate in advance every particular privacy issue raised by future technologies. So the evolving balance between the government's need to intrude on the private lives of individuals in the service of its public safety mission and the requirement to maintain individual liberty has been maintained over time by providing a degree of transparency in the use of new technologies, along with accountability to rules assured by judicial and legislative oversight. As new technologies and investigative techniques come into use, courts and legislatures have the opportunity to review these advances and make assessments of their privacy impact, guided by constitutional and public policy foundations. When new privacy risks arise or when the government powers are judged to have been extended beyond the boundaries established through the democratic process, corrective action can be taken. In order for this dynamic equilibrium of privacy and public safety to be maintained, however, transparency of the investigative process and accountability to the rule of law are essential. This appendix presents both the substantive constitutional foundations of privacy rights necessary for evaluating new technology, along with a consideration of transparency, accountability, and oversight mechanisms necessary to keep counterterrorism activities within view of the democratic process.

## G.1 SUBSTANTIVE PRIVACY RULES

In general, substantive privacy rules involve restrictions on access to and use of personal information by the government. Such restrictions are a means of limiting the power of government and private-sector institutions. For example, in the spirit of the bedrock constitutional principle of limited government, the Fourth Amendment defines limits on government power by establishing individual rights against certain intrusions. It protects privacy not only because Americans value individual liberty as an end in and of itself, but also because their collective political, cultural, and social flourishing depends on it. To this end, privacy protections generally take the form of boundaries between individuals and institutions (or sometimes other individuals). These boundaries may limit the information that is collected (in the case of wiretapping or other types of surveillance), how that information is handled (the fair information practices that seek care and openness in the management of personal information described in Box G.1), or rules governing the ultimate use of information (such as prohibitions on the use of certain health information for making employment decisions).

Today, a variety of new technologies put pressure on existing boundaries between individuals and large institutions. New surveillance and analysis technologies used in the service of counterterrorism goals are effective precisely because they give investigators new capabilities that erode the boundaries previously established between individuals and governments. For example, data mining techniques operating over large collections of information, each element of which is not particularly revealing, may yield detailed profiles of individuals, and location-aware sensor networks allow collection of tracking information on large numbers of individuals when most of them are not actually suspected of any crime at the time of data collection. New identification documents (including driver's licenses and passports) will collect biometric information in digital form on most of the population, marking the first time the digital images of the faces of the population will be available for law enforcement use. All of these technologies are susceptible to a wide variety of different uses, with widely varying intrusiveness.

### G.1.1 Privacy Challenges Posed by Advanced Surveillance and Data Mining

Many of the privacy questions facing the information age society are

## BOX G.1
## Fair Information Practices

Fair information practices are standards of practice required to ensure that entities that collect and use personal information provide adequate privacy protection for that information. These practices include notice to and awareness of individuals with personal information that such information is being collected, providing them with choices about how their personal information may be used, enabling them to review the data collected about them in a timely and inexpensive way and to contest that data's accuracy and completeness, taking steps to ensure that their personal information is accurate and secure, and providing them with mechanisms for redress if these principles are violated.

Fair information practices were first articulated in a comprehensive manner in the U.S. Department of Health, Education and Welfare's 1973 report *Records, Computers and the Rights of Citizens.*[1] This report was the first to introduce the Code of Fair Information Practices, which has proven influential in subsequent years in shaping the information practices of numerous private and governmental institutions and is still well accepted as the gold standard for privacy protection.[2]

From their origin in 1973, fair information practices "became the dominant U.S. approach to information privacy protection for the next three decades."[3] Their five principles not only became the common thread running through various bits of sectoral regulation developed in the United States, but also they were reproduced, with significant extension, in the guidelines developed by the Organization for Economic Co-operation and Development (OECD). These principles are extended in the OECD guidelines, which govern "the protection of privacy and transborder flows of personal data" and include eight principles that have come to be understood as "minimum standards . . . for the protection of privacy and individual liberties."[4] The OECD guidelines also include a statement about the degree to which data controllers should be accountable for their actions. This generally means that there are costs associated with the failure of a data manager to enable the realization of these principles.

---

[1] U.S. Department of Health, Education, and Welfare, *Records, Computers and the Rights of Citizens*, Report of the Secretary's Advisory Committee on Automated Personal Data Systems, MIT Press, Cambridge, Mass., 1973.

[2] Fair information principles are a staple of the privacy literature. See, for example, the extended discussion of these principles in D. Solove, M. Rotenberg, and P. Schwartz, *Information Privacy Law*, Aspen Publishers, New York N.Y., 2006; A. Westin, "Social and political dimensions of privacy," *Journal of Social Issues* 59(2):431-453, 2003; H. Nissenbaum, "Privacy as contextual integrity," *Washington Law Review* 79(1):119-158, February 2004; and an extended discussion and critique in R. Clarke, "Beyond the OECD guidelines: Privacy protection for the 21st century," available at http://www.anu.edu.au/people/Roger.Clarke/DV/PP21C.html.

[3] A. Westin, "Social and political dimensions of privacy," *Journal of Social Issues* 59(2):431-453, 2003, p. 436.

[4] M. Rotenberg, *The Privacy Law Sourcebook 2001*, Electronic Privacy Information Center, Washington, D.C., 2001, pp. 270-272.

As enunciated by the U.S. Federal Trade Commission (other formulations of fair information practices also exist),[5] the five principles of fair information practice include:

- *Notice and awareness.* Secret record systems should not exist. Individuals whose personal information is collected should be given notice of a collector's information practices before any personal information is collected and should be told that personal information is being collected about them. Without notice, an individual cannot make an informed decision as to whether and to what extent to disclose personal information. Notice should be given about the identity of the party collecting the data; how the data will be used and the potential recipients of the data; the nature of the data collected and the means by which it is collected; whether the individual may decline to provide the requested data and the consequences of a refusal to provide the requested information; and the steps taken by the collector to ensure the confidentiality, integrity, and quality of the data.

- *Choice and consent.* Individuals should be able to choose how personal information collected from them may be used, and in particular how it can be used in ways that go beyond those necessary to complete a transaction at hand. Such secondary uses can be internal to the collector's organization, or they can result in the transfer of the information to third parties. Note that genuinely informed consent is a sine qua non for observation of this principle. Individuals who provide personal information under duress or threat of penalty have not provided informed consent—and individuals who provide personal information as a requirement for receiving necessary or desirable services from monopoly providers of services have not, either.

- *Access and participation.* Individuals should be able to review in a timely and inexpensive way the data collected about them and to similarly contest those data's accuracy and completeness. Thus, means should be available to correct errors or, at the very least, to append notes of explanation or challenges that would accompany subsequent distributions of this information.

- *Integrity and security.* The personal information of individuals must be accurate and secure. To ensure data integrity, collectors must take reasonable steps, such as using only reputable sources of data and cross-referencing data against multiple sources, providing consumer access to data, and destroying untimely data or converting it to anonymous form. To provide security, collectors must take both procedural and technical measures to protect against loss and the unauthorized access, destruction, use, or disclosure of the data.

- *Enforcement and redress.* Enforcement mechanisms must exist to ensure that the fair information principles are observed in practice, and individuals must have redress mechanisms available to them if these principles are violated.

---

[5]See http://www.ftc.gov/reports/privacy3/fairinfo.htm.

SOURCE: National Research Council, *Engaging Privacy and Information Technology in an Information Age*, J. Waldo, H.S. Lin, and L. Millett, eds., The National Academies Press, Washington, D.C., 2007.

both challenging as a matter of public policy and difficult because they seem to call into question the adequacy of much of existing privacy law. A strong constitutional foundation constrains government actions and policies: the Fourth Amendment guarantee against unreasonable search and seizure. Although the U.S. Supreme Court has rejected the idea that the Fourth Amendment protects a general right to privacy, the amendment does create boundaries between the citizen and the powers of the state in certain domains. In the words of the Court,[1]

> [T]he Fourth Amendment cannot be translated into a general constitutional "right to privacy." That Amendment protects individual privacy against certain kinds of governmental intrusion, but its protections go further, and often have nothing to do with privacy at all. Other provisions of the Constitution protect personal privacy from other forms of governmental invasion. But the protection of a person's general right to privacy—his right to be let alone by other people—is, like the protection of his property and of his very life, left largely to the law of the individual States.

Yet the Court's interpretation of what types of intrusions raise constitutional questions has been flexible over time, reflecting the underlying values of the Fourth Amendment. In *Katz* and subsequent cases, the Court rejected the idea that rights conferred by the Fourth Amendment are determined by fixed physical or technical boundaries. As the Court explained:

> [T]he Fourth Amendment protects people, not places. What a person knowingly exposes to the public, even in his own home or office, is not a subject of Fourth Amendment protection. See *Lewis v. United States*, 385 U.S. 206, 210; *United States v. Lee*, 274 U.S. 559, 563. But what he seeks to preserve as private, even in an area accessible to the public, may be constitutionally protected.

The actual boundaries of the Fourth Amendment have changed over time, shaped by changing technological capabilities, social attitudes, government activities, and Supreme Court justices, as indicated by a series of Supreme Court decisions in the past century. The nature of this evolution, driven both by judicial intervention and legislative action, demonstrates an ongoing and vital role for policy makers and jurists to ensure that the values reflected in the Fourth Amendment are kept alive in the face of new technologies.

Whether a government activity is permissible under the Fourth Amendment is determined by the answer to two basic questions: (1) is the action a "search" within the meaning of the Constitution and (2) if it is a

---

[1] *Katz v. United States*, 389 U.S. 347 (1967).

search, is it "reasonable." A government action is considered a search if it crosses into some recognized private interest. What constitutes a private domain is sometimes easy to determine based on history and culture, the canonical one being a person's private home. But a dependence on history implies that there is no fixed definition of "private domain," and thus it falls to statutory law, judicial action, and executive branch action to protect privacy within whatever definition of private domain has been defined at the time.

As an example of how constitutional jurisprudence and statutory law have interacted to strike a balance between the protection of privacy and the legitimate needs of law enforcement, consider the legal framework surrounding telephone wiretaps. This framework requires communications carriers to provide law enforcement agencies (LEA) with lawful intercepts (LI) (e.g., wiretaps) on specific telephones under specific conditions once the appropriate warrants or subpoenas have been issued and presented. LI laws require that wiretaps be very specific—for an ongoing investigation, for a specific subject (individual), and often for a specific form of communications, such as a wire-line telephone and for a specific telephone number.

While some may have an image of a detective sitting in a smoky hotel room listening to a call via a wire tapped to a phone in an adjacent room, modern LI or electronic eavesdropping is automated as an integral part of the telecommunications infrastructure. When a law enforcement agency provides a communications carrier with a warrant for a specific lawful intercept, the communications carrier is obligated to intercept the content of specific communications and deliver them to the agency. Typically, this involves routing calls to a location designated by the law enforcement agency, where the information is captured and stored for later analysis using technologies designed for communications surveillance and analysis.

The first step in establishing today's legal framework was the *Katz* decision of the Supreme Court in 1967. In that decision, the Court found that a person in a telephone booth had a reasonable expectation of privacy and thus the content of phone conversations in the booth was entitled to the protections of the Fourth Amendment. In response to this decision, Congress passed the 1968 Wiretap Act (now known as Title III).

Strictly speaking, *Katz v. United States* addressed only the issue of whether the Fourth Amendment applied to telephone conversations held in a public telephone booth.[2] But the Wiretap Act imposed requirements on law enforcement officials to obtain warrants for wiretaps on conversa-

---

[2]Under previous precedent and law, warrants were required in order to tap phone conversations held in private residences.

tions regardless of where they were being held, and it laid out a variety of conditions that had to be met for a warrant to be issued—conditions that were not stipulated in the *Katz* decision.

Since the Wiretap Act was passed, Congress has enacted many laws governing lawful intercepts, including the Communications Assistance for Law Enforcement Act of 1994 (CALEA), the Pen Register/Trap and Trace Provisions of Title 18; and the Interception and Pen/Trap provisions of the Foreign Intelligence Surveillance Act, Title 50 U.S.C. Sections 1801-1845 (FISA). These laws have a historical context that reflects technology prevalent when the laws were written, and they were sometimes passed in response to a new Supreme Court decision or public demands for greater privacy acted on by the legislature.

### G.1.2 Evolution of Regulation of New Technologies

New technologies often pose challenges for courts and policy makers in deciding whether the surveillance power made available constitutes a permissible intrusion on a private interest.

Table G.1 illustrates the ways in which the law of electronic surveillance has evolved privacy protections over time. For example, early wiretapping was found not to violate the Fourth Amendment in 1928, but when the Supreme Court considered the question of telephone surveillance again in 1967, it reversed itself and found that citizens' reasonable expectation of privacy in private telephone calls meant that surveillance of telephone calls could be done only with a judicially approved warrant and ongoing supervision of a "detached, neutral magistrate." As reliance on new communications technologies continued, Congress stepped in to establish basic privacy protections and provisions for law enforcement access to electronic mail. In an important instance of proactive legislative action determined to be necessary to provide stable privacy protection for a new electronic communications medium, Congress acted on the belief that "the law must advance with the technology to ensure the continued vitality of the Fourth Amendment."[3]

With the advent of the World Wide Web, congressional action extended privacy protections to web and e-mail access transaction logs, along with clear procedures for legitimate law enforcement access with judicial supervision.

A more recent consideration of a new technology—use of infrared scanning of a person's home for the purpose of detecting high-heat plant grow lights (indicating a possible indoor marijuana farm)—was found to

---

[3]Senate Judiciary Committee Report on the Electronic Communications Privacy Act of 1986 (S. 2575), Report 99-541, 99th Congress, 2d Session, 1986, p. 5.

TABLE G.1  Historical Evolution of Regulation of Electronic Surveillance

|       | Communications Technology | Law | Regulatory Approach |
|-------|---------------------------|-----|---------------------|
| 1890s | Telegraph | State wiretapping crimes | Criminal prohibition |
| 1934 | Telephone | Communications Act of 1934 | Criminal prohibition and inadmissibility of evidence |
| 1967 | Telephone and bugging equipment | Fourth Amendment | Inadmissibility of evidence and civil liability |
| 1968 | Telephone and bugging equipment | Wiretap Act | Criminal prohibition, civil liability, and inadmissibility |
| 1978 | Telephone and bugging equipment | Foreign Intelligence Surveillance Act | Criminal prohibition, civil liability, and inadmissibility |
| 1986 | E-mail and Internet communications | Electronic Communications Privacy Act | Criminal prohibition and civil liability |
| 2001 | Telephone and Internet | USA Patriot Act | Criminal prohibition and civil liability |

be a violation of the Fourth Amendment. Justice Antonin Scalia found[4] that this transgressed the inviolability of the home, even though the police officers using the infrared detector did not actually enter the person's home.

Still another issue arising recently is the changing relevance of political boundaries to communications. Previous communications technologies and the laws that addressed them often reflected political boundaries. In today's communications technologies, political boundaries may be impossible to determine. This is significant, since a significant portion of the world's communications traffic is routed through U.S. switches, raising questions on the propriety of electronic eavesdropping on communications whose origins and destinations cannot be determined. In this case, recent legislation addresses this point,[5] allowing the warrant-

---

[4]"At the very core of the Fourth Amendment stands the right of a man to retreat into his own home and there be free from unreasonable governmental intrusion. With few exceptions, the question whether a warrantless search of a home is reasonable and hence constitutional must be answered no." (Kyllo, Scalia for the Court, internal quotation marks and citations omitted) *Kyllo v. United States.* 533 U.S. 27 (2001) (Scalia, J.).

[5]Protect America Act of 2007, Public Law 110-55, August 5, 2007.

less monitoring of communications that pass through the United States as long as both parties to the communications are reasonably believed to be located outside it.

As the above examples illustrate, the regulation of new technologies is affected by both by the evolution of the Supreme Court's Fourth Amendment jurisprudence on what constitutes a search and by policy choices within those boundaries made by the legislative and executive branches. Supreme Court jurisprudence on this point has evolved to limit government intrusion into certain "protected areas" because the founding fathers could not have anticipated the current and forthcoming advances in technology and practices. Furthermore, legislative and executive branch policy makers have expanded the scope of protections to ensure that constitutional values of limited government and protection of individual liberty are protected notwithstanding technological advances.

Legal analysts have a variety of views on whether regulation of new technologies should be driven by policy decisions or by the Fourth Amendment. For example, one view is that legislatures have considerable institutional advantages that enable the legislative privacy rules regulating new technologies to be more balanced, comprehensive, and effective than judicially created rules, and that the courts should adopt only modest formulations of Fourth Amendment protections in deference to these advantages.[6] Another view is that constitutionally derived protections of privacy in the face of new technologies are, by definition, more enduring and thus less subject to the often poorly justified actions of legislatures and executives, who may be acting in the heat of the moment after a terrorist incident.[7] Of course, in practice, regulation of new technologies has been influenced by both policy decisions and the Fourth Amendment.

It would be a mistake to infer from this brief history that every new surveillance technology is greeted automatically by courts and legislatures with a fixed, linear expansion of privacy protection. In some cases, long periods of time go by before a given technology receives clear privacy consideration. And it is certainly not possible to establish a priori clear measures of how much privacy protection ought to be brought to bear on new surveillance capabilities. What this history reveals is that careful consideration of the privacy impact of new surveillance powers has generally resulted in a measure of privacy protection that gives citizens confidence, while at the same time preserving apparently adequate

---

[6]See O.S. Kerr, "The Fourth Amendment and new technologies: Constitutional myths and the case for caution," *Michigan Law Review* 102(5):801-888, 2004.

[7]See, for example, P.M. Regan, *Legislating Privacy: Technology, Social Values, and Public Policy*, University of North Carolina Press, Chapel Hill, N.C., 1995, pp. 221-227.

access to data for law enforcement engaged in legitimate investigative activity.

### G.1.3 New Surveillance Techniques That Raise Privacy Questions Unaddressed by Constitutional or Statutory Privacy Rules

A number of the government counterterrorism investigative techniques with real privacy implications are likely to fall outside the boundary of what the Supreme Court today considers to be a search. Inasmuch as all levels of government are now seeking to use the most advanced, effective technologies to detect and apprehend terrorist threats, it is not surprising that many of these new technologies and techniques will have intrusive power not previously considered by either courts or legislators. The committee heard considerable testimony on the use of data mining for the purpose of identifying potential terrorist behavior. In many cases, the bulk of the information used in such data mining operations is collected from commercial data vendors and public records, such as property records, voting rolls, and other local and state databases. Access to these data is available with little or no privacy protection and little or no third-party supervision.[8] Furthermore, while data mining activity may be subject to procedural regulation under the federal Privacy Act (see Appendix F), it is not subject to any substantive statutory limitations whatsoever. It will be up to Congress to consider the appropriate limits on the use of data mining and other new privacy-invasive techniques.

### G.1.4 New Approaches to Privacy Protection: Collection Limitation Versus Use Limitation

There is growing agreement that regulation of large-scale analysis of personal information, such as data mining, will have to rely on usage limitations rather than merely collection limitations.[9] Historically, privacy

---

[8]See *Smith v. Maryland*, 442 U.S. 735 (1979) (finding no reasonable expectation of privacy transactional records of phone numbers dialed because they were "disclosed" voluntarily and duly recorded by the phone company in the ordinary course of business); and *United States v. Miller*, 425 U.S. 435 (1976) (finding no Fourth Amendment interest in banking records, since they are not confidential communications and are voluntarily presented to the bank). See also J.X. Dempsey and L.M. Flint, "Commercial data and national security," *George Washington Law Review* 72(6):6, August 2004.

[9]See the reports from the Markle Foundation Task Force on National Security in the Information Age: *Mobilizing Information to Prevent Terrorism: Accelerating Development of a Trusted Information Sharing Environment*, 3rd Report, July 13, 2006; *Creating a Trusted Network for Homeland Security*, 2nd Report, December 2, 2003; and *Protecting America's Freedom in the Information Age*, 1st Report, October 7, 2002. Available at http://www.markletaskforce.org/ [11/7/07].

against government intrusion has been protected by limiting what information the government can collect: voice conversations collected through wiretapping, e-mail collected through legally authorized access to stored data, etc. But today, as the data mining discussion in Appendix H illustrates, the greatest potential for privacy intrusion may come from analysis of data that are accessible to government investigators with little or no restriction and little or no oversight. The result is that powerful investigative techniques with significant privacy impact proceed in full compliance with existing law, but with significant unanswered privacy questions and associated concerns about data quality. However, attempts to limit collection of or access to the data that feed data mining activities may create significant burdens on legitimate investigative activity without producing any real privacy benefit. In many cases, the data in question have already been collected and access to them, under the third-party business records doctrine, will be readily granted with few strings attached.

The privacy impact of new analytic techniques that merit regulation is not access to any individual element of personal information, but rather to the overall use of a large quantity of individually innocuous items of personal information. As the debate over airline passenger screening systems has shown,[10] the main objections to proposed profiling systems are in the potential for "mission creep" and the risk of inaccurate data being used against innocent citizens.

The challenge before policy makers is how to craft appropriate privacy regulation that achieves the historic but dynamic balance between privacy protection and important public safety priorities. Establishing clear usage limitations along with traditional procedural oversight safeguards on new data analysis techniques—including but not limited to data mining—would ensure that the most powerful new investigative techniques are available against the serious threats to national security. At the same time, given the substantial new and untested power that these techniques could confer on domestic law enforcement, their use in the nonnational security arena would be limited.

## G.2 PROCEDURAL PRIVACY RULES AND THE NEED FOR OVERSIGHT

The establishment of law and regulation in any given domain is an articulation of public concerns and values in that domain. But if law and

---

[10]See Department of Homeland Security, *Notice to Establish System of Records, Secure Flight Test Records,* 69 Fed. Reg. 57,345 (Sept. 24, 2004), available at http://edocket.access.gpo.gov/2004/04-21479.htm, and comments (link to criticism publications available at http://www.epic.org/privacy/airtravel/secureflight.html).

regulation are to have any substantive or tangible impact on behavior, mechanisms are necessary for ensuring that the targets of such law and regulation behave accordingly.

In the context of the Fourth Amendment, such mechanisms are provided through third-party review of government intrusions on private domains. When properly implemented, such mechanisms provide a significant measure of accountability to ensure that these intrusions are not abused.

### G.2.1 Oversight Mechanisms of the U.S. Government

The U.S. government is based on a three-way system of internal checks and balances that was designed to limit power and ensure reviews across the executive, legislative, and judicial branches. For example, the president nominates Supreme Court and Circuit and District Court judges and the Congress votes to confirm them. The president also proposes budgets, which Congress must approve. For national security issues, the Federal Bureau of Investigations (FBI) conducts investigations inside the United States that are subject to oversight by the Federal Intelligence Security Court (established by the Foreign Intelligence Surveillance Act).[11] In addition, Congress oversees the activities of the Department of Justice, which is responsible for the FBI.[12]

Congressional oversight of executive branch departments and agencies are especially potent and controversial because congressional committees have the power of subpoena and they control budget allocations. In addition, congressional committee meetings generate intense press and public interest, especially when they are investigating failures, corruption, or other malfeasance. This form of oversight, called "fighting fires," is contrasted with scheduled regular reviews, called "patrolling streets."[13]

Some politicians are attracted to fighting fires because of the high visibility, but the patrolling streets model can be successful in preventing

---

[11]J. Berman and L. Flint, "Guiding lights: Intelligence oversight and control for the challenge of terrorism," *Criminal Justice Ethics,* Winter/Spring 2003. Available at http://www.cdt.org/publications/030300guidinglights.pdf [11/7/07].

[12]C.J. Bennett and C.D. Raab, *The Governance of Privacy: Policy Instruments in Global Perspective.* MIT Press, Cambridge, Mass., 2006.

[13]M.D. McCubbins and T. Schwartz, "Congressional oversight overlooked: police patrols versus fire alarms," *American Journal of Political Science* 28(1):165-179, February 1984; A. Lupia and M.D. McCubbins, "Designing bureaucratic accountability," *Law and Contemporary Problems* 57(1):91-126, 1994; A. Lupia and M.D. McCubbins, "Learning from oversight: Fire alarms and police patrols reconstructed," *Journal of Law, Economics and Organization* 10(1):96-125, 1994; and H. Hopenhayn and S. Lohmann, "Fire-alarm signals and the political oversight of regulatory agencies," *Journal of Law, Economics and Organization* 12(1):196-213, 1996.

problems, although it takes more effort. Critics of congressional oversight suggest that at times members may not be sufficiently informed to ask the right questions or appreciate the complexities of the agencies they review.

Executive branch agencies often create independent oversight boards to review internal activities so as to improve performance and generate public trust. Examples include the National Aeronautic and Space Administration's Shuttle Oversight Board,[14] the Department of Energy's Performance Assurance Program Independent Oversight,[15] the Food and Drug Administration's (FDA) Drug Safety Board,[16] and the Nuclear Regulatory Commission's Oversight Committee.[17] In addition, agencies often have internal oversight committees and inspectors general who monitor compliance with policy.

The judicial branch is often the arbiter of government claims that invasions of individual privacy are needed to advance other national interests. For example, warrants that allow physical searches and orders that allow wiretapping are often issued by the judicial branch upon the showing of probable cause. In this way, the courts handle many cases of potential privacy violations by federal, state, and local police or other government agencies.

Federal agencies also conduct oversight of parts of the commercial sector to ensure adherence to legal requirements and consumer protection. Examples include the Federal Reserve's regulation of banking practices and the FDA's work on pharmaceutical testing and production. Government agencies can also be the source of trusted investigations, such as the work of the National Transportation Safety Board in studying plane crashes.

Other mechanisms to detect problems in government and other organizations are sometimes applied. Some organizations include ombudsmen whose role is to constantly review practices and respond to internal or external concerns. Another strategy that has legal protection in U.S. is whistle-blowing. Government employees who report illegal or improper

---

[14]See NASA, *Standing Review Board Handbook*, August 1, 2007; available at http://fpd.gsfc.nasa.gov/NPR71205D/SRB_Handbook.pdf.

[15]See U.S. Department of Energy, *Independent Oversight and Performance Assurance Program*, DOE O 470.2B, October 31, 2002; available at http://hss.energy.gov/IndepOversight/guidedocs/o4702b/470-2b.html.

[16]See information on the Food and Drug Administration's Drug Safety Oversight Board at U.S. Food and Drug Administration, "FDA Improvements in Drug Safety Monitoring," FDA Fact Sheet, February 15, 2005; available at http://www.fda.gov/oc/factsheets/drugsafety.html.

[17]U.S. Senate, Committee on Environment and Public Works, Subcommittee on Clean Air and Nuclear Safety.

activity are protected by several laws, especially the Whistleblower Protection Act (5 U.S.C. § 1221(e)).

These mechanisms for oversight vary in the extent to which they are (and are perceived to be) independent. Independence (along with the necessary authority) is a key dimension of oversight because of the suspicion—often warranted—that oversight controlled or influenced by the entity being overseen is not meaningful and that problems revealed by nonindependent oversight will be concealed or improperly minimized. Independent oversight mechanisms also generally have greater ability to bring fresh and unbiased perspectives to an organization that is caught up in its day-to-day work.

### G.2.2 A Framework for Independent Oversight

The rich variety of independent oversight strategies makes it difficult to compare them and recognize missing features. A framework for understanding the organizational structures and operating methods would therefore help identify best practices and sources of successful outcomes. Some clarity can be gained by taking a "who, when, how, what" approach, as described by these components:

Who:   Ensure independence.
When:  Choose time for review.
How:   Set power to investigate.
What:  Raise impact of results.

Administrators will need to tune the process to fit each situation, but these components can serve as a starting point. First, some definitions: the independent oversight board members are referred to as *members*, and their goal is to review the operation of an *organization* led by *administrators* who supervise *employees*.

### Ensure Independence

Attaining the right level of independence means that the independent oversight board members are distant enough from the organization and employees so that their judgments are free from personal sympathy, coercion, bias, or conflicts of interest. However, they need to be close enough to be familiar with the organization and its operation. Members need to be knowledgeable about the domain of work and experienced at doing reviews of other organizations. In corporate audits, the independent accounting firms have helpful expertise in that they review multiple corporations, so they are familiar with standard and risky practices in each industry. Many analysts believe that the fraudulent business practices that led to the collapse of Enron could have been flagged by its accounting firm, Arthur Anderson, but they failed to do so because of their lack of independence.

Independent oversight board members should be trusted individuals whose credibility is also respected because of their career accomplishments. In addition to distance, experience, trust, and credibility, another issue tied to independence is transparency. While some activities may need to be kept private, the process should be visible enough.

### Match Nature of Review to Appropriate Stage in Event Trajectory

Oversight typically occurs at three points: before, during, and after some activity. These forms of oversight can be regarded as relevant to planning (approval of a proposed activity), execution (monitoring of activity while it is happening), and retrospective review (review of a completed activity) (see Figure G.1.)

- *Planning oversight* occurs when a specific activity has been planned but before any work begins. For example, the FISA court reviews approximately 2,000 plans for surveillance by the FBI each year; rejections are extremely rare. Another government example is the Defense Base Closure and Realignment Commission, which reviews Department of Defense decisions. Academic examples include the approval of plans for medical experiments by institutional review boards. Planning oversight also includes review boards that are convened to help make critical decisions, such as the launch of National Aeronautics and Space Administration missions, the opening of natural preserves to oil drilling, or acceptance of papers for publication in scientific journals.

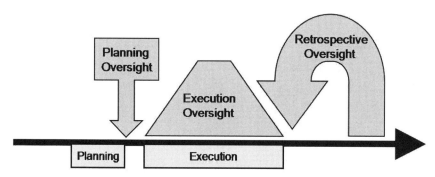

FIGURE G.1 Planning oversight is a check on plans, execution oversight is continuous review, and retrospective oversight reviews past performance.

- *Execution oversight* is the continuous oversight (patrolling streets) of a process, such as meat packing, pharmaceutical manufacture, or banking. Such processes can be labor-intensive and boring but require continuous vigilance. Independence is a challenge in these circumstances, since the oversight members may work closely with employees on a daily basis, thereby becoming personally familiar with them. The Federal Reserve Board has strict rules about personal contacts of its regulators with bank employees.

- *Retrospective oversight* occurs when the review covers previous organizational operations (fighting fires is one form) to validate performance and provide guidance for future performance. Corporate audits by independent accountants typically review a fiscal year and produce a report within 30-90 days. Audit reports must be filed with the Securities and Exchange Commission (SEC) and become public. University tenure committees are retrospective reviews, in that the members review career accomplishments of junior academics. University accreditation committees typically deal with retrospective as well as planning oversight; for example, they may review the past five years and plans for the next five years. In the U.S. government, the inspector general is an internal reviewer but sometimes functions to review other agencies or departments.

## Provide Authority to Investigate

Independent oversight boards typically receive written reports and live presentations, but in many cases they can ask questions of individuals or request further information. In some cases, they have subpoena power to require delivery of further information. A greater power to investigate raises the importance of an oversight committee and increases its perceived independence. The time limits on an independent oversight also influence its efficacy. A short review of a day or two for a review may not be enough to uncover problems, while long reviews can be a burden on organizations.

The forms of investigation vary widely, from simply reading of internal reports to extensive interviews with administrators and employees. Deeper investigations could assess organizational impact on others, such as customers, travelers, visa applicants, etc., by personal interview, survey questionnaire, or data collection (e.g., monitoring water quality).

## Disseminate Results of Oversight

Independent oversight boards typically produce a printed report, and its distribution is critical to its impact. If the only recipients are the administrators being reviewed, then there is a chance that the report will be ignored. If the recipients include employees, other stakeholders, journalists, and wider circles of the interested public, then the impact could be greater.

Independent oversight boards may also present their results verbally in private or public forums to the employees and administrators being reviewed. Some discussion may be allowed, and revised reports may be made. Such presentations can help ensure that the report is well understood and that appropriate clarifications are made, and recommendations for change contained in such reports are more likely to be implemented if they are made public.

Reports can also be made public and permanently available, as in SEC filings. A further possibility is that reports may include timetables for implementing changes and a review process to ensure that recommendations are followed.

Since there may be disagreements among independent oversight board members, a minority report may be included to allow strongly felt concerns to be raised by a subset of the members. Such minority reports, as in Supreme Court decisions, allow public exposure of alternate views that may be useful in future discussions.

### G.2.3  Applying Independent Oversight for Government Agencies to Protect Privacy

The U.S. Department of Homeland Security (DHS) has a difficult job that includes ensuring transportation safety, protecting national infrastructure, investigating terror threats, and many other tasks. For these and other purposes, DHS conducts extensive surveillance, which may invade the privacy of U.S. residents. DHS makes several efforts to assess its performance and provide internal and independent oversight. By statute, DHS has a chief privacy officer (Hugo Teufell III, appointed in July 2006), a Privacy Office, and a 20-member Data Privacy and Integrity Advisory Board (http://www.dhs.gov/xabout/structure/editorial_0510.shtm). The mission statement of the DHS Privacy Office is "to minimize the impact on the individual's privacy, particularly the individual's personal information and dignity." It remains to be seen whether the advisory board acts more as an internal review committee or a truly independent oversight board.[18]

Within DHS, the Citizenship and Immigration Services has an

---

[18]M. Rotenberg. *The Sui Generis Privacy Agency: How the United States Institutionalized Privacy Oversight After 9-11*, September 2006. Available at http://epic.org/epic/ssrn-id933690.pdf.

ombudsman office to help individuals and employers in resolving problems. They make an annual report to Congress and submit recommendations for internal improvements.

Other agencies, such as the FBI (in the Department of Justice) must request review of planned investigations by the FISA court, but they have rarely been turned down.[19] There does not seem to have been a retrospective review mechanism for the FISA court, or a retrospective review by the FISA court of FBI performance.

The president's Privacy and Civil Liberties Oversight Board (http://privacyboard.gov/) held its first meeting with six members in March 2006. This board could be helpful in generating public trust, but concerns about its independence and efficacy were raised after its first public presentation in December 2006. If this board can promote planning, execution, and retrospective oversight, it could emerge as a positive influence on many government agencies.

Public concern about warrantless domestic surveillance has become a controversial topic. A federal judge in Michigan found in July 2006 that government surveillance required review by a FISA court. After fighting this decision, the current administration agreed to FISA court oversight for at least some of their intelligence operations, but as this report is being written, the ultimate outcome of the relevant legislative proposals is unclear.

The traditional reliance on judicial review for privacy protection remains an effective process for dealing with evolving technologies and normative expectations. The judiciary's role in protecting the legal and privacy rights of citizens is effective because judicial decisions are a form of independent oversight that is widely respected. Furthermore, the rights it protects are established by the Constitution, which all branches of government are sworn to uphold.

Independent oversight is potentially very helpful for continuous improvement of government operations, especially when dealing with the complex issues of privacy protection. There are many forms of independent oversight and many strategies for carrying it out. Some government agencies conduct responsible independent oversight programs, but critics question their efficacy and independence. More troubling to critics are attempts to avoid, delay, or weaken independent oversight practices that are in place. Public discussion of independent oversight could help

---

[19]Electronic Privacy Information Center, Foreign Intelligence Surveillance Act Orders 1979-2007, updated May 8, 2008. Available at http://epic.org/privacy/wiretap/stats/fisa_stats.html. Some analysts interpret this fact to suggest that the FISA application process is more or less pro forma and does not provide a meaningful check on government power in this area, while others suggest that applications are done with particular care because the applicants know the applications will be carefully scrutinized.

resolve these differences and raise trust in government efforts to protect privacy.

### G.2.4  Collateral Benefits of Oversight

Ensuring compliance with policy is not the only benefit afforded by oversight. Indeed, administrators of government agencies face enormous challenges, not only from external pressures based on public concern over privacy, but also from internal struggles about how to motivate high performance while adhering to legal requirements and staying within budget.

Management strategies for achieving excellence in government agencies, corporations, and universities include many forms of internal review, measurement, and evaluation and a variety of strategies for external review. External reviews from consultants, advisory boards, or boards of visitors are designed to bring fresh perspectives that promote continuous improvement, while generating good will and respect from external stakeholders.

A well-designed oversight process can support the goal of continuous improvement and guide administrators in making organizational change, while raising public trust for an organization. Although many forms of oversight have been applied in corporate settings, the main approach is the board of directors. Such boards may be a weak form of oversight as they often mix internal with external participants who are less than independent. A stronger form of independent oversight and advice may come from external consultants or review panels that are convened for specific decisions or projects, but even stronger forms are possible.

For example, in the United States, corporate boards of directors are required to include an audit committee that is responsible for monitoring the external financial reporting process and related risks. An important role of the audit committee is to commission an external audit from an independent accounting firm, which is required annually for every publicly traded U.S. corporation by the SEC. These external audits are major events that provide independent oversight for financial matters with public reports to the SEC that become available to investors. In response to recent failures of independent oversight such as in the Enron and Worldcom bankruptcies, the Sarbanes-Oxley Act (2002) has substantially strengthened the rules.[20]

---

[20]J.C Thibodeau and D. Freier, *Auditing After Sarbanes-Oxley*, McGraw-Hill/Irwin, New York, N.Y., 2006.

# H

# Data Mining and Information Fusion

This appendix addresses the science and technology of data mining and information fusion and their utility in a counterterrorism context. The use of these techniques for counterterrorist purposes has substantial implications for personal privacy and freedom. While technical and procedural measures offer some opportunities for reducing the negative impacts, there is a real tension between the use of data mining for this purpose and the resulting impact on personal privacy, as well as other consequences from false positive identification. These privacy implications are primarily addressed in other parts of this report.

## H.1 THE NEED FOR AUTOMATED TECHNIQUES FOR DATA ANALYSIS

In the past 20 years, the amount of data retained by both business and government has grown to an extraordinary extent, mainly due to the recent, rapid increase in the availability of electronic storage and in computer processing speed, as well as the opportunities and competitiveness that access to information provides. Moreover, the concept of data or information has also broadened. Information that is retained for analytic purposes is no longer confined to quantitative measurements, but also includes (digitized) photographs, telephone call and e-mail content, and representations of web travels. This new view of what constitutes information that one would like to retain is inherently linked to a broader set of questions to which mathematical modeling has now been profitably applied. For example, handwritten text can now be considered to be data, and progress in automatic interpretation of handwritten text has already reached the point that over 80 percent of handwritten addresses are automatically read and sorted by the U.S. Postal Service every day. A problem of another type on which substantial progress has also been made is how to represent the information in a photograph efficiently in digital form, since every photograph has considerable redundancy in terms of information content. It is now possible to automatically detect and locate faces in digital images and, in some restricted cases, to identify the face by matching it against a database.

This new world of greatly increased data collection and novel approaches to data representation and mathematical modeling have been accompanied by the development of powerful database technologies that provide easier access to these massive amounts of collected data. These include technologies for dealing with various nonstandard data structures, including representing networks between units of interest and tools for handling the newer forms of information touched on above. A question not addressed here—but of considerable importance and a difficult challenge for the agencies responsible for counterterrorism in the United States—is how best to represent massive amounts of very disparate kinds of data in linked databases so that all relevant data elements that relate to a specific query can be easily and simultaneously accessed, contrasted, and compared.

Even with these new database management tools, the retention of data is still outpacing its effective use in many areas of application. The common concern expressed is that people are "drowning in data but starving for knowledge" (Fayyad and Uthurusamy[1] refer to this phenomenon as "data tombs"). This might be the result of several disconnects, such as collecting the wrong data, collecting data with insufficient quality, not framing the problem correctly, not developing the proper mathematical models, or not having or using an effective database management and query system. Although these problems do arise, in general, more and more areas of application are discovering novel ways in which mathematical modeling, using large amounts and new kinds of information, can address difficult problems.

Various related fields, referred to as knowledge discovery in databases (KDD), data mining, pattern recognition, machine learning, and information or data fusion (and their various synonyms, such as knowledge extraction and information discovery) are under rapid development and providing new and newly modified tools, such as neural networks,

---

[1] U. Fayyad and R. Uthurusamy, "Evolving data mining into solutions for insights," *Communications of the ACM* 45(3):28-31, 2002.

support vector machines, genetic algorithms, classification and regression trees, Bayesian networks, and hidden Markov models, to make better use of this explosion of information.

While there has been some overrepresentation of the gains in certain applications, these techniques have enjoyed impressive successes in many different areas.[2] Data mining and related analytical tools are now used extensively to expand existing business and identify new business opportunities, to identify and prevent customer churn, to identify prospective customers, to spot trends and patterns for managing supply and demand, to identify communications and information systems faults, and to optimize business operations and performance. Some specific examples include:

- In image classification, SKICAT outperformed humans and traditional computational techniques in classifying images from sky surveys comprising 3 terabytes ($10^{12}$ bytes) of image data.
- In marketing, American Express reported a 10-15 percent increase in credit card use through the application of marketing using data mining techniques.
- In investment, LBS Capital Management uses expert systems, neural nets, and genetic algorithms to manage portfolios totaling $600 million, outperforming the broad stock market.
- In fraud detection, PRISM systems are used for monitoring credit card fraud; more generally, data mining techniques have been dramatically successful in preventing billions of dollars of losses from credit card and telecommunications fraud.
- In manufacturing, CASSIOPEE diagnosed and predicted problems for the Boeing 737, receiving the European first prize for innovative application.
- In telecommunications, TASA uses a novel framework for locating frequently occurring alarm episodes from the alarm stream, improving the ability to prune, group, and develop new rules.
- In the area of data cleaning, the MERGE-PURGE system was successfully applied to the identification of welfare claims for the State of Washington.
- In the area of Internet search, data mining tools have been used to improve search tools that assist in locating items of interest based on a user profile.

Under their broadest definitions, data mining techniques include a

---

[2]U. Fayyad, G.P. Shapiro, and P. Smyth, "From data mining to knowledge discovery in databases," *AI Magazine* 17(3):37-54, 1996.

diverse set of tools for mathematical modeling, going by such names as knowledge discovery, machine learning, pattern recognition, and information fusion. The data on which these techniques operate may or may not be personally identifiable information, and indeed they may not be associated with individuals at all, although of course privacy issues are implicated when such information is or can be linked to individuals.

Knowledge discovery is a term, somewhat broader than that of data mining, which denotes the entire process of using unprocessed data to generate information that is easy to use in a decision-making context. Machine learning is the study of computer algorithms that often form the core of data mining applications. Pattern recognition refers to a class of data mining approaches that are often applied to sensor data, such as digital photographs, radiological images, sonar data, etc.

Finally, data and information fusion are data mining methods that combine information from disparate sources (often so much so that it is difficult to define a formal probabilistic model to assist in summarizing the information). Information fusion seeks to increase the value of disparate but related information above and beyond the value of the individual pieces of information ("obtaining reliable indications from unreliable indicators").

Because data mining has been useful to decision making in many diverse problem domains, it is natural and important to consider the extent to which such methodologies have utility in counterterrorism efforts, even if there is considerable uncertainty regarding the problems to which data mining can be productively applied.

One issue is whether and to what extent data mining can be effectively used to identify people (or events) that are suspicious with respect to possible engagement in activities related to terrorism; that is, whether various data sources can be used with various data mining algorithms to help select people or events that intelligence agents working in counterterrorism would be interested in investigating further. Data mining algorithms are proposed as being able to effectively rank people and events from those of greatest interest, with the potential to dramatically reduce the cases that intelligence agents have to examine.

Of course, human beings would be still required both to set the thresholds that delineate which people would receive further review and which would not (presumably dependent on available resources) and to check the cases that were selected for further inspection prior to any actions. That is, human experts would still decide, probably on an individual basis, which cases were worthy of further investigation.

A second issue is the possibility that data mining has additional uses beyond identifying and ranking candidate people and events for intelligence agents. Specifically, data mining algorithms might also be used as components of a data-supported counterterrorist system, helping to perform specific functions that intelligence agents find useful, such as helping to detect aliases, or combining all records concerning a given individual and his or her network of associates, or clustering events by certain patterns of interest, or logging all investigations into an individual's activity history. Data mining could even help with such tasks as screening baggage or containers. Such tools may not specifically rank people as being of interest or not of interest, but they could contribute to those assessments as part of a human-computer system. This appendix considers these possible roles in an examination of what is currently known about data mining and its potential for contributing to the counterterrorism effort.

An important related question is the issue of evaluating candidate techniques to judge their effectiveness prior to use. Evaluation is essential, first, because it can help to identify which among several contending methods should be implemented and whether they are sufficiently accurate to warrant deployment. Second, it is also useful to continually assess methods after they have been fielded to reflect external dynamics and to enable the methods to be tuned to optimize performance. Also, assuming that these new techniques can provide important benefits in counterterrorist applications, it is important to ask about the extent to which their application might have negative effects on privacy and civil liberties and how such negative effects might be ameliorated. This topic is the focus of Appendix L.

## H.2  PREPARING THE DATA TO BE MINED

It is well known by those engaged in implementing data mining methods that a large fraction of the energy expended in using these methods goes into the initial treatment of the various input data files so that the data are in a form consistent with the intended use (data correction and cleaning, as described in Section C.1.2). The goal here is not to provide a comprehensive list of the issues that arise in these efforts, but simply to mention some of the common hurdles that arise prior to the use of data mining techniques so that the entire process is better understood.

The following discussion focuses on databases containing personal information (information about many specific individuals), but much of the discussion is true for more general databases.

Several common data deficiencies need prior treatment:

- *Reliable linkages.* Often several databases can be used to provide information on overlapping sets of individuals, and in these cases it is extremely useful to identify which data entries are for the same individuals across the various databases. This is a surprisingly difficult and error-prone process due to a variety of complications: (1) identification numbers (e.g., Social Security numbers, SSNs) are infrequently represented in databases, and when they are, they are sometimes incorrect (SSNs, in particular, have deficiencies as a matching tool, since in some cases more than one person has the same SSN, and in others people have more than one SSN, not to mention the data files that attribute the wrong SSNs to people). (2) There are often several ways of representing names, addresses, and other characteristics (e.g., use of nicknames and maiden names). (3) Errors are made in representing names and other characteristics (e.g., misspelled names, switching first and last names). (4) Matching on a small number of characteristics, such as name and birth date, may not uniquely identify individuals. (5) People's characteristics can change over time (e.g., people get married, move, and get new jobs). Furthermore, deduplication—that is, identifying when people have been represented more than once on the same database—is hampered by the same deficiencies that complicate record linkage.

Herzog et al. point out the myriad challenges faced in conducting record linkage.[3] They point out that the ability to correctly link records is surprisingly low, given the above listed difficulties. (This is especially the case for people with common names.) The prevalence of errors for names, addresses, and other characteristics in public and commercial data files greatly increases the chances of records either being improperly linked or improperly left unlinked. Furthermore, given the size of the files in question, record linkage generally makes use of blocking variables to reduce the population in which matches are sought. Errors in such blocking variables can therefore result in two records for the same individual never being compared. Given that data mining algorithms use as a fundamental input whether the joint activities of an individual or group of individuals are of interest or not, the possibility that these joint activities are actually for different people (or that activities that are joint are not viewed as joint since the individuals are considered to be separate people) is a crucial limitation to the analysis.

• *Appropriate database structure.* The use of appropriate database management tools can greatly expedite various data mining methods. For example, the search for all telephone numbers that have either called a particular number or been called by that number can be carried out orders of magnitude faster when the database has been structured to facilitate such a search. The choice of the appropriate database framework can therefore be crucially important. Included in this is the ability to link

---

[3]T.N. Herzog, F.J. Scheuren, and W.E. Winkler, *Data Quality and Record Linkage Techniques*, Springer Science+Business Media, New York, N.Y., 2007

relevant data entries, to "drill down" to subsets of the data using various characteristics, and to answer various preidentified queries of interest.

• *Treatment of missing data.* Nonresponse (not to mention undercoverage) is a ubiquitous feature of large databases. Missing characteristics can also result from the application of editing routines that search for joint values for variables that are extremely unlikely, which if found are therefore deleted. (A canonical example is a male who reports being pregnant.) Many data mining techniques either require or greatly benefit from the use of data sets with no missing values. To create a data file with the missing values filled in, imputation techniques are used, which collectively provide the resulting database with reasonable properties, with the assumption that the missing data are missing at random. (Missing at random means that the distribution of the missing information is not dependent on unobserved characteristics. In other words, missing values have the same joint distribution as the nonmissing values, given other nonmissing values available in the database.) If the missing data are not missing at random, the resulting bias in any subsequent analysis may be difficult to address. The generation of high-quality imputations is extremely involved for massive data sets, especially those with a complicated relational structure.

• *Equating of variable definitions.* Very often, when merging data from various disparate sources, one finds information for characteristics that are similar, but not identical, in terms of their definition. This can result from time dynamics (such as similar characteristics that have different reference periods), differences in local administration, geographic differences, and differences in the units of data collection. (An example of differences in variable definitions is different diagnostic codes for hospitals in different states.) Prior to any linkage or other combination of information, such differences have to be dealt with so that the characteristics are made to be comparable from one person or unit of data collection to the next.

• *Overcoming different computing environments.* Merging data from different computer platforms is a long-standing difficulty, since it is still common to find data files in substantially different formats (including some data not available electronically). While automatic translation from one format to another is becoming much more common, there still remain incompatible formats that can greatly complicate the merging of data bases.

• *Data quality.* Deficiencies in data quality are generally very difficult to overcome. Not only can there be nonresponse and data linkage problems as indicated above, but also there can be misresponse due to a number of problems, including measurement error and dated responses. (For example, misdialing a phone number might cause one to become classified as a person of interest.) Sometimes use of multiple sources of data can provide opportunities for verification of information and can be used to update information that is not current. Also, while not a data problem per se, sometimes data (that might be of high quality) have little predictive power for modeling the response of interest. For example, data on current news magazine subscriptions might be extremely accurate, but they might also provide little help in discriminating those engaged in terrorist activities.

## H.3 SUBJECT-BASED DATA MINING AS AN EXTENSION OF STANDARD INVESTIGATIVE TECHNIQUES

This appendix primarily concerns the extent to which state-of-the-art data mining techniques, by combining information in relatively sophisticated ways, may be capable of helping police and intelligence officers reduce the threat from terrorism. However, it is useful to point out that there are applications of data mining, sometimes called subject-based data mining,[4] that are simply straightforward extensions of long-standing police and intelligence work, which through the benefits of automation can be greatly expedited and broadened in comparison to former practices, thereby providing important assistance in the fight against terrorism. Although the extent to which these more routine uses of data have already been implemented is not fully known, there is evidence of widespread use both federally and in local police departments.

For example, once an individual is under strong suspicion of participating in some kind of terrorist activity, it is standard practice to examine that individual's financial dealings, social networks, and comings and goings to identify coconspirators, for direct surveillance, etc. Data mining can expedite much of this by providing such information as (1) the names of individuals who have been in e-mail and telephone contact with the person of interest in some recent time period, (2) alternate residences, (3) an individual's financial withdrawals and deposits, (4) people that have had financial dealings with that individual, and (5) recent places of travel.

Furthermore, the activity referred to as drilling down—that is, examining that subset of a dataset that satisfies certain constraints—can also be used to help with typical police and intelligence work. For example, knowing several characteristics of an individual of interest, such as a

---

[4]J. Jonas and J. Harper, "Effective counterterrorism and the limited role of predictive data mining," pp. 1-12 in *Policy Analysis*, No. 584, CATO Institute, Washington, D.C., December 11, 2006.

description of their automobile, a partial license plate, and/or partial fingerprints, might be used to provide a much smaller subset of possible suspects for further investigation.

The productivity and utility of a subject-based approach to data mining depends entirely on the rules used to make inferences about subjects of interest. For example, if the rules for examining the recent places to which an individual has traveled are unrelated to the rules for flagging the national origin of large financial transactions, inferences about activities being worthy of further investigation may be less useful than if these rules are related. Counterterrorism experts thus have the central role in determining the content of the applicable rules, and most experts can make up lists of patterns of behavior that they would find worrisome and therefore worthy of further investigation. For example, these might include the acquisition of such materials as toxins, biological agents, guns, or components of explosives (when their occupations do not involve their use) by a community of individuals in regular contact with each other. Implemented properly, rule-based systems could be very useful for reducing the workload of intelligence analysts by helping them to focus on subjects worthy of further investigation.

The committee recognizes that when some of the variables in question refer to personal characteristics rather than behavior, issues of racial, religious, and other kinds of stereotyping immediately arise. The committee is silent on whether and under what circumstances personal characteristics do have predictive value, but even if they do, policy considerations may suggest that they not be used anyway. In such a situation, policy makers would have to decide whether the value for counterterrorism added by using them would be large enough to override the privacy and civil liberties interests that might be implicated through such use.

## H.4 PATTERN-BASED DATA MINING TECHNIQUES AS ILLUSTRATIONS OF MORE SOPHISTICATED APPROACHES

Originating in various subdisciplines of computer science, statistics, and operations research, a class of relevant data mining techniques for counterterrorist application includes (1) those that might be used to identify combinations of variables that are associated with terrorist activities and (2) those that might identify anomalous patterns that experts would anticipate would have a higher likelihood of being linked to terrorist activities. The identification of combinations of variables that are associated with terrorist activities essentially requires a training set—which is a set of data representing the characteristics of people (or other units) of interest and those not of interest, so that the patterns that best discriminate between these two groups can be discerned.[5] This use of a training set is referred to as supervised learning.

The creation of a training set requires the existence of ground truth. That is, for a supervised learning application to learn to distinguish X (i.e., things or people or activities of interest) from not-X (i.e., things or people or activities not of interest), the training set must contain a significant number of examples of both X and not-X.

For an example, consider airport baggage inspections. Here, supervised learning techniques can provide an improvement over rule-based expert systems by making use of feedback loops using training sets to refine algorithms through continued use and evaluation. Machines that use various types of sensing to "look" inside baggage for weapons and explosives can be trained over time to discriminate between suspicious bags and nonsuspicious ones. It might be possible, given the large volume of training data that can be collected from many airports, that they might be trained over time to demonstrate greater proficiency than human inspectors.

The inputs to such a procedure could include the types of bags, the arrangement of items inside the bags, the images recorded when the bags are sensed, and information about the traveler. Useful training sets should be very easy to produce in this application for two reasons. First, many people (sometimes inadvertently) pack forbidden items in carry-on luggage, thereby providing many varied instances of data from which the system could learn. Second, ground truth is available, in the sense that bags selected for further inspection can be objectively determined to contain forbidden items or not. (It would be useful, in such an application, to randomly select bags that were viewed as uninteresting for inspection to measure the false negative rate.) Furthermore, if necessary, a larger number of examples of forbidden articles can be introduced artificially—this process would increase the number of examples from which an algorithm might learn to recognize such items.[6]

The requirement in supervised learning methods that a training set must contain a significant number of labeled examples of both X and not-X places certain limitations on their use. In the context of distinguishing between terrorist and nonterrorist activity, because of the relative infrequency of terrorist activity, only a few instances can be included in a training set, and thus learning to discriminate between

---

[5]There is a slightly different definition of a training set when the goal is estimation instead of classification.

[6]However, performance would improve only with respect to information contained in images of the bag—because such seeding would necessarily be carried out by a nonrandom set of the population, it would not be possible to improve performance with respect to information about bag owners.

normal activity and preterrorist activity through use of a labeled training set will be extremely challenging. Moreover, even a labeled training set can miss unprecedented types of attacks, since the ground truth they contain (whether or not an attack occurred) is historical rather than forward-looking.

By contrast, a search for anomalous patterns is an example of unsupervised learning, which is often based on examples for which no labels are available. The definition of anomalous behavior that is relevant to terrorist activity is rather fuzzy and unclear, although it can be separated into two distinct types. First, behavior of an individual or household can be distinctly different from its own historical behavior, although such differences may not (indeed, most often will not) relate specifically to terrorist behavior. For example, credit card use or patterns of telephone calls can be distinctly different from those observed for the same individual or individuals in the past. This is referred to as signature-based anomaly detection. Second, behavior can be distinctly different cross-sectionally; that is, an individual or household's behavior can be distinctly different from that of other comparable individuals or households. Unsupervised learning seeks to identify anomalous patterns, some of which might indicate novel forms of terrorist activity. Candidate patterns must be checked against and validated by expert judgment.

As an example, consider the simultaneous booking of seats on an aircraft of a group of unrelated individuals from the same foreign country without a return ticket. A statistical model could be developed to estimate how often this pattern would occur assuming no terrorism and therefore how anomalous this circumstance was. If it turned out that such a pattern was extremely common, possibly no further action would be taken. However, if this were an extremely rare occurrence, and assuming that intelligence analysts viewed this pattern as suspicious, further investigation could be warranted.

A more recent class of data mining techniques, which are still under development, use relational databases as input.[7] Relational databases represent linkages between units of analysis, and in a counterterrorism context the key example is social networks. Social networks are people who regularly communicate with each other, for example, by telephone or e-mail, and who might be acting in concert. Certainly, if one could produce a large relational database of individuals known to be in communication, it would be useful. One could then identify situations similar to those in which each member acquired an uninteresting amount of some chemical, but in which the total amount over all communicating individu-

[7]E. Segal, D. Pe'er, A. Regev, D. Koller, and N. Friedman, "Learning module networks," *Journal of Machine Learning Research* 6(Apr):557-588, 2005.

als was capable of doing considerable harm. Of course, the vast majority of the networks would be entirely innocent, and only a few would be worthy of further investigation. However, having the potential for such assessments could be useful.

A key question then is, how useful is pattern-based data mining likely to be in counterterrorism? Without more empirical experience, it is difficult to make strong assertions, but some things are relatively clear. When training sets are available, as in the case of baggage inspection, pattern-based data mining techniques are very likely to provide substantial benefits. At this point, it is not known how prevalent such applications are likely to be, but an effort should be made to identify such situations, given the strong tools available in such cases. Also, when there is a specific initiating person(s) or event(s) that is known to be of interest, as argued in the previous section, subject-based techniques are certain to be very useful in helping those working in counterintelligence to expeditiously find other people and events of interest.

In the absence of training sets, and for the situation in which there are no initiating persons or events to provide initial foci of investigation, the benefits obtained from the use of pattern-based data mining techniques for counterterrorism are likely to be minimal. The reason is that ordinary people often engage in anomalous activities. Many people have the experience of having been temporarily restricted from making credit card purchases because their recent transactions have been viewed as being atypical. People travel to places they haven't been before, make larger withdrawals of funds than they have before, buy things they haven't bought before, and they call and e-mail people whom they have not called or e-mailed before.

A basic result from multivariate statistical analysis is that, when more characteristics are considered simultaneously, it is more likely for such joint events to be unusual relative to the remainder of the data. So, if the simultaneous actions of travel, communications, purchases, movement of funds, and so on are considered jointly, it is more likely that a joint set of characteristics will be viewed as anomalous. Therefore, searches for anomalous activities, without being trained and without using some linkage to a ground truth assessment of whether the activity is or is not terrorist-related, are much more likely to focus on innocent activity rather than activity related to terrorism.

Data mining tools can also be useful to intelligence analysts if they can reduce the time it currently takes them to carry out their current duties, as long as their accuracies are not less than those of the analysts. (Of course, if the analysts are unable to do a good job, because of data inadequacies for example, automated data mining tools will also result in a bad job, only faster. In such a case, the tools can't hurt, but spending money to acquire them may not be the best use of limited resources.)

As an illustration, consider a suite of data mining tools that facilitates the detection of aliases, record linkages concerning a given individual and his or her network of associates, identification of cluster of related events by certain patterns of interest, and indexed audio/images/video from surveillance monitors. Add to this suite data mining tools that performed as well as a very good analyst in identifying patterns of interest but did so more quickly. Such a suite could improve the productivity of an analyst significantly by allowing him or her to spend less time on "grunt work" and to spend more time on cases that did warrant further investigation. Note also that these activities are likely not to require training sets for their development.

It is not the goal of this appendix to include a description of the objectives or operations of the leading data mining techniques. (Excellent tutorials exist for most of the important methods, and software is typically readily available. Also, a number of recent texts provide excellent descriptions of the majority of the current data mining techniques.[8]) Some of the prominent techniques are listed in Box H.1.

Different techniques have different attributes, which make any given technique more or less suitable in a given application. These attributes include whether or the extent to which a given technique:

1. *Is scalable.* Scalability indicates whether the technique will run efficiently on very large data sets. Scalability is important because some data sets are far too large for an inefficient technique to process in any reasonable length of time.

2. *Easily incorporates privacy protections.* If so, it will be possible to incorporate into the methodology algorithms that provide reasonable protections against disclosures.

3. *Is easily interpretable.* An easily interpretable technique is one for which the general predictive model underlying the technique can be communicated to analysts without specific training.

4. *Is able to handle missing data.* Some techniques are better than others at handling data sets with missing values.

5. *Has effective performance with low-quality data* (i.e., with a small fraction of data having widely discrepant values).

6. *Has effective performance in the face of erroneous record linkages.* The

---

[8]T. Hastie, R. Tibshirani, and J. Friedman, *The Elements of Statistical Learning; Data Mining, Inference and Prediction,* Springer-Verlag, New York, N.Y., 2001; C. Bishop, *Pattern Recognition and Machine Learning,* Springer-Verlag, New York, N.Y., 2006; T. Mitchell, *Machine Learning,* McGraw Hill, Columbus, Ohio, 1997.

---

**BOX H.1**
**Common Data Mining Techniques**

Logistic regression
Regression trees
Bagging
Hidden Markov models
Boosting
Genetic algorithms
Bayesian networks
Cluster analysis
Neural networks
Genetic algorithms
Classification trees
Nearest neighbor estimation
Support vector machines
Supervised learning
Random forests
Recursive partitioning

---

issue arises because record linkages often result in data with such values (10 percent or more of the data may have such values), and some techniques do not perform reliably when applied to such data.

7. *Is resistant to gaming.* Resistance to gaming indicates whether an adversary can take countermeasures to reduce the effectiveness of the method.

## H.5 THE EVALUATION OF DATA MINING TECHNIQUES

It is crucially important that analysts planning to use a data mining algorithm for counterterrorism have some objective understanding of its performance, both prior to use and continually updated while in use. Evaluation provides the basis for (1) an understanding of the quality of the assessments provided, which is particularly important when those assessments are to be used in conjunction with other sources of information; (2) a quantitative way of judging the trade-offs between the benefits derived from the use of an algorithm and its associated costs, especially including a decrease in privacy, and particularly when those trade-offs justify its use; and (3) determining when a competing algorithm should be adopted in replacement or determining when a modification should be made. Evaluation of data mining techniques can be particularly difficult in certain counterterrorism applications, for several reasons discussed below.

### H.5.1 The Essential Difficulties of Evaluation

Evaluation of data mining methods can be carried out in two very general ways. First, internal validation can be used: an algorithm is examined step-by-step, assessing the likelihood of any assumptions obtaining, the quality of input data, the validity of any statistical models, etc. Sensitivity analyses are used in internal validation to examine the impact of divergences from the ideal. Second, external validation compares the predictions to ground truth for situations in which ground truth is available. External validation is very strongly preferred, since it is a direct assessment of the value of a data mining tool.

As mentioned above, data mining algorithms could play a number of very disparate supplementary roles in counterterrorism. In some cases, evaluation might be obvious, such as when the data mining tool performs the same function currently performed by intelligence agents, but much faster. An example might that of logging all investigations into an individual's activity history.

However, in situations in which a pattern-based data mining algorithm is being used to discriminate between people or events of interest and those not of interest, a training set is not only extremely important for developing the data mining algorithm, but it is also nearly essential for carrying out an evaluation of such an algorithm when it has completed development. The difficulties in developing such algorithms therefore translate to difficulties in their evaluation.

As far as the committee knows, there are no data sets available that represent the activities of a diverse group of people including both terrorists (i.e., people of interest and worthy of further investigation) and non-terrorists (i.e., those not of interest) and also where they are correctly identified as such in the database. Also, since the development of procedures used to discriminate between two populations is greatly facilitated when there are substantial numbers of both types represented in the training set, the rarity of terrorist events, and more broadly the rarity of people of interest, complicates both the development and the evaluation of data mining techniques for counterterrorism.

Even if a procedure could be evaluated on a current training set, there is always the possibility that terrorists could adjust (game) their procedures to avoid detection once a methodology is implemented.[9]

---

[9]For example, a wide variety of countermeasures to polygraph use are well-known. See, for example, National Research Council, *The Polygraph and Lie Detection*, The National Academies Press, Washington, D.C., 2003.

Even without gaming, other dynamics might impact the effectiveness of a methodology over time. So not only is there a need for evaluation, but also there is a need for constant reevaluation.

To address this situation, evaluation must be carried out as an iterative process, in which techniques are initially implemented on a research basis, followed by a period of continuous evaluation and testing. Then, only those procedures that have demonstrated their utility would be formally deployed, and, after deployment, procedures would be continuously evaluated both to monitor their performance given the dynamic nature of the threat and to tune procedures to increase their effectiveness.

The importance of evaluation here is difficult to overstate, since the use of ineffective data mining procedures represents a threefold cost. First, there is the potentially enormous cost of using a less effective algorithm for identifying terrorists and possibly not preventing an attack. Second, there is the serious impact each additional data mining procedure has on the freedoms and privacy of U.S. citizens. Third, investigating false leads from ineffective data mining procedures may waste substantial resources, reducing the energies that can be addressed to real threats. For these reasons, evaluation plays an important role in the committee's framework. It is therefore vitally important that procedures be comprehensively evaluated both in development and if implemented, throughout the history of their use, and further that implementation be contingent on a careful assessment of a technique's effectiveness, as well as its costs in terms of impact on privacy and its required resources for continued use. Furthermore, it is crucial, given the finite resources and the costs to privacy, that poorly performing procedures be removed from development or from use as soon as possible.

## H.5.2 Evaluation Considerations

Some progress in the evaluation of data mining techniques for counterterrorism can be made without the use of training sets. In the dichotomous supervised learning case, in which one is using data mining to discriminate between terrorist activities and nonterrorist activities, two types of errors that can be made are false positives and false negatives.

While some data mining techniques do separate the cases into those of interest and those not of interest, most data mining techniques only rank-order the cases from those of least interest to those of greatest interest, without specifying where a line should be drawn between the two groups. However, in practice, the intelligence and police agencies are likely to draw a line at some point, based on the results of the data mining algorithm, and some people will be further investigated (which may mean having an analyst look over the data and okay or not okay further investigation) and some people will not be investigated. Therefore, for evaluation purposes, it makes sense to proceed as if there are false positives and false negatives that are the direct result of the application of data mining methods.

Even without a training set, the assessment of the false positive rate for a procedure is in some sense straightforward, because if a procedure identifies a number of people as being of interest, one can further investigate (a sample of) such people and determine whether they were, in fact, of interest. However, this procedure is clearly resource-intensive.

The assessment of the false negative rate is considerably more difficult than for the false positive rate. A number of ideas might be suggested to produce a type of training set for use in evaluation:

• Have intelligence and police officers look at data on (likely) tens of thousands of individuals, identifying some as worthy of further investigation, with the remainder not of interest. However, the likely result is that the training set constructed in this way will not contain very many people of interest given the rarity of terrorist activity.

• To deal with the lack of identified people of interest, one could relax the definition of "person of interest" to include people with less direct links to terrorist activity. This will boost the number of people of interest in the training set. However, the obvious problem is that the resulting data mining procedure will then be oriented to identify many more false positive cases.

• Another way of increasing the number of identified people of interest is to introduce synthetic data that represent fictitious people worth further investigation.

Any of these ideas will require assessments of which cases are and are not of interest, which will require more resource-intensive use of analysts to make the assessments. Once such a training set is created, algorithms can then be run on the data to determine the ability of a procedure to correctly discriminate between those of interest and those not of interest.

The goal is that, over time, the data mining procedures trained on such data would mimic what the intelligence officers would do if they could process millions of data records. The downside is that the data mining algorithm is then limited to mimicry and does not have the capacity to anticipate new terrorism patterns that might elude intelligence experts.

A variety of approaches can be used for evaluating data mining methodologies. Many possibilities for evaluation exist in addition to the ones described below, including various forms of sensitivity analysis and measuring the performance of an algorithm using mixtures of real and synthetic data sets.

## Cross-Validation

One should not evaluate a data mining routine on the same training set that was used to develop the procedure, since the routine will then be overfit to that particular data set and therefore assessment of its performance on that training set will be optimistically biased. Cross-validation is one approach that can be used to counter this bias.

Cross-validation denotes an approach in which the available data are first separated into a training subset and a test subset. The system is trained using only the training subset, and the resulting trained procedure is then evaluated on the held-out test set of data. One typical procedure, called $k$-fold cross-validation, involves randomly splitting the training sample into $k$ equal-sized subsets, and then training a procedure using all but the $i$th subset, evaluating that procedure by using it to predict the response of interest for the cases on the set-aside $i$th subset. This process is repeated so that each of the $k$ subsets is used as the test set on one of the folds, and the evaluated accuracy over these $k$ repetitions is averaged.

While cross-validation is strongly recommended as an evaluation tool, it has two limitations. First, as mentioned above, for any supervised learning technique, since a training set is typically not representative of time dynamics, cross-validation does not evaluate a procedure's value for future data sets. Second, using this technique, one is evaluating each procedure as a single entity. However, the data mining procedures will be used as elements of a portfolio approach to counterterrorism. Therefore, what is desired is not how a procedure performs in isolation, but what a procedure adds to an existing group of techniques. In that sense, novelty may be much more valuable than correspondence with an extremely useful methodology that is already implemented.

Finally, cross-validation is most readily applied to data sets in which the specified subsets have no relationships with each other. It is likely that cross-validation can also be applied to more complicated data structures, such as networks, but additional research may be needed to determine the best way to do this.

## Face Validity

Another evaluation tool is face validity. Generally speaking, procedures that produce sensible outputs in response to given, often extreme inputs (often best-case and worst-case scenarios) are said to have gained face validity. In addition, input data for fictitious individuals that are designed to provoke an investigation given current procedures, and which are subsequently ranked as being of high interest using a particular data mining algorithm, provide some degree of face validity for that algorithm and procedure. The same is true for fictitious inputs for cases that would be of no interest to counterterrorism analysts for further investigation. Another way that a data mining procedure used for counterterrorism would gain face validity would be if counterterrorism analysts found the results of the procedure useful in their work.

So if a data mining routine provides rankings of interest for people (or other units of analysis) that an analyst finds saves time in deciding where to focus investigative attention, that outcome is a good starting point for indicating the potential value of the algorithm.

However, achieving face validity is a very limited form of evaluation. It does not help to tune or optimize a procedure, it is by its nature a small sample assessment, and experts in the field might have difficulty agreeing on whether a particular approach has face validity or objectively comparing several competing techniques. However, face validity is, at the least, a necessary hurdle for a methodology to overcome prior to fielding.

The committee suggests that the results of expert judgment should be retained in some fashion and incorporated into the data mining procedures in use over time so that their subsequent use reflects this input. This can be done in several ways, but the basic idea is that cases that experts view differently from the data mining procedure—for example, a person clearly of interest who receives a low ranking by the data mining procedure—should result in modifications to the procedure to avoid repeating that error in the future. To support this, not only should experts examine cases identified as of interest to discover false positives, but also a sample of those identified as not of interest should be reviewed in order to have some possibility, admittedly remote, of discovering false negatives. The evaluation and improvement of data mining procedures for counterterrorism needs to be an iterative process.

Finally, one could use face validity as a method for evaluating competing algorithms. One could conduct an experiment in which investigators are given leads from two competing data mining algorithms, denoted A and B to blind the comparison. At the end of the experiment, the experts involved in the experiment could be asked whether they preferred the leads from A or B.

## Gaming and Countermeasures

Another topic that needs to be considered in evaluating data mining procedures for use in counterterrorism is the extent to which these procedures can be gamed. That is, if someone has some general knowledge of the procedures being used, could their behavior be adjusted to reduce the effectiveness of the data mining technique (or to completely defeat the algorithm)?

Of course, specific knowledge of the precise procedures (and the specific parameter values) being used would be enormously valuable, although nearly impossible to obtain. What is more likely is that there would be a general understanding of what is being carried out. Certainly, there would be advantages to the typical actions those engaged in illegal activities already take to mask their identities, such as the use of false identifications and aliases, frequent changes of residences, etc.

However, our broad expectation is that some of the patterns that would be focused on through use of data mining would be difficult to mask. Therefore, while some gaming of the routines used would be effective, having a sufficiently diverse portfolio of algorithms might, over time, provide alternate avenues toward the discovery of terrorists engaged in many different kinds of terrorist activities. A general statement is that it is not whether a procedure can or cannot be gamed, but how relatively easily a procedure can be gamed relative to other competing ones, what is the impact on the procedure's effectiveness, and how the opportunities for gaming can be reduced. Keeping the procedures and the input data sources secret reduces the opportunity for gaming, though at the same time it runs counter to the public's right to know what the government is doing that may compromise personal privacy and other rights. Finding the appropriate middle ground is difficult.

The issue of how an adversary might take countermeasures against any data mining system or data collection effort raises an important policy issue regarding the costs and benefits of greater transparency into these systems and efforts (i.e., more public knowledge about the nature of the data being collected and how the systems work). As noted above, costs could include an increased risk of adversary circumvention of these systems and efforts and perhaps also strong negative reactions of citizens attempting to stop the loss of privacy and confidentiality. However, greater transparency is likely to result in increasing trust in government and some relief that the threat of terrorism was possibly being reduced.

## Reducing Bias in Evaluation

A central issue in research and development is evaluation. It is easy to propose techniques, and vendors and other interested parties propose purportedly new techniques all the time. Government agencies like the U.S. Department of Homeland Security (DHS) and the National Security Agency (NSA) will acquire data mining algorithms for use in counterterrorism in two ways: from outside developers (contractors) and from algorithm developers within the agency. But both internal and external developers are likely to have biases and vested interests in the outcome of any evaluation they may have conducted to judge the performance of an algorithm that they have developed. Thus, before deployment and operational use, such techniques must be as carefully and comprehensively evaluated as possible using the best available evaluation techniques and methods. Many such techniques and methods are used in sophisticated commercial applications.

For these reasons, independent checks on the evaluation work of developers are necessary to minimize the possibility of bias, regardless of whether proprietary claims are asserted. Thus, those conducting the checks should have as much information as necessary to conduct the reviews involved (e.g., full access to descriptions of the algorithms, results of previous evaluations, and descriptions of adjustments that have been made in response to earlier evaluations) and work as independently as possible from the developers.

Evaluators can also build on the foundations provided by preliminary or internal evaluations, since a great deal can be learned about the performance of a system through its performance throughout development. Developers often view these as proprietary, so if DHS or NSA is at the early stage of requesting proposals for development of such techniques, the sharing of such information must be specified in the contract prior to the beginning of work.

Finally, it is also important to subject work in this area to peer evaluation to the extent possible consistent with the needs to protect classified information. Engagement of the best talent and expertise available and solicitation of their contributions as input to the decision making process are important. Such expertise is generally needed to make critical judgments about vendor claims concerning new technological solutions, and it is essential toward deploying effective measures to security problems. Possible mechanisms to support such contributions include interagency professional agreements, sabbatical arrangements for academics, consulting agreements, and external advisory groups.

## H.6 EXPERT JUDGMENT AND ITS ROLE IN DATA MINING

The importance of responsible expert judgment in various aspects of data mining, from research and development to field deployment, cannot be overstated. Expert judgment (of individuals with different background and experiences) is critical both in operations and in development.

From an operational standpoint, human beings are required to interpret the results of a data mining application. As noted above, data mining generally does not identify cases of interest. Instead, data mining rank-orders cases from those of no interest to those of great interest. But it is a matter of human judgment to set thresholds (e.g., those above a certain specified line are of interest, and those below a certain, different, specified line are not of interest) and to determine exceptions (e.g., closer examination of person X who ranked above the threshold indicates that in fact he is not of interest). That is, human experts must decide, probably on an individual basis, which cases are worthy of further investigation or other action. Therefore, there is a need to consider the operator and the data mining algorithms as a sociotechnical system, as well as a need to determine how operators and the data mining technology can best work together.

As an example of a sociotechnical issue, consider a frequently held belief in the infallibility of a computer. Although in principle a human expert may be required to validate and check a computer's conclusions or rank orderings, in practice it is all too easy for the human—especially a young and inexperienced one—to play it safe by accepting at face value a machine-generated conclusion. Procedures and incentives must be developed to shape the human's behavior so that she or he is neither too trusting nor too skeptical of the computer's output.

From a development standpoint, human judgment and expertise play critical roles in shaping how a given system works. In addition to the above-mentioned role for experts in deciding which cases should be further investigated, expert assessments also have other important roles:

• Deciding which variables are discriminating and the values of these variables that indicate whether a given case is of interest or not. For example, a variable ("item purchased") and an amount may be associated with a credit card transaction, some purchases and amounts should be indicated as being of interest and some not, and this is probably best determined by experts. Additional work is needed to determine which input data sets contain potentially relevant information.

• Deciding on criteria to separate anomalous patterns (i.e., patterns that are unusual in some sense) into those that are and are not potentially threatening and indicative of terrorist activity.

• Deciding on the specific form of the algorithm that is evaluated for use. (For example, should one use a transformed or untransformed version of a predictor in a logistic regression model?)

• Improving the robustness of data mining routines against gaming and steps taken to "fly under the radar." For example, a routine may be adjusted to account for an individual making many small purchases over an extended period of time by making that effectively equal to a large one-time purchase.

These multiple and significant roles for expert judgment remain even with the best of data mining technologies. Over time, it may be that more of this expertise can be represented in the portfolio of techniques used in an automated way, but there will always be substantial deficiencies that will require expert oversight to address.

## H.7 ISSUES CONCERNING THE DATA AVAILABLE FOR USE WITH DATA MINING AND THE IMPLICATIONS FOR COUNTERTERRORISM AND PRIVACY

It is generally the case that the effectiveness of a data mining algorithm is much more dependent on the predictive power of the data collected for use than on the precise form of the algorithm. For example, it typically does not matter that much, in discriminating between two populations, whether one uses logistic regression, a classification tree, a neural net, a support vector machine, or discriminant analysis. Priority should therefore be given to obtaining data of sufficient quality and in sufficient quantity to have predictive value in the fight against terrorism.

The first step is to ensure that the data are of high quality, especially when they are to be linked. When derived from record linkages, data tend to assume the worst accuracies in the original data sets rather than the best. Inaccurate data, regardless of quantity, will not produce good or useful results in this counterterrorism context.

A second step is to ensure that the amount of data is adequate—although as a general rule, the collection of more data on people's activities, movements, communications, financial dealings, etc., results in greater opportunities for a loss of privacy and the misuse of the information. Portions of the committee's framework provide for best practices to minimize the damage done to privacy when information is collected on individuals, but ultimately, a policy still needs to be identified that specifies how much additional data should be used for obtaining better results.

Insight into the specifics of the trade-off can be obtained through the use of synthetic data for the population at large (i.e., the haystack within which terrorist needles are hiding) without compromising privacy. At the outset, researchers would use as much synthetic data as they were able to generate in order to assess the effectiveness of a given data mining technique. Then, by removing databases one by one from the scope of the analysis, they would be able to determine the magnitude of the negative impact of such removal. With this analysis in hand, policy makers would have a basis on which to make decisions about the trade-off between accuracy and privacy.

## H.8 DATA MINING COMPONENTS IN AN
## INFORMATION-BASED COUNTERTERRORIST SYSTEM

It is too limiting a perspective to view data mining algorithms only as stand-alone procedures and not to view them as potentially components of a data-supported counterterrorist system. Consider, for example, that data mining techniques have played an essential role as various components of the algorithm that comprises an Internet search engine.

A search engine, at the user level, is not a data mining system, but instead a database with a natural query language. However, the component processes of populating this database, ranking the results, and making the query language more robust are all carried out through the essential use of data mining algorithms. These component processes include (1) spell correction, (2) demoting Web sites that are trying various techniques to inflate their "page rank," (3) identifying Web sites with duplicate content, (4) clustering web pages by concept or similarity of central topic, (5) modifying ranking functions based on the history of users' click sequences, and (6) indexing images and video.

Without these and other features, implemented partially in response to efforts to game search engines, search results would be nearly useless compared with their current value. But as these features have been added over the years, they have increased the value of search engines enormously over their initial implementations, and today search engines are an indispensable part of an individual's online experience.

In a somewhat similar way, one can imagine a search engine, in a general sense of the term, that was designed and optimized for counterterrorist applications. Such a system could, among other things: (a) generalize/specialize the detection of aliases and/or address the ambiguity in foreign names, (b) combine all records concerning a given individual and his or her network of associates, (c) cluster related events by certain patterns of interest and other topics (such as the acquisition of materials and expertise useful for the development of explosives, toxins, and biological agents), (d) log all investigations into an individual's activity history and develop ratings of people as to their degree of interest, and (e) index audio/images/video from surveillance monitors.

All of these are typical data mining applications that do not depend on the existence of training data, and they would seem to be critical components in any counterterrorism system that is designed to collect, organize, and make available for query information on individuals and other units of interest for possibly further data collection, investigation, and analysis. Therefore, data mining might provide many component processes of what would ideally be a large counterterrorism system, with human analysts and investigators playing an essential role alongside specific data mining tools.

Over time, as more data are acquired and different sources of data are found to be more or less useful, as attempts at gaming are continuously monitored and addressed, as various additional unforeseen complexities arise and are addressed, a system could conceivably be developed that could provide substantial assistance in reducing the risk from terrorism. Few of the necessary components of this idealized system currently exist, and therefore this is not something that could be implemented quickly. However, in the committee's view, the threat from terrorism is very likely to persist, and therefore the committee is in support of a fully supported research and development program with the goal of examining the potential effectiveness of such a system.

It is important to point out that each of the above component applications is quite non-trivial. For example, part (b) "combine all records concerning a given individual and his or her network of associates" would be an extremely complicated tool to develop in a way that would be easy to access and use.

And it is useful to point out that when viewing data mining applications as part of a system, their role and therefore their evaluation changes. For example, consider a data mining algorithm that was extremely good at identifying patterns of behavior that are *not* indicative of terrorist activity but was not nearly as effective at identifying patterns that are. Such a component process could be useful as a filter, reducing the workload of investigators, and thereby freeing up resources to devote to a smaller group of individuals of potential interest. This algorithm would fail as a stand-alone tool, but as part of a system, it might perform a useful function.

Development of such a system would certainly be extremely challenging, and success in reducing the threat from terrorism would be a significant achievement. Therefore, research and development of such an approach requires the direct involvement of data mining experts of the first rank. What is needed is not simply the modification of commercial off-the-shelf techniques developed for various business applications, but a dedicated collaborative research effort involving both data miners and intelligence analysts with the goal of developing what are currently non-existent techniques and tools.

## H.9 INFORMATION FUSION

Another class of data mining techniques, referred to as "information fusion," might be useful in counterterrorism. Information fusion refers to a class of methods for combining information from disparate sources in order to make inferences that may not be possible from a single source. One possible, more limited application to counterterrorism is matching people using a variety of sources of information, including address, name, and date of birth, as well as fingerprints, retinal scans, and other biometric information. A broader application of information fusion is identifying patterns that are jointly indicative of terrorist activity.

With respect to the narrower application of person matching, there are different ways of aggregating information to measure the degree to which the personal information matches. One can develop (a) distance metrics using sums of distances using the measured quantities themselves, (b) sums of measures of the assessment of the degree of match for each characteristic, and (c) voting rules that aggregate over whether or not there is a match for each characteristic. There may be advantages in different applications to combining information at different levels of the decision process. (A common approach to joining information at level (a) is through use of the Fellegi-Sunter algorithm.) The committee thinks that information fusion might prove helpful in this limited application. However, the problems mentioned above concerning the difficulties of record linkage will greatly reduce the effectiveness of many information fusion algorithms that are used to assist in person matching.

Regarding the broader application, consider the problem of identifying whether there is a terrorist threat from the following disparate sources of information: recent meetings of known terrorists, greater than usual movement of funds from countries known to harbor terrorists, and greater than usual purchases of explosives in the United States. Information fusion uses such techniques as the Kalman filter and Bayesian networks to learn how to optimally join disparate pieces of information at different levels of the decision process, by either combining individual data elements or combining higher level assessments for the decision at hand, in order to make improved decisions in comparison to more informal use of the disparate information.

Clearly, information fusion directly addresses an obvious need that arises repeatedly in the attempt to use various data sources and types of data for counterterrorism. Intelligence agencies will have surveillance photographs, information on monetary transactions, information on the purchase of dangerous materials, communications of people with suspected terrorists, movements of suspected people into and out of the country, and so on, all of which will need to be combined in some way to make decisions as to whether to initiate further and more intrusive investigations.

To proceed, information fusion for these broader applications typically requires estimates of a number of parameters, such as conditional probabilities, that model how to link the evidence received at various levels of the decision process to the phenomenon of interest. An example might be the probability that a terrorist act is planned in country B in the next three months, given a monetary movement of more than X dollars from a bank in country A to one in country B in the last six months and the purchase in the last two months of more than the usual amounts of explosives of a certain type and greater than usual air travel in the last two months of individuals from country A to country B. Clearly, a conditional probability like this would be enormously useful to have, but how could one estimate it? It is possible that this conditional probability could be expressed as an arithmetic function of simpler conditional probabilities under some conditional independence assumptions, but then there is the problem of validating those assumptions to link those more primitive conditional probabilities to the desired conditional probability.

More fundamentally, information fusion for the broader problem of counterterrorism requires a structure that expresses the forms in which information is received and how it should be combined. At this time, especially given the great infrequency of terrorist events, it will be extremely difficult to validate either the above assumptions or the overall structure proposed for use. Therefore, while information fusion is likely to be useful for some limited problems, it does not currently seem likely to be productive for the broad problem of identifying people and events of interest.

## H.10  AN OPERATIONAL NOTE

The success of any data mining enterprise depends on the availability of relevant data in the universe of data being mined and the ability of the data mining algorithms being used to identify patterns of interest.

In the first instance (availability of data), the operational security skills of the would-be terrorists are the determining factor as to whether data is informative. For terrorists planning high-end attacks (e.g., nuclear explosions involving tens or hundreds of thousands of deaths), the means and planning needed for carrying out a successful attack are complex indeed. On one hand, almost by definition, a terrorist group that could carry out such an attack would have a considerable level of sophistication, and it would take great care to minimize its database tracks. Thus, for attacks at the high end, those intending to carry out such attacks may be better able to reduce the evidence of their activities. On the other hand, the complicated planning necessary for these attacks might provide greater opportunity for data mining to succeed. The trade-off in this case is difficult to evaluate.

In the second instance, regarding the identification of patterns of interest against a noisy background, the primary issue is the fact that the means to carry out small-scale terrorist attacks (e.g., attacks that might result in a few to a few dozen deaths) are easily available. Though not a terrorist, in 2007 the Virginia Tech shooter, for example, killed a few dozen individuals with guns purchased over the counter at a gun store.

---

**BOX H.2**
**An Illustrative Compromise in Operational**
**Security from a Terrorist Perspective**

A conversation between a U.S. person and an unknown individual in Pakistan is intercepted. The call was initiated in the Detroit area from a pay phone using a prepaid phone card. The conversation was conducted in the Arabic language. The initiator is informing the recipient of the upcoming "marriage" of the initiator's brother in a few weeks. The initiator makes reference to the "marriage" of the "dead infidel" some years ago and says this "marriage" will be "similar but bigger." The recipient cautions the initiator about talking on the telephone and terminates the call abruptly.

The intelligence analyst's interpretation of this conversation is that "marriage" is open code for martyrdom. Interrogation of another source indicates that the association of "marriage" and "dead infidel" is a reference to the Oklahoma City bombing. It is the analyst's assessment that a major ANFO or ANNM attack on the continental United States is imminent. Red team analysis concludes that large quantities of ammonium nitrate can be untraceably acquired by making cash purchases that are geographically and temporally distributed.

A "tip" such as this phone conversation might well trigger a major ad hoc data mining exercise through previously unsearched databases, such as those of home improvement and gardening suppliers.

---

Moreover, the planning needed to carry out such an attack is fairly minimal, especially if the terrorist is willing to die. Thus, those intending to carry out relatively small-scale attacks might in principle leave a relevant database track, but the difficult (and for practical purposes, probably insoluble) problem would be the ability to identify that track and infer terrorist actions against a much larger background of innocuous activity.

For practical purposes, then, data mining tools may be most useful against the intermediate scale of terrorist attack (say, car or truck bombs using conventional explosives that might cause many tens or hundreds of deaths). Moreover, as a practical matter, terrorists must face the possibility of unknown leakages—telltale signs that a terrorist group may not know they are leaving, or human intelligence tips that cue counterterrorism authorities about what to look for (Box H.2)—and likelihood of such leakages can be increased by a comprehensive effort that aggressively seeks relevant intelligence information from all sources. This point further underscores the importance of seeing data mining as one element of a comprehensive counterterrorist effort.

# H.11 ASSESSMENT OF DATA MINING FOR COUNTERTERRORISM

Past successes in applying data mining techniques in many diverse domains have interested various government agencies in exploring the extent to which data mining could play a useful role in counterterrorism. On one hand, this track record alone is not an unreasonable basis for interest in exploring, through research and development, the potential applicability of data mining for this purpose. On the other hand, the operational differences between the counterterrorism application and other domains in which data mining has proven its value are significant, and the intellectual burden that researchers must surmount in order to demonstrate the utility of data mining for counterterrorism is high.

As an illustration of these differences, consider first the use of data mining for credit scoring. Credit scoring, as described in Hand and in Lambert,[10] makes use of the history of financial transactions, current debts, income, and accumulated wealth for a given individual, as well as for similar individuals, to develop models of how people behave who are likely to default on a loan, and those who are not likely. Such histories are extensive and have been collected for many years.

Training sets are developed that contain the above information on people who have been approved for loans who later paid in full and also those who were approved for loans and who later defaulted. Training sets are sometimes augmented by data on a sample of those who would not have been approved for a loan but who were granted one nonetheless, and whether or not they later defaulted on the loan. Training sets in this application can be used to develop very predictive models that discriminate well between those for whom additional loans would be both a good and a bad decision on the part of the credit granting institution.

The utility of training sets in this application benefits from the prevalence of the failure to repay loans. While there is a great interest in reducing the number of bad loans to the extent possible, missing a small percentage of bad loans is not a catastrophe. Therefore, false negatives are to be avoided, but a few bad loans are acceptable. While there is a substantial effort to game the process of awarding credit, it has been possible to discover ways to adjust the models that are used to discriminate between good and bad loan applications to retain their utility. Finally, while applications for credit from those new to the database are problem-

---

[10]D.J. Hand and W.E. Henley, "Statistical classification methods in consumer credit scoring: A review," *Journal of the Royal Statistical Society, Series A* 160(3):523-541, 1997; also D. Lambert, "What Use is Statistics for Massive Data?," Bell Labs/Lucent Technologies, Murray Hill, N.J., unpublished paper, 2000.

atic, it has also been possible to develop models that can be used for initial loan applicants to handle those without a credit history.[11]

By contrast, consider the contrasting problem of implementing a "no-fly" list. Although the details of actual programs remain secret, enough is known in the public domain to identify key differences between this problem and that of credit scoring. Some data on behavior relevant to potential terrorist activity (or more likely past activity) are available, but they are very incomplete, and the predictive power of the data collected and the patterns viewed as being related to terrorist activity is quite low. (For example, it is known that an individual with a name that is similar to that of a person on a terrorist watch list is cause for suspicion and additional screening.) Labeled training sets for supervised learning methods cannot be developed because the number of people that have attempted to initiate attacks on aircraft and other terrorist activity is extremely small. Furthermore, gaming—for example, the use of aliases and false documentation, including passports—is difficult to adjust to. Finally, as in credit scoring, there is a need for a process to deal with individuals for whom no data are available, but in this application there seems to be much less value in "borrowing information" from other people.

Given these differences, it is not surprising that the base technologies in each example have compiled vastly different track records: data mining for credit scoring is widely acknowledged as an extremely successful application of data mining, while the various no-fly programs (e.g., CAPPS II) have been severely criticized for their high rate of false positives.[12] Box H.3 describes the largely unsuccessful German experience with counterterrorist profiling based on personal characteristics and backgrounds.

At a minimum, subject-based data mining (Section H.3) is clearly relevant and useful. This type of data mining—for example, structured searches for identifying those in regular contact with known terrorists

---

[11]This description ignores some complexities. All loans are not of equal dollar amount, so making a number of mistakes on a group of loan decisions is not well summarized by the number of mistakes made, that is, the amount loaned in error is also useful to know. Furthermore, it may be profitable to let in some poor loans if more profit is made collectively through the group of loans. Also, there is a selection problem, in that typically it is not known for those rejected for a loan whether that decision was appropriate or not. Finally, external circumstances can change, for example, an economic recession can occur, which may impact the effectiveness of the models used.

[12]Implementing the no-fly list also illustrates the importance of human intervention. In most cases, individuals flagged for further screening are indeed allowed to board aircraft, although they may miss their flight or suffer further inconvenience or harm. The reason they are allowed to do so is because the data mining technology has flagged them as likely risks, but the additional (human-based) screening efforts, though time-consuming, have determined that the individual in question is not likely to be a risk.

## BOX H.3
## The German Experience with Profiling

In the aftermath of the September 11, 2001, terrorist attacks on the United States, German law enforcement authorities sought to explore the possibilities of using large-scale statistical profiling of entire sectors of the population with the purpose of identifying potential terrorists. An initial profile was developed, largely based on the social characteristics of the known perpetrators of 9/11 (male, 18-40 years old, current or former student, Islamic, legal resident in Germany, and originating from one of a list of 26 Muslim countries). This profile was scanned against the registers of residents' registration offices, universities, and the Central Foreigners' Register to identify individuals matching the defined profile—an exercise that resulted in approximately 32,000 entries.

Individuals in this database were then checked against another database of about 4 million individuals identified as possibly having the relevant knowledge to carrying out a terrorist attack, or who had familiarity with places that could constitute possible terrorist targets. This included, for example, individuals with a pilot's license (or attending a course to obtain it), members of sporting aviation associations, as well as employees of airports, nuclear power plants, chemical plants, the rail service, laboratories and other research institutes, as well as students of the German language at the Goethe Institutes.

The comparison of these two databases yielded 1,689 individuals as potential "sleepers." These individuals were investigated at greater length by the German police, but after one year not one sleeper had been identified. Seven individuals suspected of being members of a terrorist cell in Hamburg were arrested, but they did not fit the statistical profile.

In the entire profiling exercise, data were collected and analyzed on about 8.3 million individuals—with a null result to show for it. The exercise was terminated after about 18 months (in summer 2003) and the databases deleted. (In April 2006, the German Federal Constitutional Court declared the then-terminated exercise unconstitutional.)

SOURCE: Adapted from Giovanni Capoccia, "Institutional Change and Constitutional Tradition: Responses to 9/11 in Germany," in Martha Crenshaw (ed.), *The Consequences of Counterterrorist Policies in Democracies*, New York, Russell Sage, forthcoming.

or identifying those, possibly as part of a group, who are collecting large quantities of toxins, biological agents, explosive material, or military equipment—might well identify individuals of interest that warrant further investigation, especially if their professional and personal lives indicate that they have no need for such material. (Such searches could also result in a large number of false positives that would require human judgment to dispose of.) Such searches are within the purview of law enforcement and intelligence analysts today, and it would be surprising if such searches were not being conducted today as extensions of standard investigative techniques.

These approaches have been criticized because they are relevant primarily to future events that have a nontrivial similarity to past events, thus providing little leverage in anticipating terrorist activities that are qualitatively different from those carried out in the past. But even if this criticism is valid (and only research and experience will provide such indications), there is definite and important benefit in being able to reduce the risk from known forms of terrorist activity. Forcing terrorists to use new approaches implies new training regimes, new operational difficulties, and new resource requirements—all of which complicate their own planning and reduce the likelihood of successful execution.

The jury is still out on whether pattern-based data mining algorithms produced without the benefits of machine learning will be similarly useful, and in particular whether such techniques could be useful in discovering more subtle, novel patterns of behavior as being indicative of the planning of a terrorist event that would have been unrecognized a priori as such by intelligence analysts. Jonas and Harper (2006) refer to this kind of data mining as "pattern-based" data mining.[13] The distinction between subject-based and pattern-based data mining is important. Subject-based data mining is focused on terrorist activities that are either precedented (because analysts have some retrospective understanding of them) or anticipated (because analysts have some basis for understanding the precursors to such activities), while pattern-based data mining is focused on future terrorist activities that are unanticipated and unprecedented (that is, activities that analysts are not able to predict or anticipate).

Subject-based techniques have the advantage of being based on strongly predictive models. For example, being a close associate of someone suspected of terrorist activity and having similar connections to persons or groups of interest are strong predictors that a given person will also be of interest for further investigation. By contrast, pattern-based techniques, in the absence of a training set, are likely to have substantially less predictive power than the subject-based patterns chosen by counterintelligence experts based on their experience—and consequently a very large false positive rate. (Indeed, one might expect such an outcome, since pattern-based techniques, by definition, seek to discover anomalous patterns that are not a priori associated with terrorist activity and therefore have no historical precedents to support them. Pattern-based techniques

---

[13]J. Jonas and J. Harper, "Effective counterterrorism and the limited role of predictive data mining," pp. 1-12 in *Policy Analysis, No. 584*, CATO Institute, Washington, D.C., December 11, 2006.

are also, at their roots, tools for identifying correlations, and as such they do not provide insight into why a particular pattern may arise.)

Jonas and Harper (2006) identify three factors that are likely to have a bearing on the utility of data mining for counterterrorist purposes:

- The ability to identify subtle and complex data patterns indicating likely terrorist activity,
- The construction of training sets that facilitate the discovery of indicative patterns not previously recognized by intelligence analysts, and
- The high false positive rates that are likely to result from the problems in the first two bullets.

A number of approaches can be taken to possibly address this argument. For example, as mentioned above, it may be possible to develop training sets by broadening the definition of what patterns of behavior are of interest for further investigation, although that raises the false positive rate. Also, it may be possible to reduce the rate of false positives to a manageable percentage by using a judicious mix of human analysis and different automated tools. However, this is likely to be very resource intensive. The committee does not know whether there are a large number of useful behavioral profiles or patterns that are indicative of terrorist activity.

In addition to these issues, a variety of practical considerations are relevant, including the paucity of data, the often-poor quality of primary data, and errors arising from linkage between records. (Section H.2 discusses additional issues in more detail.)

# I

# Illustrative Government Data Mining Programs and Activity

Several federal agencies have sought to use data mining to reduce the risk of terrorism, including the Department of Defense (DOD), the Department of Homeland Security (DHS), the Department of Justice (DOJ), and the National Security Agency (NSA). Some of the data mining programs have been withdrawn; some are in operation; some have changed substantially in scope, purpose, and practice since they were launched; and others are still in development. This appendix briefly describes a number of the programs, their stated goals, and their current status (as far as is known publicly).[1]

The programs described vary widely in scope, purpose, and sophistication. Some are research efforts focused on the fundamental science of data mining; others are intended as efforts to create general toolsets and developer toolkits that could be tailored to meet various requirements. Most of the programs constitute specific deployments of one or more forms of data mining technology intended to achieve particular

---

[1] A 2004 U.S. Government Accountability Office (GAO) report provided a comprehensive survey of data mining systems and activities in federal agencies up to that time. See GAO, *Data Mining: Federal Efforts Cover a Wide Range of Uses*, GAO-04-548, GAO, Washington, D.C., May 2004. Other primary resources: J.W. Seifert, *Data Mining and Homeland Security: An Overview*, RL31798, Congressional Research Service, Washington, D.C., updated June 5, 2007; U.S. Department of Homeland Security (DHS), "Data Mining Report: DHS Privacy Office Response to House Report 108-774," DHS, Washington, D.C., July 6, 2006; DHS Office of Inspector General, "Survey of DHS Data Mining Activities," OIG-06-56, DHS, Washington, D.C., August 2006.

operational goals. The programs vary widely in sophistication of the technologies used to achieve operational goals; they also vary widely in the sources of data used (such as government data, proprietary information from industry groups, and data from private data aggregators) and in the forms of the data (such as structured and unstructured). The array of subject matter of the projects is broad: they cover law enforcement, terrorism prevention and pre-emption, immigration, customs and border control, financial transactions, and international trade. Indeed, the combination of the variety of applications and the variety of definitions of what constitutes data mining make any overall assessment of data mining programs difficult.

The scientific basis of many of these programs is uncertain or at least not publicly known. For example, it is not clear whether any of the programs have been subject to independent expert review of performance. This appendix is intended to be primarily descriptive, and the mention of a given program should not be taken as an endorsement of its underlying scientific basis.

## I.1  TOTAL/TERRORISM INFORMATION AWARENESS (TIA)

Status: *Withdrawn* as such, but see Appendix J for a description.

## I.2  COMPUTER-ASSISTED PASSENGER PRESCREENING SYSTEM II (CAPPS II) AND SECURE FLIGHT

Status: CAPPS II *abandoned;* Secure Flight *planned for deployment in 2008.*

In creating the Transportation Security Administration (TSA), Congress directed that it implement a program to match airline passengers against a terrorist watch list. CAPPS II was intended to fulfill that directive. It was defined as a prescreening system whose purpose was to enable TSA to assess and authenticate travelers' identities and perform a risk assessment to detect persons who may pose a terrorist-related threat. However, it went beyond the narrow directive of checking passenger information against a terrorist watch list and included, for instance, assessment of criminal threats. According to the DHS fact sheet on the program, CAPPS II was to be an integral part of its layered approach to security, ensuring that travelers who are known or potential threats to aviation are stopped before they or their baggage board an aircraft.[2] It

---

[2]U.S. Department of Homeland Security, "Fact Sheet: CAPPS II at a Glance," February 13, 2004, available at http://www.dhs.gov/xnews/releases/press_release_0347.shtm.

was meant to be a rule-based system that used information provided by the passenger (name, address, telephone number, and date of birth) when purchasing an airline ticket to determine whether the passenger required additional screening or should be prohibited from boarding.

CAPPS II would have examined both commercial and government databases to assess the risk posed by passengers. In an effort to address privacy and security concerns surrounding the program, DHS issued a press release about what it called myths and facts about CAPPS II.[3] For instance, it stated that retention of data collected would be limited—that all data collected and created would be destroyed shortly after the completion of a traveler's itinerary. It also said that no data mining techniques would be used to profile and track citizens, although assessment would have extended beyond checking against lists and would have included examining a wide array of databases. A study by GAO in 2004 found that TSA was sufficiently addressing only one of eight key issues related to implementing CAPPS II.[4] The study found that accuracy of data, stress testing, abuse prevention, prevention of unauthorized access, policies for operation and use, privacy concerns, and a redress process were not fully addressed by CAPPS II. Despite efforts to allay concerns, CAPPS II was abandoned in 2004. It was replaced in August 2004 with a new program called Secure Flight.

Secure Flight is designed to fulfill the Congressional directive while attempting to address a number of concerns raised by CAPPS II. For instance, unlike CAPPS II, Secure Flight makes TSA responsible for cross-checking passenger flight information with classified terrorist lists rather than allowing such checking to be done by contracted vendors. Although the possibility of using commercial databases to check for threats is still included, the use of commercial data is now precluded.[5] Other differences between CAPPS II and Secure Flight include limiting screening to checking for terrorism threats, not criminal offenses (although this was initially included), and using only historical data during testing phases. TSA states that the mission of Secure Flight is "to enhance the security of domestic commercial air travel within the United States through the

---

[3]U.S. Department of Homeland Security, "CAPPS II: Myths and facts," February 13, 2004, available at http://www.dhs.gov/xnews/releases/press_release_0348.shtm.

[4]U.S. Government Accountability Office (GAO), *Aviation Security: Computer-Assisted Passenger Prescreening System Faces Significant Implementation Challenges*, GAO-04-385, GAO, Washington, D.C., February 2004.

[5]U.S. Transportation Security Administration, "Secure Flight: Privacy Protection," available at http://www.tsa.gov/what_we_do/layers/secureflight/secureflight_privacy.shtm.

use of improved watch list matching."[6] According to TSA, when implemented, Secure Flight would:

- Decrease the chance of compromising watch-list data by centralizing use of comprehensive watch lists.
- Provide earlier identification of potential threats, allowing for expedited notification of law-enforcement and threat-management personnel.
- Provide a fair, equitable, and consistent matching process among all aircraft operators.
- Offer consistent application of an expedited and integrated redress process for passengers misidentified as posing a threat.

However, Secure Flight has continued to raise concerns about privacy, abuse, and security. A 2006 GAO study of the program found that although TSA had made some progress in managing risks associated with developing and deploying Secure Flight, substantial challenges remained.[7] After publication of the study report, TSA announced that it would reassess the program and make changes to address concerns raised in the report. The 2006 DHS Privacy Office report on data mining did not include an assessment of Secure Flight; it stated that searches or matches are done with a known name or subject and thus did not meet the definition of data mining used in the report.[8] In a prepared statement before the Senate Committee on Commerce, Science, and Transportation in January 2007, the TSA administrator noted progress in addressing those concerns and the intention to make the program operational by some time in 2008.[9] Most recently, DHS Secretary Michael Chertoff announced that Secure Flight would no longer include data mining and would restrict information collected about passengers to full name and, optionally, date of birth and sex. Chertoff stated that Secure Flight will not collect commercial data, assign risk scores, or attempt to predict behavior, as was

---

[6]U.S. Transportation Security Administration, "Secure Flight: Layers of Security," available at http://www.tsa.gov/what_we_do/layers/secureflight/index.shtm.

[7]U.S. Government Accountability Office (GAO), *Aviation Security: Significant Management Challenges May Adversely Affect Implementation of the Transportation Security Administration's Secure Flight Program,* GAO-06-374T, GAO, Washington, D.C., February 9, 2006.

[8]U.S. Department of Homeland Security (DHS), "Data Mining Report: DHS Privacy Office Response to House Report 108-774," July 6, 2006; DHS Office of Inspector General, *Survey of DHS Data Mining Activities,* OIG-06-56, DHS, Washington, D.C., August 2006, p. 20, footnote 25.

[9]Prepared Statement of Kip Hawley, Assistant Secretary of the Transportation Security Administration Before the U.S. Senate Committee on Commerce, Science and Transportation, January 17, 2007, available at http://www.tsa.gov/press/speeches/air_cargo_testimony.shtm.

envisioned in earlier versions of the program.[10] The information provided will be compared with a terrorist watch list.

## I.3 MULTISTATE ANTI-TERRORISM INFORMATION EXCHANGE (MATRIX)

Status: *Pilot program ended; no follow-on program started.*

This program was an effort to support information-sharing and collaboration among law-enforcement agencies.[11] It was run as a pilot project administered by the Institute for Intergovernmental Research for DHS and DOJ.[12] MATRIX involved collaborative information-sharing between public, private, and nonprofit institutions. A Congressional Research Service (CRS) report described MATRIX as a project that "leverages advanced computer/information management capabilities to more quickly access, share, and analyze public records to help law enforcement generate leads, expedite investigations, and possibly prevent terrorist attacks."[13] The MATRIX system was developed and operated by a private Florida-based company, and the Florida Department of Law Enforcement controlled access to the program and was responsible for the security of the data.[14] Although "terrorism" is part of the program name, the primary focus appears to have been on law enforcement and criminal investigation. Until the system was redesigned, participating states were required to transfer state-owned data to a private company.[15] The core function of the system was the Factual Analysis Criminal Threat Solution (FACTS) application used to query disparate data sources by using available investigative information, such as a portion of a vehicle license number, to combine records dynamically to identify people of potential interest. According

---

[10]Michael J. Sniffen, "Feds off simpler flight screening plan," *Associated Press*, August 9, 2007.

[11]U.S. Department of Homeland Security (DHS), "MATRIX Report: DHS Privacy Office Report to the Public Concerning the Multistate Anti-Terrorism Information Exchange," DHS, Washington, D.C., December 2006, p. 1.

[12]The Institute for Intergovernmental Research (IIR) is a Florida-based nonprofit research and training organization specializing in law enforcement, juvenile justice, criminal justice, and homeland security. See http://www.iir.com/default.htm.

[13]W.J. Krouse, *The Multi-State Anti-Terrorism Information Exchange (MATRIX) Pilot Project*, RL32536, U.S. Congressional Research Service (CRS), Washington, D.C., August 18, 2004, p. 1, italics original. Note that the official Web site for MATRIX program cited in this CRS report is no longer available.

[14]Ibid., p. 2. The company, Seisint, was acquired by Reed Elsevier subsidiary LexisNexis in July 2004.

[15]J. Rood, "Controversial data-mining project finds ways around privacy laws," *CQ Homeland Security—Intelligence*, July 23, 2004, p. 1.

to the CRS report, FACTS included crime-mapping, association-charting, lineup and photograph montage applications, and extensive query capabilities.[16] Data sources available in the system included both those traditionally available to law enforcement—such as criminal history, corrections-department information, driver's license, and motor-vehicle data—and nontraditional ones, such as:[17]

- Pilot licenses issued by the Federal Aviation Administration,
- Aircraft ownership,
- Property ownership,
- U.S. Coast Guard vessel registrations,
- State sexual-offender lists,
- Corporate filings,
- Uniform Commercial Code filings or business liens,
- Bankruptcy filings, and
- State-issued professional licenses.

Concerns were raised that the data would be combined with private data, such as credit history, airline reservations, and telephone logs; but the MATRIX Web site stated those would not be included. The system initially included a scoring system called High Terrorist Factor (HTF) that identified people who might be considered high-risk, although it was later claimed that the HTF element of the system had been eliminated.[18]

The pilot program ended in April 2005. Legal, privacy, security, and technical concerns about requirements to transfer state-owned data to MATRIX administrators and continuing costs associated with using the system prompted several states that initially participated or planned to participate in MATRIX to withdraw.[19] By March 2004, 11 of the 16 states that originally expressed interest in participating had withdrawn from the program. In an attempt to address some of the concerns, the architecture was changed to allow distributed access to data in such a way that no

---

[16]Krouse, op. cit., p. 4.

[17]Data sources are identified in the Congressional Research Service report (Krouse, op. cit., p. 6) as referenced from the official MATRIX program Web site, http://www.matrix-at.org, which is no longer available.

[18] B. Bergstein, "Database firm gave feds terror suspects: 'Matrix' developer turned over 120,000 names," Associated Press, May 20, 2004, available at http://www.msnbc.msn.com/id/5020795/.

[19]See, for instance, Georgia Department of Motor Vehicle Safety, "Department of Motor Vehicle Safety's Participation in MATIX," September 29, 2003; New York State Police, Letter to Chairman of MATRIX, March 9, 2004; Texas Department of Public Safety, Letter to Chair, Project MATRIX, May 21, 2003. Those documents and additional information on state involvement are available from the American Civil Liberties Union (ACLU) Web site at http://www.aclu.org/privacy/spying/15701res20050308.html.

data transfers from state-controlled systems would be required to share data.[20] The CRS report concluded that "it remains uncertain whether the MATRIX pilot project is currently designed to assess and address privacy and civil liberty concerns."[21] For instance, there appears to have been no comprehensive plan to put safeguards and policies in place to avoid potential abuses of the system, such as monitoring of activities of social activists or undermining of political activities.[22]

## I.4 ABLE DANGER

Status: *Terminated in January 2001.*

This classified program established in October 1999 by the U.S. Special Operations Command and ended by 2001 called for the use of data mining tools to gather information on terrorists from government databases and from open sources, such as the World Wide Web.[23] The program used link analysis to identify underlying connections between people.[24] Analysis would then be used to create operational plans designed to disrupt, capture, and destroy terrorist cells. Link analysis is a form of network analysis that uses graph theory to identify patterns and measure the nature of a network. The related social-network analysis is now considered a critical tool in sociology, organizational studies, and information sciences. Cohesion, betweenness, centrality, clustering coefficient, density, and path length are some of the measures used in network analysis to model and quantify connections. The combination of complex mathematics and the enormous volumes of data required to gain an accurate and complete picture of a network make the use of information technology critical if useful analysis is to be performed on a large scale. Several network-mapping software packages are available commercially. Applications include fraud detection, relevance ratings in Internet search engines, and epidemiology.

---

[20]Rood, op. cit., p. 1.

[21]Krouse, op. cit., p. 10.

[22]Ibid., p. 8.

[23]U.S. Department of Defense (DOD) Office of the Inspector General, "Report of Investigation: Alleged Misconduct by Senior DOD Officials Concerning the Able Danger Program and Lieutenant Colonel Anthony A. Shaffer, U.S. Army Reserve," Case Number H05L97905217, DOD, Washington, D.C., September 18, 2006.

[24]"Link analysis" was an informal term used to describe the analysis of connections between individuals rather than any kind of formal "record linkage" between database records.

Able Danger was focused specifically on mapping and analyzing relationships within and with Al Qaeda. The program became public in 2005 after claims made by Rep. Curt Weldon that Able Danger had identified 9/11 hijacker Mohammad Atta before the attack surfaced in the mass media. A member of Able Danger, Anthony Shaffer, later identified himself as the source of Weldon's information.[25] He further claimed that intelligence discovered as part of Able Danger was not passed on to Federal Bureau of Investigation (FBI) and other civilian officials. Shaffer said a key element of Able Danger was the purchase of data from information brokers that identified visits by individuals to specific mosques.[26] That information was combined with other data to identify patterns and potential relationships among alleged terrorists. Claims made by Shaffer were refuted in a report written by the DOD inspector general.[27] The report showed examples of the types of charts produced by link analysis.[28] It characterized Able Danger operations as initially an effort to gain familiarity with state-of-the-art analytical tools and capabilities and eventually to apply link analysis to a collection of data from other agencies and from public Web sites to understand Al Qaeda infrastructure and develop a strategy for attacking it.[29] The program was then terminated, having achieved its goal of developing a (still-classified) "campaign plan" that "formed the basis for follow-on intelligence gathering efforts."[30] An investigation by the Senate Select Committee on Intelligence concluded that Able Danger had not identified any of the 9/11 hijackers before September 11, 2001.[31]

No follow-on intelligence effort using link-analysis techniques developed by Able Danger has been publicly acknowledged. However, the existence of a program known as Able Providence supported through the Office of Naval Intelligence, which would reconstitute and improve

---

[25]Cable News Network, "Officer: 9/11 panel didn't receive key information," August 17, 2005, available at http://www.cnn.com/2005/POLITICS/08/17/sept.11.hijackers.

[26]J. Goodwin, "Inside Able Danger—The secret birth, extraordinary life and untimely death of a U.S. military intelligence program," Government Security News, September 5, 2005, available at http://www.gsnmagazine.com/cms/lib/410.pdf.

[27]U.S. Department of Defense Office of the Inspector General, "Report of Investigation: Alleged Misconduct by Senior DOD Officials Concerning the Able Danger Program and Lieutenant Colonel Anthony A. Shaffer, U.S. Army Reserve," Case Number H05L97905217, September 18, 2006.

[28]Ibid., pp. 8-9.

[29]Ibid., p. 14.

[30]Ibid.

[31]G. Miller, "Alarming 9/11 claim is baseless, panel says," Los Angeles Times, December 24, 2006.

on Able Danger, was reported by Weldon in testimony to the U.S. Senate Committee on the Judiciary as part of its investigation.[32]

## I.5 ANALYSIS, DISSEMINATION, VISUALIZATION, INSIGHT, AND SEMANTIC ENHANCEMENT (ADVISE)

Status: *Under development (some deployments decommissioned).*

This program, being developed by DHS, was intended to help to detect potentially threatening activities by using link analysis of large amounts of data and producing graphic visualizations of identified linkage patterns. It was one of the most ambitious data mining efforts being pursued by DHS. ADVISE was conceived as a data mining toolset and development kit on which applications could be built for deployment to address specific needs. An assessment of ADVISE was not included in the 2006 DHS Privacy Office report on data mining, because it was considered a tool or technology and not a specific implementation of data mining.[33] That position was noted in a GAO report on the program that questioned the decision not to include a privacy assessment of the program, given that "the tool's intended uses include applications involving personal information, and the E-Government Act, as well as related Office of Management and Budget and DHS guidance, emphasize the need to assess privacy risks early in systems development."[34]

The GAO report identified the program's intended benefit as helping to "detect activities that threaten the United States by facilitating the analysis of large amounts of data that otherwise would be very difficult to review," noting that the tools developed as part of ADVISE are intended to accommodate both structured and unstructured data.[35] The report concluded that ADVISE raised a number of privacy concerns and that although DHS had added security controls related to ADVISE, it had failed to assess privacy risks, including erroneous associations of people, misidentification of people, and repurposing of data collected for other

---

[32]Representative Curt Weldon in testimony to the United States Senate Committee on the Judiciary, September 21, 2005, available at http://judiciary.senate.gov/testimony.cfm?id=1606&wit_id=4667. See also P. Wait, "Data-mining offensive in the works," *Government Computer News*, October 10, 2005.

[33]U.S. Department of Homeland Security (DHS), "Data Mining Report: DHS Privacy Office Response to House Report 108-774," DHS, Washington, D.C., July 6, 2006; DHS Office of Inspector General, "Survey of DHS Data Mining Activities," OIG-06-56, DHS, Washington, D.C., August 2006, p. 20, footnote 25.

[34]U.S. Government Accountability Office (GAO), *Data Mining: Early Attention to Privacy in Developing a Key DHS Program Could Reduce Risks*, GAO-07-293, GAO, Washington, D.C., February 2007, p. 3.

[35]Ibid., p. 3.

purposes.[36] It called on DHS "to conduct a privacy impact assessment of the ADVISE tool and implement privacy controls as needed to mitigate any identified risks."[37]

DHS responded to the GAO report, saying that it was in the process of developing a privacy impact assessment tailored to the unique character of the ADVISE program (as a tool kit). A later DHS Privacy Office report did review the ADVISE program and drew a careful distinction between ADVISE as a technology framework and ADVISE deployments.[38] The report first reviewed the technology framework in light of privacy compliance requirements of the DHS Privacy Office described in the report.[39] In light of those requirements, it then assessed six planned deployments of ADVISE:[40]

- *Interagency Center for Applied Homeland Security Technology (ICAHST).* ICAHST evaluates promising homeland-security technologies for DHS and other government stakeholders in the homeland-security technology community.
- *All-Weapons of Mass Effect (All-WME).* Originally begun by the Department of Energy, All-WME used classified message traffic collected by the national laboratories' field intelligence elements to analyze information related to foreign groups and organizations involved in WME material flows and illicit trafficking. Deployment has been discontinued.
- *Biodefense Knowledge Management System.* This was a series of three deployment initiatives planned by the Biodefense Knowledge Center with the overall goal of identifying better methods for assisting DHS analysts in identifying and characterizing biological threats posed by terrorists. All the deployments have ended, and there are no plans for future deployments.
- *Remote Threat Alerting System (RTAS).* RTAS sought to determine whether the ADVISE technology framework could assist DHS Customs and Border Protection (CBP) in identifying anomalous shipments on the basis of cargo type and originating country. All RTAS activities ended in September 2006.
- *Immigration and Customs Enforcement Demonstration (ICE Demo).* This deployment was operated by the DHS Science and Technology Director-

---

[36]Ibid., p. 18.

[37]Ibid., from "Highlights: What GAO Recommends." See also p. 23.

[38]U.S. Department of Homeland Security, "DHS Privacy Office Review of the Analysis, Dissemination, Visualization, Insight and Semantic Enhancement (ADVISE) Program," DHS, Washington, D.C., July 11, 2007. Page 2 discusses and defines these terms.

[39]Ibid., pp. 3-5.

[40]Ibid. Definitions and descriptions of these programs are drawn from the report beginning on p. 7.

ate and Lawrence Livermore National Laboratory to determine whether the ADVISE technology framework could assist DHS Immigration and Customs Enforcement (ICE) in using existing ICE data better. All activity related to this deployment of ADVISE has ended.

- *Threat Vulnerability Integration System (TVIS).* TVIS used a series of data sets to identify opportunities to test the capability of the ADVISE technology framework to help analysts in the DHS Office of Intelligence and Analysis. Early pilot deployment phases have been followed by subsequent pilot deployment phases.

The report found that some of the deployments did use personally identifiable information without conducting privacy impact assessments.[41] It also recommended short- and long-term actions to address the problems. In particular, it recommended actions that would integrate privacy compliance requirements into project development processes, echoing recommendations made in the GAO report on the program.[42] DHS ended the program in September 2007, citing the availability of commercial products to provide similar functions at much lower cost.[43]

### I.6 AUTOMATED TARGETING SYSTEM (ATS)

Status: *In use.*

This program is used by CBP, part of DHS, to screen cargo and travelers entering and leaving the United States by foot, car, airplane, ship, and rail. ATS aassess risks by using data mining and data-analysis techniques. The risk assessment and links to information on which the assessment is based are stored in the ATS for up to 40 years.[44] The assessment is based on combining and analyzing data from several existing sources of information—including the Automated Commercial System, the Automated Commercial Environment System, the Advance Passenger Information System, and the Treasury Enforcement Communications System—and from people crossing the U.S. land border known (the Passenger Name Record).

ATS compares a traveler's name with a list of known and suspected

---

[41]Ibid., p. 1.

[42]U.S. Government Accountability Office (GAO), *Data Mining: Early Attention to Privacy in Developing a Key DHS Program Could Reduce Risks,* GAO-07-293, GAO, Washington, D.C., February 2007. See especially p. 24.

[43]M.J. Sniffen, "DHS Ends Criticized Data-Mining Program," *Washington Post,* September 5, 2007.

[44]U.S. Department of Homeland Security, "Notice of Privacy Act system of records," *Federal Register* 71(212): 64543-64546, November 2, 2006.

terrorists. It also performs link analysis, checking, for example, the telephone number associated with an airline reservation against telephone numbers used by known terrorists.[45] Such checking has been credited by DHS with preventing entry of suspected terrorists and with identifying criminal activity,[46] but concerns about high numbers of false alarms, efficacy of the risk assessment, lack of a remediation process, and ability of the agency to protect and secure collected data properly have been raised by some in the technical community and by civil-liberties groups.[47]

## I.7 THE ELECTRONIC SURVEILLANCE PROGRAM

Status: *Continuing subject to oversight by the Foreign Intelligence Surveillance Court.*

This program, also called the Terrorist Surveillance Program, involves the collection and analysis of domestic telephone-call information with the goal of targeting the communications of Al Qaeda and related terrorist groups and affiliated individuals. Details about the program remain secret, but as part of the program the president authorized NSA to eavesdrop on communications of people in the United States without obtaining a warrant when there is "reasonable basis to conclude that one party to the communication is a member of Al Qaeda."[48]

Existence of the program first surfaced in a *New York Times* article published in December 2005.[49] Questions as to the legality of the program led a federal judge to declare the program unconstitutional and illegal and to order that it be suspended. That ruling was overturned on appeal on narrow grounds regarding the standing of the litigants rather than the legality of the program.[50] In a letter to the Senate Committee on the

---

[45]Remarks of Stewart Baker, Assistant Secretary for Policy, Department of Homeland Security at the Center for Strategic and International Studies, Washington, D.C., December 19, 2006.

[46]Ibid. Baker noted the use of ATS to identify a child-smuggling ring. CBP officers who examined ATS data noticed that a woman with children had not taken them with her on the outbound flight; this led to further investigation.

[47]See, for instance, B. Schneier, "On my mind: They're watching," *Forbes*, January 8, 2007; Electronic Privacy Information Center, Automated Targeting System, http://www.epic.org/privacy/travel/ats/default.html.

[48]Press briefing by Attorney General Alberto Gonzales and General Michael Hayden, Principal Deputy Director for National Intelligence, December 19, 2005, available at http://www.whitehouse.gov/news/releases/2005/12/20051219-1.html.

[49]J. Risen and E. Lichtblau, "Bush lets U.S. spy on callers without courts," *New York Times*, December 16, 2005.

[50]A. Goldstein, "Lawsuit against wiretaps rejected," *The Washington Post*, July 7, 2007, p. A1.

Judiciary on January 17, 2007, Attorney General Alberto Gonzales stated that the program would not be reauthorized by the president although the surveillance program would continue subject to oversight by the Foreign Intelligence Surveillance Court (FISC).

Although the legality of the program has been the primary focus of the press, it is unclear to what extent data mining technology is used as part of the program. Some press reports suggest that such technology is used extensively to collect and analyze data from sources that include telephone and Internet communication, going well beyond keyword searches to use link analysis to uncover hidden relationships among data points.[51] The adequacy of the FISC to address technology advances, such as data mining and traffic-analysis techniques, has also been called into question.[52]

As this report is being written (June 2008), changes in the Foreign Intelligence Surveillance Act are being contemplated by Congress. The final disposition of the changes is not yet known.

## I.8 NOVEL INTELLIGENCE FROM MASSIVE DATA (NIMD) PROGRAM

Status: *In progress.*

NIMD is a research and development program funded by the Disruptive Technology Office,[53] which is part of the Office of the Director of National Intelligence. The program, which has many similarities to the Total/Terrorism Information Awareness program, is focused on the development of data mining and analysis tools to be used in working with massive data. According to a "Call for 2005 Challenge Workshop Proposals," "NIMD aims to preempt strategic surprise by addressing root causes of analytic errors related to bias, assumptions, and premature attachment to a single hypothesis."[54] Two key challenges are identified: data triage to support decision-making and real-time analysis of petabytes of data and practical knowledge representation to improve machine processing and

---

[51]E. Lichtblau and J. Risen, "Spy agency mined vast data trove, official report," *New York Times,* December 23, 2005; S. Harris, "NSA spy program hinges on state-of-the-art technology," *National Journal,* January 20, 2006.

[52]See, for instance, K.A. Taipale, "Whispering wires and warrantless wiretaps: Data mining and foreign intelligence surveillance," *NYU Review of Law and Security,* Issue 7, Supplemental Bulletin on Law and Security, Spring 2006.

[53]The Disruptive Technology Office was previously known as the Advanced Research and Development Activity (ARDA).

[54]Advanced Research Development Activity, "Call for 2005 Challenge Workshop Proposals," available at http://nrrc.mitre.org/arda_explorprog2005_cfp.pdf.

data-sharing among disparate agencies and technologies. The challenge identifies five focus areas for NIMD research with the overarching goal of building "smart software assistants and devil's advocates that help analysts deal with information overload, detect early indicators of strategic surprise, and avoid analytic errors": "modeling analysts and analytic processes, capturing and reusing prior and tacit knowledge, generating and managing hypotheses, organizing/structuring massive data (mostly unstructured text), and human interaction with information."

Advocacy groups and some members of Congress have expressed concerns that at least some of the research done as part of the TIA program has continued under NIMD.[55] In contrast with TIA, Congress stipulated that technologies developed under the program are to be used only for military or foreign intelligence purposes against non-U.S. citizens.

### I.9 ENTERPRISE DATA WAREHOUSE (EDW)

Status: *Operational since 2000 and in use.*

This system collects data from CBP transactional systems and subdivides them into data sets for analysis.[56] The data sets are referred to as data marts. Their creation is predicated on the need for a specific grouping and configuration of selected data.[57] EDW acquires and combines data from several customs and other federal databases to perform statistical and trend analysis to look for patterns, for instance, to determine the impact of an enforcement action or rule change.[58] EDW uses commercial off-the-shelf technology for its analysis.[59] EDW data are treated as read-

---

[55]"U.S. still minding terror data," *Associated Press*, Washington, D.C., February 23, 2004; M. Williams, "The Total Information Awareness Project lives on," *Technology Review*, April 26, 2006.

[56]U.S. Department of Homeland Security (DHS), "Data Mining Report: DHS Privacy Office Response to House Report 108-774," July 6, 2006, pp. 20-21.

[57]An explanation of the distinction between a data warehouse and a data mart is provided as a footnote in DHS Office of Inspector General, "Survey of DHS Data Mining Activities," OIG-06-56, August 2006, p. 11.

[58]See U.S. Customs and Border Protection, "U.S. Customs data warehousing," available at http://www.cbp.gov/xp/cgov/trade/automated/automated_systems/data_warehousing. xml; databases used as sources include Automated Commercial System (ACS), Automated Commercial Environment (ACE), Treasury Enforcement Communications System (TECS), Administrative and Financial Systems, the Automated Export System. See U.S. Customs, "Enterprise Data Warehouse: Where it stands, where it's heading," U.S. Customs Today, August 2000, available at http://www.cbp.gov/custoday/aug2000/dwartic4.htm.

[59]U.S. Department of Homeland Security (DHS), "Data Mining Report: DHS Privacy Office Response to House Report 108-774," DHS, Washington, D.C., July 6, 2006, p. 21.

only; all changes occur in source systems propagated to it periodically (every 24 hours).[60]

## I.10 LAW ENFORCEMENT ANALYTIC DATA SYSTEM (NETLEADS)

Status: *In use.*

This program facilitates ICE law-enforcement activities and intelligence analysis capabilities through the use of searches and pattern recognition based on multiple data sources.[61] As with EDW, NETLEADS uses data marts. Link analysis is used to show relationships, such as associations with known criminals. Information analyzed includes criminal-alien information and terrorism, smuggling, and criminal-case information derived from federal and state government law-enforcement and intelligence agencies' data sources and commercial sources.[62] The technology includes timeline analysis, which allows comparisons of relationships at different times. Trend analysis across multiple cases can also be performed in the context of particular investigations and intelligence operations.

## I.11 ICE PATTERN ANALYSIS AND INFORMATION COLLECTION SYSTEM (ICEPIC)

Status: *Operating as pilot program as of July 2006; planned to enter full-scale operation in fiscal year 2008.*[63]

Whereas the NETLEADS focus is on law enforcement, ICEPIC focuses on the goal of disrupting and preventing terrorism.[64] Link analysis is performed to uncover nonobvious associations between individuals and organizations to generate counterterrorism leads. Data for analysis is drawn from DHS sources and from databases maintained by the Department of State, DOJ, and the Social Security Administration. ICEPIC uses technology from IBM called Non-obvious Relationships Awareness (NORA) to perform the analysis.[65] ICEPIC, NETLEADS, and two other systems—the Data Analysis and Research for Trade Transparency System

---

[60]Ibid.

[61]Ibid., pp. 21-24.

[62]Ibid., pp. 22-23.

[63]Immigration and Customs Enforcement Fact Sheet, http://www.ice.gov/pi/news/factsheets/icepic.htm.

[64]U.S. Department of Homeland Security Office of Inspector General, "Survey of DHS Data Mining Activities," OIG-06-56, DHS, Washington, D.C., August 2006, p. 11.

[65]U.S. Department of Homeland Security (DHS), "Data Mining Report: DHS Privacy Office Response to House Report 108-774," DHS, Washington, D.C., July 6, 2006, pp. 24-26.

(DARTTS) and the Crew Vetting System (CVS)—all use association, the process of discovering two or more variable that are related, as part of the analysis.[66]

## I.12 INTELLIGENCE AND INFORMATION FUSION (I2F)

Status: *In development.*

Using commercial off-the-shelf systems, this program uses tools for searching, link analysis, entity resolution, geospatial analysis, and temporal analysis to provide intelligence analysts with an ability to view, query, and analyze information from multiple data sources.[67] The program is focused on aiding in discovery and tracking of terrorism threats to people and infrastructure. With three other DHS programs—Numerical Integrated Processing System (NIPS), Questioned Identification Documents (QID), and Tactical Information Sharing System (TISS)—I2F uses collaboration processes that support application of cross-organizational expertise and visualization processes that aid in presentation of analysis results.[68] Data may be drawn from both government and commercial sources.

## I.13 FRAUD DETECTION AND NATIONAL SECURITY DATA SYSTEM (FDNS-DS)

Status: *In use but without analytical tools to support data mining; support for data mining capabilities not expected for at least 2 years.*

This program (formerly the Fraud Tracking System) is used to track immigration-related fraud, public-safety referrals to ICE, and national-security concerns discovered during background checks.[69] In its present form, FDNS-DS is a case-management system with no analytical or data mining tools. It is planned to add those capabilities to allow identification of fraudulent schemes.

---

[66]U.S. Department of Homeland Security Office of Inspector General, "Survey of DHS Data Mining Activities," OIG-06-56, DHS, Washington, D.C., August 2006, pp. 9-11.

[67]U.S. Department of Homeland Security (DHS), "Data Mining Report: DHS Privacy Office Response to House Report 108-774," DHS, Washington, D.C., July 6, 2006, p. 26.

[68]U.S. Department of Homeland Security Office of Inspector General, "Survey of DHS Data Mining Activities," OIG-06-56, DHS, Washington, D.C., August 2006, p. 13.

[69]U.S. Department of Homeland Security (DHS), "Data Mining Report: DHS Privacy Office Response to House Report 108-774," DHS, Washington, D.C., July 6, 2006, p. 27.

## I.14 NATIONAL IMMIGRATION INFORMATION SHARING OFFICE (NIISO)

Status: *In use without data mining tools; pilot project that includes data mining capabilities being planned.*

This program is responsible for fulfilling requests for immigration-related information from other DHS components and law-enforcement and intelligence agencies.[70] The program does not include any data mining tools and techniques, relying instead on manual searches based on specific requests to supply information to authorized requesting agencies. Plans to add such analytical capabilities are being developed. Data for analysis would include data collected by immigration services, publicly available information, and data from commercial aggregators.[71]

## I.15 FINANCIAL CRIMES ENFORCEMENT NETWORK (FinCEN) AND BSA DIRECT

Status: *FinCEN in use; BSA Direct withdrawn.*

FinCEN applies data mining and analysis technology to data from a number of sources related to financial transactions to identify cases of money-laundering and other financial elements of criminal and terrorist activity. The goal of FinCEN is to promote information-sharing among law-enforcement, regulatory, and financial institutions.[72] FinCEN is responsible for administering the Bank Secrecy Act (BSA). As part of that responsibility, it uses data mining technology to analyze data collected on the basis of requirements of BSA and to identify suspicious activity tied to terrorists and organized crime.

In 2004, FinCEN began a program called BSA Direct intended to provide law-enforcement agencies with access to BSA data and to data mining capabilities similar to those available to FinCEN.[73] BSA Direct was permanently halted in July 2006 after cost overruns and technical implementation and deployment difficulties.[74]

---

[70]U.S. Department of Homeland Security (DHS), "Data Mining Report: DHS Privacy Office Response to House Report 108-774," DHS, Washington, D.C., July 6, 2006, p. 28.

[71]Ibid.

[72]See the FinCEN Web site at http://www.fincen.gov/af_faqs.html for further details on its mission.

[73]Statement of Robert W. Werner before the House Committee on Government Reform Subcommittee on Criminal Justice, Drug Policy, and Human Resources, May 11, 2004, p. 3, available at http://www.fincen.gov/wernertestimonyfinal051104.pdf.

[74]FinCEN, "FinCEN Halts BSA Direct Retrieval and Sharing Project," July 13, 2006, available at http://www.fincen.gov/bsa_direct_nr.html.

## I.16  DEPARTMENT OF JUSTICE PROGRAMS
## INVOLVING PATTERN-BASED DATA MINING

Status: *All programs under development or in use.*

Responding to requirements of the USA PATRIOT Improvement and Reauthorization Act of 2005,[75] DOJ submitted a report to the Senate Committee on the Judiciary that identified seven programs that constitute pattern-based data mining as defined in the act.[76] The report carefully scoped what was considered pattern-based data mining on the basis of the definition of the act to determine which programs it was required to report on.[77] For each program identified, the report provides a description, plans for use, efficacy, potential privacy and civil-liberties impact, legal and regulatory foundation, and privacy- and accuracy-protection policies.[78] The report notes that the scope of the programs and the detail provided vary widely. The following is a summary of the programs drawn from the DOJ report.[79]

- *System-to-Assess-Risk (STAR) Initiative.* Focused on extending the capabilities of the Foreign Terrorist Tracking Task Force (FTTTF), this program is a risk-assessment software system that is meant to help analysts to set priorities among persons of possible investigative interest. Data used by STAR are drawn from the FTTTF data mart, an existing data repository "containing data from U.S. Government and proprietary sources (e.g., travel data from the Airlines Reporting Corporation) as well as access to publicly available data from commercial data sources (such as ChoicePoint)."[80] STAR is under development.
- *Identity Theft Intelligence Initiative.* This program extracts data from the Federal Trade Commission's Identity Theft Clearinghouse and compares them with FBI data from case complaints of identity theft and with suspicious financial transactions filed with FinCEN. Further comparisons are made with data from private data aggregators, such as Lexis-Nexis, Accurint, and Autotrack. On the basis of the results of the analysis, FBI creates a knowledge base to evaluate identity-theft types, identify

---

[75]U.S. Pub. L. No. 109-177, Sec. 126.

[76]U.S. Department of Justice, "Report on 'Data-mining' Activities Pursuant to Section 126 of the USA PATRIOT Improvement and Reauthorization Act of 2005," July 9, 2007, available at http://www.epic.org/privacy/fusion/doj-dataming.pdf.

[77]Ibid., pp. 1-6.

[78]The report includes a review of only six of the seven initiatives identified, saying that a supplemental report on the seventh initiative will be provided at a later date.

[79]Ibid., pp. 7-30.

[80]Ibid., p. 8. ChoicePoint is a private data aggregator; see http://www.choicepoint.com/index.html.

identity-theft rings through subject relationships, and send leads to field offices. The program has been operational since 2003.

• *Health Care Fraud Initiative.* This program is used by FBI analysts to research and investigate health-care providers. The program draws data from Medicare "summary billing records extracted from the Centers for Medicare and Medicaid Services (CMS), supported by the CMS Fraud Investigative Database, Searchpoint [the Drug Enforcement Administration's pharmaceutical-claims database], and the National Health Care Anti-Fraud Association Special Investigative Resource and Intelligence System (private insurance data)."[81] The program has been in use since 2003.

• *Internet Pharmacy Fraud Initiative.* This program's aim is to search consumer complaints (made to the Food and Drug Administration and Internet Fraud Complaint Center) involving alleged fraud by Internet pharmacies to develop common threads indicative of fraud by such pharmacies. Data on Internet pharmacies available from open-source aggregators are also incorporated into the analysis. The program began in December 2005 and is operational.

• *Housing Fraud Initiative.* This program run by the FBI uses public-source data containing buyer, seller, lender, and broker identities and property addresses purchased from ChoicePoint to uncover fraudulent housing purchases. All analysis is done by FBI analysts manually (that is, not aided by computer programs) to identify connections between individuals and potentially fraudulent real-estate transactions. The program first became operational in 1999 and continues to be extended by Choice-Point as new real estate transaction information becomes available.

• *Automobile Accident Insurance Fraud Initiative.* This program run by FBI was designed to identify and analyze information regarding automobile-insurance fraud schemes. Data sources include formatted reports of potential fraudulent claims for insurance reimbursement as identified and prepared by the insurance industry's National Insurance Crime Bureau, FBI case-reporting data, commercial data aggregators, and health-care insurance claims information from the Department of Health and Human Services (DHHS) and the chiropractic industry. The program is being run as a pilot program in use by only one FBI field office. No target date has been set for national deployment.

In addition to the programs identified as meeting the definition of pattern-based data mining used by the DOJ report, several programs were identified as potentially meeting other definitions of data mining. That report does not provide details about the programs, but it includes brief

---

[81]Ibid., p. 20.

sketches of them. The programs identified as "advanced analytical tools that do not meet the definition in Section 126" and included in the DOJ report are as follows: [82]

- Drug Enforcement Administration (DEA) initiatives:

—*SearchPoint*. DEA project that uses prescription data from insurance and cash transactions obtained commercially from ChoicePoint, included the prescribing official (practitioner), the dispensing agent (pharmacy, clinic, hospital, and so on), and the name and quantity of the controlled substance (drug information) to conduct queries about practitioners, pharmacies, and controlled substances to identify the volume and type of controlled substances being subscribed and dispensed.

—*Automation of Reports of Consolidated Orders System (ARCOS)*. DEA uses data collected from manufacturers and distributors of controlled substances and stored in the ARCOS database to monitor the flow of the controlled substances from their point of manufacture through commercial distribution channels to point of sale or distribution at the dispensing or retail level (hospitals, retail pharmacies, practitioners, and teaching institutions).

—*Drug Theft Loss (DTL) Database*. This is similar to ARCOS, but the data source is all DEA controlled-substance registrants (including practitioners and pharmacies).

—*Online Investigative Project (OIP)*. OIP enables DEA to scan the Internet in search of illegal Internet pharmacies. The tool searches for terms that might indicate illegal pharmacy activity.

- Bureau of Alcohol, Tobacco, Firearms, and Explosives initiatives:

—*Bomb Arson Tracking System (BATS)*. BATS enables law-enforcement agencies to share information related to bomb and arson investigations and incidents. The source of information is the various law-enforcement agencies. Possible queries via BATS include similarities of components, targets, or methods. BATS can be used, for example, to make connections between multiple incidents with the same suspect.

—*GangNet*. This system is used to track gang members, gangs, and gang incidents in a granular fashion. It enables sharing of information among law-enforcement agencies. It can also be used to identify trends, relationships, patterns, and demographics of gangs.

- Federal Bureau of Investigation initiative:

—*Durable Medical Equipment (DME) Initiative*. DME is designed to help in setting investigative priorities on the basis of analysis of suspicious claims submitted by DME providers by contractors for CMS. Data

---

[82]Ibid., pp. 31-35. Descriptions are drawn from the report.

sources include complaint reports from the CMS and DHHS Inspector General's office and FBI databases.

- Other DOJ activities:

—*Organized Crime and Drug Enforcement Task Force (OCDETF) Fusion Center.* OCDETF maintains a data warehouse named Compass that contains relevant drug and related financial intelligence information from numerous law-enforcement organizations. As stated in the report, "the goal of the data warehouse is to use cross-case analysis tools to transform multi-agency information into actionable intelligence in order to support major investigations across the globe."[83]

—*Investigative Data Warehouse (IDW).* Managed by FBI, this warehouse enables investigators to perform efficient distributed searches of data sources across FBI. IDW provides analysts with the capability to examine relationships between people, places, communication devices, organizations, financial transactions, and case-related information.

—*Internet Crime Complaint Center (IC3).* A partnership between FBI and the National White Collar Crime Center (NW3C), IC3 is focused on cybercrime. It provides a reporting mechanism for suspected violations. Reports are entered into the IC3 database, which can then be queried to discover common characteristics of complaints.

—*Computer Analysis and Response Team (CART) Family of Systems.* This is a set of tools used to support computer forensics work. CART maintains a database of information collected from criminal investigations. Data can be searched for similarities among confiscated computer hard drives.

Before publication of the report, many of the programs were either unknown publicly or had unclear scopes and purposes. Commenting on the DOJ report shortly after its delivery to the Senate Committee on the Judiciary, Senator Patrick Leahy commented that "this report raises more questions than it answers and demonstrates just how dramatically the Bush administration has expanded the use of this technology, often in secret, to collect and sift through Americans' most sensitive personal information," and said that the report provided "an important and all too rare ray of sunshine on the Department's data mining activities and provides Congress with an opportunity to conduct meaningful oversight of this powerful technological tool."[84]

---

[83]Ibid., p. 34.

[84]Comment of Senator Patrick Leahy, Chairman, Senate Judiciary Committee on Department of Justice's Data Mining Report, July 10, 2007; see http://leahy.senate.gov/press/200707/071007c.html.

# J

# The Total/Terrorist Information Awareness Program

## J.1  A BRIEF HISTORY[1]

In 2002, in the wake of the September 11, 2001, attacks, the Defense Advanced Research Projects Agency (DARPA) of the U.S. Department of Defense (DOD) launched a research and development effort known as the Total Information Awareness (TIA) program. Later renamed the Terrorism Information Awareness program, TIA was a research and development program intended to counter terrorism through prevention by developing and integrating information analysis, collaboration, and decision-support tools with language-translation, data-searching, pattern-recognition, and privacy-protection technologies.[2] The program included the development of a prototype system/network to provide an environment for integrating technologies developed in the program and as a testbed for conducting experiments. Five threads for research investigation were to be pursued: secure collaborative problem-solving among disparate agencies and institutions, structured information-searching and pattern recognition based

---

[1]This description of the TIA program is based on unclassified, public sources that are presumed to be authoritative because of their origin (for example, Department of Defense documents and speeches by senior program officials). Recognizing that some aspects of the program were protected by classification, the committee believes that this description is accurate but possibly incomplete.

[2]Defense Advanced Research Programs Agency (DARPA), "Report to Congress Regarding the Terrorism Information Awareness Program: In response to Consolidated Appropriations Resolution, 2003, Pub. L. No. 108-7, Division M, § 111(b)," DARPA, Arlington, Va., May 20, 2003.

on information from a wide array of data sources, social-network analysis tools to understand linkages and organizational structures, data-sharing in support of decision-making, and language-translation and information-visualization tools. A technical description of the system stressed the importance of using real data and real operational settings that were complex and huge.[3]

The TIA program sought to pursue important research questions, such as how data mining techniques might be used in national-security investigations and how technological approaches might be able to ameliorate the privacy impact of such analysis. For example, in a speech given in August 2002, John Poindexter said that[4]

> IAO [Information Awareness Office] programs are focused on making Total Information Awareness—TIA—real. This is a high level, visionary, functional view of the world-wide system—somewhat over simplified. One of the significant new data sources that needs to be mined to discover and track terrorists is the transaction space. If terrorist organizations are going to plan and execute attacks against the United States, their people must engage in transactions and they will leave signatures in this information space. This is a list of transaction categories, and it is meant to be inclusive. Currently, terrorists are able to move freely throughout the world, to hide when necessary, to find sponsorship and support, and to operate in small, independent cells, and to strike infrequently, exploiting weapons of mass effects and media response to influence governments. We are painfully aware of some of the tactics that they employ. This low-intensity/low-density form of warfare has an information signature. We must be able to pick this signal out of the noise. Certain agencies and apologists talk about connecting the dots, but one of the problems is to know which dots to connect. The relevant information extracted from this data must be made available in large-scale repositories with enhanced semantic content for easy analysis to accomplish this task. The transactional data will supplement our more conventional intelligence collection.

Nevertheless, authoritative information about the threats of interest to the TIA program is scarce. In some accounts, TIA was focused on a generalized terrorist threat. In other informed accounts, TIA was premised on the notion of protecting a small number of high-value targets in the United States, and a program of selective hardening of those targets

---

[3]Defense Advanced Research Programs Agency (DARPA), *Total Information Awareness Program System Description Document*, version 1.1, DARPA, Arlington, Va., July 19, 2002.

[4]J. Poindexter, Overview of the Information Awareness Office, Remarks prepared for DARPATech 2002 Conference, Anaheim, Calif., August 2, 2002, available at http://www.fas.org/irp/agency/dod/poindexter.html.

would force terrorists to carry out attacks along particular lines, thus limiting the threats of interest and concern to TIA technology.

The TIA program was cast broadly as one that would "integrate advanced collaborative and decision support tools; language translation; and data search, pattern recognition, and privacy protection technologies into an experimental prototype network focused on combating terrorism through better analysis and decision making."[5] Regarding data-searching and pattern recognition, research was premised on the idea that

> ... terrorist planning activities or a likely terrorist attack could be uncovered by searching for indications of terrorist activities in vast quantities of transaction data. Terrorists must engage in certain transactions to coordinate and conduct attacks against Americans, and these transactions form patterns that may be detectable. Initial thoughts are to connect these transactions (e.g., applications for passports, visas, work permits, and drivers' licenses; automotive rentals; and purchases of airline ticket and chemicals) with events, such as arrests or suspicious activities.[6]

As described in the DOD TIA report, "These transactions would form a pattern that may be discernable in certain databases to which the U.S Government would have lawful access. Specific patterns would be identified that are related to potential terrorist planning."[7]

Furthermore, the program would focus on analyzing nontargeted transaction and event data en masse rather than on collecting information on specific individuals and trying to understand what they were doing. The intent of the program was to develop technology that could discern event and transaction patterns of interest and *then* identify individuals of interest on the basis of the events and transactions in which they participated. Once such individuals were identified, they could be investigated or surveilled in accordance with normal and ordinary law-enforcement and counterterrorism procedures.

The driving example that motivated TIA was the set of activities of the 9/11 terrorists who attacked the World Trade Center. In retrospect, it was discovered that they had taken actions that together could be seen

---

[5]Defense Advanced Research Programs Agency (DARPA), "Report to Congress Regarding the Terrorism Information Awareness Program: In response to Consolidated Appropriations Resolution, 2003, Pub. L. No. 108-7, Division M, § 111(b)," DARPA, Arlington, Va., May 20, 2003.

[6]DARPA. *Defense Advanced Research Projects Agency's Information Awareness Office and Terrorism Information Awareness Project.* Available at http://www.taipale.org/references/iaotia.pdf.

[7]Defense Advanced Research Programs Agency (DARPA), "Report to Congress Regarding the Terrorism Information Awareness Program: In response to Consolidated Appropriations Resolution, 2003, Pub. L. No. 108-7, Division M, § 111(b)," May 20, 2003, p. 14.

as predictors of the attack even if no single action was unlawful. Among those actions were flight training (with an interest in level flight but not in takeoff and landing), the late purchase of one-way air tickets with cash, foreign deposits into banking accounts, and telephone records that could be seen to have connected the terrorists. If the actions could have been correlated before the fact, presumably in some automated fashion, suspicions might have been aroused in time to foil the incident before it happened.

Because the TIA program was focused on transaction and event data that were already being collected and resident in various databases, privacy implications generally associated with the collection of data per se did not arise. But the databases were generally privately held, and many privacy questions arose because the government would need access to the data that they contained. The databases also might have contained the digital signatures of most Americans as they conducted their everyday lives, and this gave rise to many concerns about their vast scope.

After a short period of intense public controversy, Congress took action on the TIA program in 2003. Section 8131 of H.R. 2658, the Department of Defense Appropriations Act of 2004, specified that

> (a) Notwithstanding any other provision of law, none of the funds appropriated or otherwise made available in this or any other Act may be obligated for the Terrorism Information Awareness Program: Provided, That this limitation shall not apply to the program hereby authorized for processing, analysis, and collaboration tools for counterterrorism foreign intelligence, as described in the Classified Annex accompanying the Department of Defense Appropriations Act, 2004, for which funds are expressly provided in the National Foreign Intelligence Program for counterterrorism foreign intelligence purposes.

> (b) None of the funds provided for processing, analysis, and collaboration tools for counterterrorism foreign intelligence shall be available for deployment or implementation except for:

> (1) lawful military operations of the United States conducted outside the United States; or

> (2) lawful foreign intelligence activities conducted wholly overseas, or wholly against non-United States citizens.

> (c) In this section, the term "Terrorism Information Awareness Program" means the program known either as Terrorism Information Awareness or Total Information Awareness, or any successor program, funded by the Defense Advanced Research Projects Agency, or any other Department or element of the Federal Government, including the individual components of such Program developed by the Defense Advanced Research Projects Agency.

It is safe to say that the issues raised by the TIA program have not been resolved in any fundamental sense. Though the program itself was terminated, much of the research under it was moved from DARPA to another group, which builds technologies primarily for the National Security Agency, according to documents obtained by the *National Journal* and to intelligence sources familiar with the move. The names of key projects were changed, apparently to conceal their identities, but their funding remained intact, often under the same contracts.[8]

The immediate result, therefore, of congressional intervention was to drive the development and deployment of data mining at DOD from public view, relieve it of the statutory restrictions that had previously applied to it, block funding for research into privacy-enhancing technologies, and attenuate the policy debate over the appropriate roles and limits of data mining. Law and technology scholar K.A. Taipale wrote:[9]

> At first hailed as a "victory" for civil liberties, it has become increasingly apparent that the defunding [of TIA] is likely to be a pyrrhic victory. . . . Not proceeding with a focused government research and development project (in which Congressional oversight and a public debate could determine appropriate rules and procedures for use of these technologies and, importantly, ensure the development of privacy protecting technical features to support such policies) is likely to result in little security and, ultimately, brittle privacy protection. . . . Indeed, following the demise of IAO and TIA, it has become clear that similar data aggregation and automated analysis projects exist throughout various agencies and departments not subject to easy review.

Thus, many other data mining activities supported today by the U.S. government continue to raise the same issues as did the TIA program: the potential utility of large-scale databases containing personal information for counterterrorism and law-enforcement purposes and the potential privacy impact of the use of such databases by law-enforcement and national-security authorities.

## J.2 A TECHNICAL PERSPECTIVE ON TIA'S APPROACH TO PROTECTING PRIVACY

As noted above, managers of the TIA program understood that their approach to identifying terrorists before they acted had major privacy implications. To address privacy issues in TIA and similar programs, such

---

[8]S. Harris, "TIA lives on," *National Journal*, February 23, 2006, available at http://nationaljournal.com/about/njweekly/stories/2006/0223nj1.htm#.

[9]K.A. Taipale, "Data mining and domestic security: Connecting the dots to make sense of data," *Columbia Science and Technology Law Review* 5(2):1-83, 2003.

as MATRIX, Tygar[10] and others have advocated the use of what has come to be called selected revelation, involving something like the risk-utility tradeoff in statistical disclosure limitation. Sweeney[11] used the term to describe an approach to disclosure limitation that allows data to be shared for surveillance purposes "with a sliding scale of identifiability, where the level of anonymity matches scientific and evidentiary need." That corresponds to a monotonically increasing threshold for maximum tolerable risk in the risk-utility confidentiality-map framework previously described in Duncan et al.[12] Some related ideas emanate from the computer-science literature, but most authors attempt to demand a stringent level of privacy, carefully defined, and to restrict access by adding noise and limitations on the numbers of queries allowed (e.g., see Chawla et al.[13]).

The TIA privacy report suggests that[14]

> selective revelation [involves] putting a security barrier between the private data and the analyst, and controlling what information can flow across that barrier to the analyst. The analyst injects a query that uses the private data to determine a result, which is a high-level sanitized description of the query result. That result must not leak any private information to the analyst. Selective revelation must accommodate multiple data sources, all of which lie behind the (conceptual) security bar-

---

[10]J.D. Tygar, "Privacy Architectures," presentation at Microsoft Research, June 18, 2003, available at http://research.microsoft.com/projects/SWSecInstitute/slides/Tygar.pdf; J.D. Tygar, "Privacy in sensor webs and distributed information systems," pp. 84-95 in *Software Security Theories and Systems*, M. Okada, B. Pierce, A. Scedrov, H. Tokuda, and A. Yonezawa, eds., Springer, New York, 2003.

[11]L. Sweeney, "Privacy-preserving surveillance using selective revelation," LIDAP Working Paper 15, Carnegie Mellon University, 2005; updated journal version is J. Yen, R. Popp, G. Cybenko, K.A. Taipale, L. Sweeney, and P. Rosenzweig, "Homeland security," *IEEE Intelligent Systems* 20(5):76-86, 2005.

[12]G.T. Duncan, S.E. Fienberg, R. Krishnan, R. Padman, and S.F. Roehrig, "Disclosure limitation methods and information loss for tabular data," pp. 135-166 in *Confidentiality, Disclosure and Data Access: Theory and Practical Applications for Statistical Agencies*, P. Doyle, J. Lane, J. Theeuwes, and L. Zayatz, eds., North-Holland, Amsterdam, 2001. See also G.T. Duncan, S.A. Keller-McNulty, and S.L. Stokes, *Database Security and Confidentiality: Examining Disclosure Risk vs. Data Utility Through the R–U Confidentiality Map*, Technical Report 142, National Institute of Statistical Sciences, Research Triangle Park, N.C., 2004; G.T. Duncan and S.L. Stokes, "Disclosure risk vs. data utility: The R–U confidentiality map as applied to topcoding," *Chance* 17(3):16-20, 2004.

[13]S.C. Chawla, C. Dwork, F. McSherry, A. Smith, and H. Wee, "Towards Privacy in Public Datatbases," in *Theory of Cryptography Conference Proceedings*, J. Kilian, ed., Lecture Notes in Computer Science, Volume 3378, Springer-Verlag, Berlin, Germany.

[14]Information Systems Advanced Technology (ISAT) panel, *Security with Privacy*, DARPA, Arlington, Va., 2002, p. 10, available at http://www.cs.berkeley.edu/~tygar/papers/ISAT-final-briefing.pdf.

rier. Private information is not made available directly to the analyst, but only through the security barrier.

One effort to implement this scheme was dubbed privacy appliances by Golle et al. and was intended to be a stand-alone device that would sit between the analyst and the private data source so that private data stayed in authorized hands.[15] The privacy controls would also be independently operated to keep them isolated from the government. According to Golle et al., the device would provide:

- *Inference control* to prevent unauthorized individuals from completing queries that would allow identification of ordinary citizens.
- *Access control* to return sensitive identifying data only to authorized users.
- *Immutable audit trails* for accountability.

Implicit in the TIA report and in the Golle et al. approach was the notion that linkages between databases behind the security barrier would use identifiable records and thus some form of multiparty computation method involving encryption techniques.

The real questions of interest in "inference control" are, What disclosure-limitation methods should be used? To which databases should they be applied? How can the "inference control" approaches be combined with the multiparty computation methods? Here is what is known in the way of answers:

- Both Sweeney and Golle et al. refer to microaggregation, known as $k$-anonymity, but with few details on how it could be used in this context. The method combines observations in groups of size $k$ and reports either the sum or the average of the group for each unit. The groups may be identified by clustering or some other statistical approach. Left unsaid is what kinds of users might perform with such aggregated data. Furthermore, neither $k$-anonymity nor any other confidentiality tool does anything to cope with the implications of the release of exactly linked files requested by "authorized users."
- Much of the statistical and operations-research literature on confidentiality fails to address the risk-utility trade-off, largely because it

---

[15]Philippe Golle et al. "Protecting Privacy in Terrorist Tracking Applications," presentation to Computers, Freedom, and Privacy 2004, available at http://www.cfp2004.org/program/materials/w-golle.ppt.

focuses primarily on privacy or on technical implementations without understanding how users wish to analyze a database.[16]

• A clear lesson from the statistical disclosure-limitation literature is that privacy protection in the form of "safe releases" from separate databases does not guarantee privacy protection for a merged database. A figure in Lunt et al.[17] demonstrates recognition of that by showing privacy appliances applied for the individual databases and then independently for the combined data.

• There have been a small number of crosswalks between the statistical disclosure-limitation literature on multiparty computation and risk-utility trade-off choices for disclosure limitation. Yang et al. provide a starting point for discussions on *k*-anonymity.[18] There are clearly a number of alternatives to *k*-anonymity and alternatives that yield "anonymized" databases of far greater statistical utility.

• The "hype" associated with the TIA approach to protection has abated, largely because TIA no longer exists as an official program. But similar programs continue to appear in different places in the federal government and no one associated with any of them has publicly addressed the privacy concerns raised here regarding the TIA approach.

When Congress stopped the funding for DARPA's TIA program in 2003, work on the privacy appliance's research and development effort at PARC Research Center was an attendant casualty. Thus, prototypes of the privacy appliance have not been made publicly available since then, nor are they likely to appear in the near future. The claims of privacy protection and selective revelation continued with MATRIX and other data warehouse systems but without an attendant research program, and the federal government continues to plan for the use of data mining techniques in other initiatives, such as the Computer Assisted Passenger Pro-

---

[16]R. Gopal, R. Garfinkel, and P. Goes, "Confidentiality via camouflage: The CVC approach to disclosure limitation when answering queries to databases," *Operations Research* 50:501-516, 2002.

[17]T. Lunt, J. Staddon, D. Balfanz, G. Durfee, T. Uribe, D. Smetters, J. Thornton, P. Aoki, B. Waters, and D. Woodruff, "Protecting Privacy in Terrorist Tracking Applications," presentation at the University of Washington/Microsoft Research/Carnegie Mellon University Software Security Summer Institute, *Software Security: How Should We Make Software Secure?* on June 15-19, 2003, available at http://research.microsoft.com/projects/SWSecInstitute/five-minute/Balfanz5.ppt.

[18]Z. Yang, S. Zhong, and R.N. Wright, "Anonymity-preserving data collection," pp. 334-343 in *Proceedings of the 11th ACM SIGKDD International Conference on Knowledge Discovery and Data Mining—KDD'05*, Association for Computing Machinery, New York, N.Y., 2005.

filing System II (CAPPS II). Similar issues arise in the use of government, medical, and private transaction data in bioterrorism surveillance.[19]

## J.3 ASSESSMENT

Section J.1 provided a brief history of the TIA program. Whatever one's views regarding the desirability or technical feasibility of the TIA program, it is clear that from a political standpoint, the program was a debacle. Indeed, after heated debate, the Senate and House appropriations committees decided to terminate funding of the program.[20] On passage of the initial funding limitation, a leading critic of the TIA program, Senator Ron Wyden, declared:

> The Senate has now said that this program will not be allowed to grow without tough Congressional oversight and accountability, and that there will be checks on the government's ability to snoop on law-abiding Americans.[21]

The irony of the TIA debate is that although the funding for the TIA program was indeed terminated, both research on and deployment of data mining systems continue at various agencies (Appendix I, "Illustrative Government Data Mining Programs and Activity"), but research on privacy-management technology did not continue, and congressional oversight of data mining technology development has waned to some degree.

The various outcomes of the TIA debate raise the question of whether the nature of the debate over the program (if not the outcome) could have been any different if policy makers had addressed in advance some of the difficult questions that the program raised. In particular, it is interesting to consider questions in the three categories articulated in the framework of Chapter 2: effectiveness, consistency with U.S. laws and values, and possible development of new laws and practices. The TIA example further illustrates how careful consideration of the privacy impact of new technologies is needed before a program seriously begins the research stage.

The threshold consideration of any privacy-sensitive technology is whether it is effective in meeting a clearly defined law-enforcement or

---

[19]See S.E. Fienberg and G. Shmueli, "Statistical issues and challenges associated with rapid detection of bio-terrorist attacks," *Statistics in Medicine* 24:513-529, 2005; L. Sweeney, "Privacy-Preserving Bio-Terrorism Surveillance," presentation at AAAI Spring Symposium, AI Technologies for Homeland Security, Stanford University, Stanford, Calif., 2005.

[20]U.S. House, Conference Report on H.R. 2658, Department of Defense Appropriations Act, (House Report 108-283), U.S. Government Printing Office, Washington, D.C., 2004.

[21]Declan McCullagh, "Senate limits Pentagon 'snooping' plan," CNET News.com, January 24, 2003. Available at http://sonyvaio-cnet.com.com/2100-1023_3-981945.html.

national-security purpose. The question of effectiveness must be assessed through rigorous testing guided by scientific standards. The TIA research program proposed an evaluation framework, but none of the results of evaluation have been made public. Some testing and evaluation may have occurred in a classified setting, but neither this committee nor the public has any knowledge of results. Research on how large-scale data-analysis techniques, including data mining, could help the intelligence community to identify potential terrorists is certainly a reasonable endeavor. Assuming that initial research justifies additional effort on the basis of scientific standards of success, the work should continue, but it must be accompanied by a clear method for assessing the reliability of the results.

Even if a proposed technology is effective, it must also be consistent with existing U.S. law and democratic values. First, one must assess whether the new technique and objective comply with law. In the case of TIA, DARPA presented to Congress a long list of laws that it would comply with and affirmed that "any deployment of TIA's search tools may occur only to the extent that such a deployment is consistent with current law." Second, inasmuch as TIA research sought to enable the deployment of very large-scale data mining over a larger universe of data than the U.S. government had previously analyzed, even compliance with then-current law would not establish consistency with democratic values.

The surveillance power that TIA proposed to put in the hands of U.S. investigators raised considerable concern among policy makers and the general public. That the program, if implemented, could be said to comply with law did not address those concerns. In fact, the program raised the concerns to a higher level and ultimately led to an effort by Congress to stop the research altogether.

TIA-style data mining was, and still is, possible because there are few restrictions on government access to third-party business records. Any individual business record (such as a travel reservation or credit-card transactions) may have relatively low privacy sensitivity when looked at in isolation; but when a large number of such transaction records are analyzed over time, a complete and intrusive picture of a person's life can emerge.

Developing the technology to derive such individual profiles was precisely the objective of the TIA program. It proposed to use such profiles in only the limited circumstances in which they indicated terrorist activity. That may be a legitimate goal and could ultimately be recognized explicitly as such by law. However, that the program was at once legal and at the same time appeared to cross boundaries not previously crossed by law-enforcement or national-security investigations gives rise to questions that must be answered.

John Poindexter, director of the DARPA office responsible for TIA, was aware of the policy questions and took notable steps to include in the technical research agenda various initiatives to build technical mechanisms that might minimize the privacy impact of the data mining capabilities being developed. In hindsight, however, a more comprehensive analysis of both the technical and larger public-policy considerations associated with the program was necessary to address Congress's concerns about privacy impact.

# K

# Behavioral-Surveillance
# Techniques and Technologies

The primary question in behavioral science as applied to the use of behavioral technologies in the antiterrorism effort is, How can detection of particular behaviors and the attendant biological activity be used to indicate current and future acts of terrorism?

## K.1 THE RATIONALE FOR BEHAVIORAL SURVEILLANCE

Some behavioral methods attempt to detect terrorist activity directly (for example, through surveillance at bridges, docks, and weapon sites). However, the focus in this appendix is on behavioral methods that are more indirect. Such methods are used to try to detect patterns of behavior that are thought to be precursors or correlates of wrongdoing (such as deception and expression of hostile emotions) or that are anomalous in particular situations (for example, identifying a person who fidgets much more and has much more facial reddening than others in a security line).

Many behavioral-detection methods monitor biological systems (such as cardiac activity, facial expressions, and voice tone) and use physiological information to draw inferences about internal psychological states (for example, "on the basis of this pattern of physiological activity, this person is likely to be engaged in deception"). In most situations, the easiest and most accurate way to determine past, current, and future behavior might be to ask the person what he or she has been doing, is doing, and plans to do. However, the terrorist's desire to avoid detection and the "cat and mouse" game that is played by terrorists and their pursuers make such a verbal mode of information-gathering highly unreliable.

Because verbal reports can be manipulated and controlled so easily, we might turn to biological systems that are less susceptible to voluntary control or that provide detectable signs when they are being manipulated. Once we move to the biological level, however, we have abandoned direct observation of terrorist behavior and moved into the realm of inference of likely behavior from more primitive and less specific sources. Biobehavioral methods can be powerful and useful, but they are intrinsically subject to three limitations:

- *Many-to-one.* Any given pattern of physiological activity can result from or correlate with a number of quite different psychological or physical states.
- *Probabilistic.* Any detected sign or pattern conveys some *likelihood* of the behavior, intent, or attitude of interest but not an absolute certainty.
- *Errors.* In addition to the highly desirable true positives and true negatives that are produced, there will be the troublesome false positives (an innocent person is thought to be guilty) and false negatives (a guilty person is thought to be innocent). Depending on the robustness of the biobehavioral techniques involved, it may be possible in the face of countermeasures for a subject to induce false negatives by manipulating his or her behavior.

In addition, even if deception or the presence of an emotion can be accurately and reliably detected, information about the reason for deception, a given emotion, or a given behavior is not available from the measurements taken. A person exhibiting nervousness may be excited about meeting someone at the airport or about being late. A person lying about his or her travel plans may be concealing an extramarital affair. A person fidgeting may be experiencing back pain. None of those persons would be the targets of counterterrorist efforts, nor should they be—and the possibility that their true motivations and intents may be revealed has definite privacy implications.

## K.2 MAJOR BEHAVIORAL-DETECTION METHODS

Most behavioral methods are based on monitoring the activity of neural systems that are thought to be difficult to control voluntarily or that reveal measurable signs when they are being controlled.

## K.2.1 Facial Expression

Facial muscles are involved in the expression and communication of emotional states. They can be activated both voluntarily and involuntarily,[1] so there is ample opportunity for a person to interfere with the expression of emotion in ways that serve personal goals. There is strong scientific evidence that different configurations of facial-muscle contractions are associated with what are often called basic emotions.[2] Those emotions include anger, contempt, disgust, fear, happiness, surprise, and sadness. There is also evidence that other emotions can be identified on the basis of patterns of movement in facial *and* bodily muscles (for example, embarrassment[3]) and that distinctions can be made between genuine felt happiness and feigned unfelt happiness according to whether a smile (produced by the zygomatic major muscles) is accompanied by the contraction of the muscles (orbicularis oculi) that circle the eyes.[4]

Facial-muscle activity can be measured accurately by careful examination of the changes in appearance that are produced as the muscles cause facial skin to be moved.[5] Trained coders working with video recordings can analyze facial expressions reliably, but it is extremely time-consuming (it can take hours to analyze a few minutes of video fully). Greatly simplified methods that focus only on the key muscle actions involved in a few emotions of interest and that are appropriate for real-time screening are being developed and tested. Some basic efforts to develop automated computer systems for analyzing facial expressions have also been undertaken,[6] but the problems inherent in adapting them for real-world, naturalistic applications are enormous.[7]

---

[1]W.E. Rinn, "The neuropsychology of facial expression: A review of the neurological and psychological mechanisms for producing facial expressions," *Psychological Bulletin* 95(1):52-77, 1984.

[2]P. Ekman, "An argument for basic emotions," *Cognition and Emotion* 6(3-4):169-200, 1992.

[3]D. Keltner, "Signs of appeasement: Evidence for the distinct displays of embarrassment, amusement, and shame," *Journal of Personality and Social Psychology* 68(3):441-454, 1995.

[4]P. Ekman and W.V. Friesen, "Felt, false and miserable smiles," *Journal of Nonverbal Behavior* 6(4):238-252, 1982.

[5]P. Ekman and W.V. Friesen, *Facial Action Coding System*, Consulting Psychologists Press, Palo Alto, Calif., 1978.

[6]J.F. Zlochower, A.J. J. Lien, and T. Kanade, "Automated face analysis by feature point tracking has high concurrent validity with manual FACS coding," *Psychophysiology* 36(1):35-43, 1999.

[7]For example, according to a German field test of facial recognition conducted in 2007, an accuracy of 60 percent was possible under optimal conditions, 30 percent on average (depending on light and other factors). See Bundeskriminalamt (BKA), *Face Recognition as a Tool for Finding Criminals: Picture-man-hunt*, Final report, BKA, Wiesbaden, Germany, February 2007. Available in German at http://www.cytrap.eu/files/EU-IST/2007/pdf/2007-07-FaceRecognitionField-Test-BKA-Germany.pdf.

There are several other ways to measure facial-muscle activity. The electrical activity of the facial muscles themselves can be measured (with electromyelography [EMG]). That requires the application of many electrodes to the face, each placed to maximize sensitivity to the action of particular muscles and minimize sensitivity to the action of other muscles. Because of the overlapping anatomy of facial muscles, their varied sizes, and their high density in some areas (such as around the mouth), the EMG method may be better suited to simple detection of emotional valence (positive or negative) and intensity than to the detection of specific emotions. Another indirect method of assessing facial-muscle activity is to measure the "heat signature" of the face associated with changes in blood flow to different facial regions.[8] That information can be read remotely by using infrared cameras; however, the spatial and temporal resolutions are problematic.

Even if a method emerged that allowed facial-muscle activity to be measured reliably, comprehensively, economically, and unobtrusively, there would be the issue of its utility in a counterterrorism effort. Necessary (but not sufficient) conditions for utility would include:

- The availability of tools that can determine the specific emotion that is being signaled if and when emotional facial expression is displayed.
- The superiority of a facial-expression–based emotion-prediction system to a system based on any other biological or physiological markers.
- The detectability of indicators that a person is attempting to conceal his or her true emotional state or to shut down facial expression entirely, such as
—Small, fleeting microexpressions of the emotion being felt.
—Tell-tale facial signs of attempted control (such as tightening of some mouth muscles).
—Signs that particular emotions are being simulated.
—Characteristic increases in cardiovascular activity mediated by the sympathetic branch of the autonomic nervous system.[9]

Because no specific facial sign is associated with committing or plan-

---

[8]See, for example D.A. Pollina and A. Ryan, *The Relationship Between Facial Skin Surface Temperature Reactivity and Traditional Polygraph Measures Used in the Psychophysiological Detection of Deception: A Preliminary Investigation*, U.S. Department of Defense Polygraph Institute, Ft. Jackson, S.C., 2002.

[9]J.J. Gross and R.W. Levenson, "Emotional suppression: Physiology, self-report, and expressive behavior," *Journal of Personality and Social Psychology* 64(6):970-986, 1993; J.J. Gross and R.W. Levenson, "Hiding feelings: The acute effects of inhibiting negative and positive emotion," *Journal of Abnormal Psychology* 106(1):95-103, 1997.

ning a terrorist act, using facial measurement in a counterterrorism effort will have to be based on some combination of the detection of facial expressions thought to indicate malevolent intent (such as signs of anger, contempt, or feigned happiness in some situations), the detection of facial expressions thought to indicate deception,[10] and the detection of facial expressions that are anomalous compared with those of other people in the same situation.

Results of research on the connection between facial expression and emotional state suggest correlations between the two. However, the suggestive findings have generally not been subject to rigorous, controlled tests of accuracy in a variety of settings that might characterize real-world application contexts.

## K.2.2 Vocalization

In addition to the linguistic information carried by the human voice, a wealth of paralinguistic information is carried in pitch, timbre, tempo, and the like and is thought to be related to a person's emotional state.[11] Those paralinguistic qualities of speech can be difficult to control voluntarily, so they are potentially useful for detecting underlying emotional states and deception. In the emotion realm, much of the promise of mapping paralinguistic qualities of vocalization onto specific emotions has yet to be realized, and the history of using paralinguistic markers in the deception realm is not very encouraging. At one time, a great deal of attention was given to the detection of deception by quantifying microtremors in the voice ("voice stress analyzers"), but this approach has failed to withstand scientific scrutiny.[12]

Why has more progress not been made in using paralinguistic qualities of speech to detect emotions and deception? There are several possible reasons. The relationships between paralinguistic qualities of speech and psychological states are much weaker than originally thought. The field has not yet identified the right characteristics to measure. And siz-

---

[10]P. Ekman and M. O'Sullivan, "From flawed self-assessment to blatant whoppers: The utility of voluntary and involuntary behavior in detecting deception," *Behavioral Sciences and the Law. Special Issue: Malingering* 24(5):673-686, 2006.

[11]K.R. Scherer, *Vocal Measurement of Emotion*, Academic Press, Inc., San Diego, Calif., 1989.

[12]See, for example, National Research Council, *The Polygraph and Lie Detection*, The National Academies Press, Washington, D.C., 2003; Mitchell S. Sommers, "Evaluating voice-based measures for detecting deception," *The Journal of Credibility Assessment and Witness Psychology* 7(2):99-107, 2006, available at http://truth.boisestate.edu/jcaawp/2006_No_2/2006_99-107.pdf; J. Masip, E. Garrido, and C. Herrero, "The detection of deception using voice stress analyzers: A critical review," *Estudios de PsicologÃa* 25(1):13-30, 2004.

able individual differences in speech need to be accounted for before interindividual consistencies will emerge. In the interim, new approaches that do not rely on paralinguistic vocalizations in isolation but rather combine them with other indicators of deception and emotion (such as facial expressions and physiological indicators) may prove useful. Is is ironic that it is fairly simple to obtain high-quality, noninvasive samples of vocalizations in real-world contexts. Moreover, cost-effective, accurate instrumentation for analyzing the acoustic properties of speech is readily available. Thus, the tools are already in place; it is just the science that is lagging.

### K.2.3 Other Muscle Activity

Technology is readily available for quantifying the extent of overall motor activity (sometimes called gross motor activity or general somatic activity). It can be done with accelerometers attached to a person (some are built into watch-like casings) or with pressure-sensitive devices (such as piezoelectric transducers) placed under standing and sitting areas. The latter can be used to track motion in multiple dimensions and thus enable characterization of patterns of pacing, fidgeting, and moving. Although clearly not specifically related to any particular emotional or psychological state, high degrees of motor activity may be noteworthy when they are anomalous in comparison with usual levels of agitation and tension.

### K.2.4 Autonomic Nervous System

The autonomic nervous system (ANS) controls the activity of the major organs, including the heart, blood vessels, kidneys, pancreas, lungs, stomach, and sweat glands. Decades of methodological development in medicine and psychophysiology have produced ways to measure a wide array of autonomic functions reliably and noninvasively. Some of the measures are direct (such as using the electrical activity of the heart muscle to determine heart rate), and some are indirect (such as estimating vascular constriction by using the reflection of infrared light to determine the amount of blood pooling in peripheral sites or using impedance methods to measure the contractile force of the left ventricle as it pumps blood from the heart to the rest of the body). Additional work has been directed toward developing methods of ambulatory monitoring that enable tracking of ANS activity in freely moving people. Remote sensing of autonomic function is still in its infancy, but some progress has been made in using variation in surface temperature to indicate patterns of blood flow.

Measures of ANS activity are essentially measures of arousal and reflect the relative activation and deactivation of various organ systems to provide the optimal milieu to support current body activity (such as sleep, digestion, aggression, and thinking). Debate has raged over the decades as to whether specific patterns of autonomic activity are associated with particular psychological states, including emotions. For emotion, the issue is whether the optimal bodily milieu for anger (ANS support for fighting) is different from that for disgust (ANS support for withdrawal and expulsion of harmful substances). Evidence in support of that kind of autonomic specificity for at least some of the basic emotions is drawn from experimentation, metaphors found in language (such as association of heat and pressure with anger or of coolness with fear), and observable signs of autonomic activity (such as crying during sadness but not during fear or gagging during disgust but not during angers). There are a number of reviews of these issues and the associated scientific evidence.[13]

Over the years, patterns of ANS activity have been mapped onto several nonemotionl states. Among the more durable of them have been the distinction between stimulus intake (elevated skin conductance plus heart rate deceleration) and stimulus rejection (elevated skin conductance plus heart rate acceleration),[14] and the more recent distinction between the cardiovascular responses to threat (moderate increases in cardiac contractility, no change or decrease in cardiac output, and no change or increase in total peripheral resistance) and to challenge (increase in cardiac contractility, increase in cardiac output, and decrease in total peripheral resistance).[15]

Regardless of the putative pattern, using the existence of any particular pattern of ANS activity by itself to infer psychological or emotional states is fraught with danger. The ANS is the slave to many masters, and any ANS pattern may reflect any of a host of nonpsychological and psychological states.

The other way in which ANS monitoring has been used extensively is to detect deception. The use of autonomic measurement in lie-detection technology has a long history in law enforcement, security screening, and personnel selection. Despite its history (which continues), most of the major scientific investigations of the validity of the polygraph have raised serious reservations. For example, an independent review of the

---

[13]R.W. Levenson, "Autonomic specificity and emotion," pp. 212-224 in *Handbook of Affective Sciences*, R.J. Davidson, K.R. Scherer, and H.H. Goldsmith, eds., Oxford University Press, New York, N.Y., 2003.

[14]J.I. Lacey, J. Kagan, B.C. Lacey, and H.A. Moss, "The visceral level: Situational determinants and behavioral correlates of autonomic response patterns," pp. 161-196 in *Expression of the Emotions in Man*, P.H. Knapp, ed., International University Press, New York, N.Y., 1963.

[15]J. Tomaka, J. Blascovich, R.M. Kelsey, and C.L. Leitten, "Subjective, physiological, and behavioral effects of threat and challenge appraisal," *Journal of Personality and Social Psychology* 65(2):248-260, 1993.

use of the polygraph commissioned by the Office of Technology Assessment concluded that[16]

> there is at present only limited scientific evidence for establishing the validity of polygraph testing. Even where the evidence seems to indicate that polygraph testing detects deceptive subjects better than chance (when using the control question technique in specific-incident criminal investigations), significant error rates are possible, and examiner and examinee differences and the use of countermeasures may further affect validity. (p. 96)

In 2003, a review by the National Research Council was similarly critical,[17] concluding that the polygraph has a better than chance but far less than perfect performance in detecting specific incidents of deception but that it is not acceptable for use in general screening and is highly vulnerable to countermeasures. In considering the use of the polygraph in antiterrorism efforts, it is important to weigh its possible utility in "guilty knowledge" situations (for example, the person being interrogated is denying knowing something that he or she knows) against the likelihood that the person will be trained in using countermeasures. Empirically, "guilty knowledge" studies indicate that the polygraph confers at best a minimal advantage in identifying such situations and suggest that guilty parties may not need to take countermeasures at all to evade detection by a polygraph.

### K.2.5 Central Nervous System

The brain is clearly the source of motivated behavior—both good and evil. Thus, measuring brain activity is appealing if the goal is to detect intentions, motives, planned behaviors, allegiances, and a host of other mental states related to terrorism and terrorist acts. The electrical activity of the brain can be measured directly with electroencephalography (EEG) and indirectly with such methods as magnetoencephalography (MEG, which detects changes in magnetic fields produced by the brain's electrical activity), positron-emission tomography (PET, which uses radioactive markers to track blood flow into the brain areas that are most active), and functional magnetic resonance imaging (fMRI, which uses strong magnetic fields to detect changes in the magnetic properties of blood flowing

---

[16]Office of Technology Assessment, *Scientific Validity of Polygraph Testing: A Research Review and Evaluation*, U.S. Congress, Office of Technology Assessment, Washington, D.C., 1983.

[17]National Research Council, *The Polygraph and Lie Detection*, The National Academies Press, Washington, D.C., 2003.

through the brain that occur when active brain areas use the oxygen carried by red blood cells).

The electrical activity of the brain can be monitored using these technologies while an individual is undertaking fairly complex behavioral activities and can sometimes be linked to particular discrete stimulus events. An overarching goal of the research using these methods has been to understand how and where in the brain such basic mental activities as error detection, conflict monitoring, emotion activation, and behavioral regulation occur. In most brain research, the focus has been more on specific cognitive processes than on specific emotions. Some patterns of brain activity can be used to predict when a person is experiencing emotion but not the particular emotion. In addition to emotional activation, some patterns indicate attempts at emotion regulation and control.[18]

Each of the existing measures of brain activity has advantages and disadvantages in temporal resolution, spatial resolution, invasiveness, susceptibility to movement artifact, methodological requirements, and expense. Much of the current excitement in the field is focused on fMRI. Viewed from the perspective of counterterrorism, fMRI presents numerous challenges: subjects must be supine in a tube for a long period (typically 15 minutes to 2 hours), temporal resolution is low, and the method is highly vulnerable to movement artifacts (movements greater than 3 mm can result in unusable images). Although the committee heard testimony about detection of deception with fMRI, the paucity of research supporting it and the considerable constraints associated with it make it difficult to imagine its having any immediate antiterrorism utility.

## K.3 ASSESSING BEHAVIORAL-SURVEILLANCE TECHNIQUES

Proponents and advocates (especially vendors) often seek to demonstrate the validity of a particular approach to behavioral surveillance or deception detection by presenting evidence that it discriminates accurately between truthfulness and deception in a particular sample of examinees. Although such evidence would be necessary to accept claims of validity, it is far from sufficient.

The 2003 National Research Council report on the polygraph and lie detection[19] provided a set of questions that guide the collection of credible evidence to support claims of validity of any proposed technique for deception detection and a set of characteristics of high-quality studies

---

[18]K.N. Ochsner and J.J. Gross, "The cognitive control of emotion," *Trends in Cognitive Sciences* 9(5):242-249, 2005.

[19]National Research Council, 2003, op. cit.

that address issues of accuracy. Those questions and characteristics are presented in Box K.1.

## K.4  BEHAVIORAL AND DATA MINING METHODS: SIMILARITIES AND DIFFERENCES

Behavioral and data mining methods have many similarities and some key differences. Perhaps most important, they face many of the same challenges and can both be evaluated in the overall framework presented in this report. These are some characteristics that are common to the two methods:

- *Probabilistic.* Data mining and behavioral surveillance seek patterns that are likely to be associated with terrorist acts. Successful methods will need to have high rates of true positives and true negatives and low rates of false positives and false negatives. Because of the low base rate of terrorism in most contexts (for example, in airport security lines), both methods will detect many acts of malfeasance that are not directly related to terrorism (for example, acts by people who have committed or are planning other crimes). The value and cost of these "true positives of another sort" must be considered in evaluating any applications of the methods.
- *Remote and secret monitoring.* Data mining and some kinds of behavioral surveillance allow information to be collected and analyzed without direct interaction with those being monitored.
- *Countermeasures.* Data mining and behavioral surveillance are vulnerable to countermeasures and disinformation.
- *Gateways to human judgment.* Data mining and behavioral surveillance may best be viewed as ways to identify situations that require follow-up investigation by skilled interviewers, analysts, and scientists.
- *Privacy.* Data mining and behavioral surveillance raise serious concerns for the protection of individual liberties and privacy.
- *Need for prior empirical demonstration.* Data mining and behavioral surveillance should deployed operationally on a wide scale only after their utility has been empirically demonstrated in the laboratory and on a limited scale in operational contexts.
- *Need for continuing evaluation.* The use of data mining and of behavioral surveillance should be accompanied by a continuing process of evaluation to establish utility, accuracy and error rates, and violation of individual privacy.

The following are some of the important differences:

- *Collection versus analysis.* Techniques for detecting deception require

**BOX K.1**
**Questions for Assessing Validity and**
**Characteristics of Accurate Studies**

*Questions for Assessing Validity*

• Does the technique have a plausible theoretical rationale, that is, a proposed psychological, physiological, or brain mechanism that is consistent with current physiological, neurobiological, and psychological knowledge?

• Does the psychological state being tested for (deception or recognition) reliably cause identifiable behavioral, physiological, or brain changes in individuals, and are these changes measured by the proposed technique?

• By what mechanisms are the states associated with deception linked to the phenomena the technique measures?

• Are optimal procedures being used to measure the particular states claimed to be associated with deception?

• By what mechanisms might a truthful response produce a false positive result with this technique? What do practitioners of the technique do to counteract or correct for such mechanisms? Is this response to the possibility of false positives reasonable considering the mechanisms involved?

• By what means could a deceptive response produce a false negative result? That is, what is the potential for effective countermeasures? What do practitioners of the technique do to counteract or correct for such phenomena? Is this response to the possibility of false negatives and effective countermeasures reasonable considering the mechanisms involved?

• Are the mechanisms purported to link deception to behavioral, physiological, or brain states and those states to the test results universal for all people who might be examined, or do they operate differently in different kinds of people or in different situations? Is it possible that measured responses do not always have the same meaning or that a test that works for some kinds of examinees or situations will fail with others?

• How do the social context and the social interactions that constitute the examination procedure affect the reliability and validity of the recordings that are obtained?

• Are there plausible alternative theoretical rationales regarding the underlying mechanisms that make competing empirical predictions about how the technique performs? What is the weight of evidence for competing theoretical rationales?

*Research Methods for Demonstrating Accuracy*

• *Randomized experimentation.* In analog studies, this means that examinees are randomly assigned to be truthful or deceptive. It is also useful to have studies in which examinees are allowed to decide whether to engage in the target behavior. Such studies gain a degree of realism for what they lose in experimental control.

• *Manipulation checks.* If a technique is claimed to measure arousal, for example, there should be independent evidence that experimental manipulations actually create different levels of arousal in the different groups.

- *Blind administration and blind evaluation of the technique,* Whoever administers and scores tests based on the technique must do so in the absence of any information on whether the examinee is truthful or deceptive.
- *Adequate sample sizes.* Most of the studies examined [in the National Research Council 2003 polygraph report] were based on relatively small sample sizes that were sometimes adequate to allow for the detection of statistically significant differences but were insufficient for accurate assessment of accuracy. Changing the results of only a few cases might dramatically affect the implications of these studies.
- *Appropriate comparison conditions and experimental controls.* These conditions and controls will vary with the technique. A suggestion of what may be involved is the idea in polygraph research of comparing a polygraph examination with a bogus polygraph examination, with neither the examiner nor the examinee knowing that the test output might be bogus.
- *Cross-validation of any exploratory data analytic solution on independent data.* Any standardized or computerized scoring system for measurements from a technique cannot be seriously considered as providing accurate detection unless it has been shown to perform well on samples of examinees different from those on whom it was developed.
- *Examinees masked to experimental hypotheses if not to experimental condition.* It is important to sort out precisely what effect is being measured. For example, the results of a countermeasures study would be more convincing if examinees were instructed to expect that the examiner is looking for the use of countermeasures, among other things, rather than being instructed explicitly that this is a study of whether countermeasures work and can be detected.
- *Standardization.* An experiment should have sufficient standardization to allow reliable replication by others and should analyze the results from all examinees. It is important to use a technique in the same way on all the examinees, which means: clear reporting of how the technique was administered; sharply limiting the examiner's discretion in administering the technique and interpreting its results; and using the technique on all examinees, not only the ones whose responses are easy to classify. If some examinees are dropped from the analysis, the reasons should be stated explicitly. This is a difficult test for a procedure to pass, but it is appropriate for policy purposes.
- *Analysis of sensitivity and specificity or their equivalents.* Data should be reported in a way that makes it possible to calculate both the sensitivity and specificity of the technique, preferably at multiple thresholds for diagnostic decision making or in a way that allows comparisons of the test results with the criterion on other than binary scales.

---

SOURCE: National Research Council, *The Polygraph and Lie Detection*, The National Academies Press, Washington, D.C., 2003, pp. 222-224.

the collection of physiological and biological data, whereas data-mining is a technique for analyzing already-collected data.

- *Degree of intrusiveness.* Traditional jurisprudence and ethics generally regard a person's body as worthy of a higher degree of protection than his or her information, residences, or possessions. Thus, techniques that require the collection of physiological and biological data (especially data relevant to one's *thoughts*) are arguably more intrusive than collection schemes directed at different kinds of personal data.

# L

# The Science and Technology
# of Privacy Protection

To the extent that there is a tension between counterterrorism efforts and protection of citizens' privacy, it is useful to understand how it may be possible to design counterterrorism information systems to minimize their impact on privacy. This appendix considers privacy protection from two complementary perspectives—privacy protection that is built into the analytical techniques themselves and privacy protection that can be engineered into an operational system. The appendix concludes with a brief illustration of how government statistical agencies have approached confidential data collection and analysis over the years. A number of techniques described here have been proposed for use in protecting privacy; none would be a panacea, and several have important weaknesses that are not well understood and that are discussed and illustrated.

## L.1 THE CYBERSECURITY DIMENSION OF PRIVACY

Respecting privacy interests necessarily means that parties that *should* not have access to personal information *do* not have such access. Security breaches are incompatible with protecting the privacy of personal information, and good cybersecurity for electronically stored personal information is a necessary (but not sufficient) condition for protecting privacy.

From a privacy standpoint, the most relevant cybersecurity technologies are encryption and access controls. Encryption obscures digitally stored information so that it cannot be read without having the key necessary to decrypt it. Access controls provide privileges of different sorts to specified users (for example, the system may grant John Doe the right to know that a file exists but not the right to view its contents, and it may give Jane Doe both rights). Access controls may also be associated with audit logs that record what files were accessed by a given user.

Because of the convergence of and similarities between communication and information technologies, the technologies face increasingly similar threats and vulnerabilities. Furthermore, addressing these threats and vulnerabilities entails similar countermeasures or protection solutions. A fundamental principle of security is that no digital resource that is in use can be absolutely secure; as long as information is accessible, it is vulnerable. Security can be increased, but the value of increased security must be weighed against the increase in cost and the decrease in accessibility.

Human error, accident, and acts of God are the dominant sources of loss and damage in information and communication systems, but the actions of hackers and criminals are also of substantial concern. Terrorists account for a small percentage of losses, financial and otherwise, but could easily exploit vulnerabilities in government and business to cause much more serious damage to the nation. Security analysts and specialists report a large growth in the number and diversity of cyberthreats[1] and vulnerabilities.[2] Despite a concurrent growth in countermeasures (that is, security technologies[3]) penetrations and losses are increasing. A data-breach chronology reports losses of 104 million records (for example, in lost laptop computers) containing personally identifiable information from January 2005 to February 2007.[4] The Department of Homeland Security National Cyber Security Division reports that over 25 new vulnerabilities were discovered each day in 2006.[5]

The state of government information security is unnecessarily weak.

---

[1]A.T. Williams, A. Hallawell, R. Mogull, J. Pescatore, N. MacDonald, J. Girard, A. Litan, L. Orans, V. Wheatman, A. Allan, P. Firstbrook, G. Young, J. Heiser, and J. Feiman, *Hype Cycle for Cyberthreats*, Gartner, Inc., Stamford, Conn., September 13, 2006.

[2]National Vulnerability Database, National Institute of Standards and Technology Computer Security Division, sponsored by the U.S. Department of Homeland Security National Cyber Security Division/U.S. Computer Emergency Readiness Team (US-CERT), available at http://nvd.nist.gov/.

[3]A.T. Williams, A. Hallawell, R. Mogull, J. Pescatore, N. MacDonald, J. Girard, A. Litan, L. Orans, V. Wheatman, A. Allan, P. Firstbrook, G. Young, J. Heiser, and J. Feiman, *Hype Cycle for Cyberthreats*, Gartner, Inc., Stamford, Conn., September 13, 2006.

[4]A Chronology of Data Breaches, Privacy Rights Clearing House.

[5]National Vulnerability Database, National Institute of Standards and Technology Computer Security Division, sponsored by the U.S. Department of Homeland Security National Cyber Security Division/U.S. Computer Emergency Readiness Team (US-CERT), available at http://nvd.nist.gov/.

For example, the U.S. Government Accountability Office (GAO) noted in March 2008 that

> [m]ajor federal agencies continue to experience significant information security control deficiencies that limit the effectiveness of their efforts to protect the confidentiality, integrity, and availability of their information and information systems. Most agencies did not implement controls to sufficiently prevent, limit, or detect access to computer networks, systems, or information. In addition, agencies did not always effectively manage the configuration of network devices to prevent unauthorized access and ensure system integrity, patch key servers and workstations in a timely manner, assign duties to different individuals or groups so that one individual did not control all aspects of a process or transaction, and maintain complete continuity of operations plans for key information systems. An underlying cause for these weaknesses is that agencies have not fully or effectively implemented agencywide information security programs. As a result, federal systems and information are at increased risk of unauthorized access to and disclosure, modification, or destruction of sensitive information, as well as inadvertent or deliberate disruption of system operations and services. Such risks are illustrated, in part, by an increasing number of security incidents experienced by federal agencies.[6]

Such performance is reflected in the public's lack of trust in government agencies' ability to protect personal information.[7] Security of government information systems is poor despite many relevant regulations and guidelines.[8] Most communication and information systems are unnecessarily vulnerable to attack because of poor security practices, and

---

[6]Statement of Gregory C. Wilshusen, GAO Director for Information Security Issues, "Information Security: Progress Reported, but Weaknesses at Federal Agencies Persist," Testimony Before the Subcommittee on Federal Financial Management, Government Information, Federal Services, and International Security, Committee on Homeland Security and Governmental Affairs, U.S. Senate, GAO-08-571T, March 12, 2008. Available at http://www.gao.gov/new.items/d08571t.pdf.

[7]L. Ponemon, *Privacy Trust Study of United States Government*, The Ponemon Institute, Traverse City, Mich., February 15, 2007.

[8]Appendix III, OMB Circular A-130, "Security of Federal Automated Information Resources," (Office of Management and Budget, Washington, D.C.) revises procedures formerly contained in Appendix III, OMB Circular No. A-130 (50 FR 52730; December 24, 1985), and incorporates requirements of the Computer Security Act of 1987 (P.L. 100-235) and responsibilities assigned in applicable national security directives. See also Federal Information Security Management Act of 2002 (FISMA), 44 U.S.C. § 3541, et seq., Title III of the E-Government Act of 2002, Public Law 107-347, 116 Stat. 2899, available at http://csrc.nist.gov/drivers/documents/FISMA-final.pdf.

the framework outlined in Chapter 2 identifies data stewardship as a critical evaluation criterion.[9]

Although cybersecurity and privacy are conceptually different, they are often conflated—with good reason—in the public's mind. Cybersecurity breaches—which occur, for example, when a hacker breaks into a government information system that contains personally identifiable information (addresses, Social Security numbers, and so on)—are naturally worrisome to the citizens who may be affected. They do not particularly care about the subtle differences between a cybersecurity breach and a loss of privacy through other means; they know only that their privacy has been (potentially) invaded and that their loss of privacy may have deleterious consequences for them. That reaction has policy significance: the government agency responsible (perhaps even the entire government) is viewed as being incapable of protecting privacy, and public confidence is undermined when it asserts that it will be a responsible steward of the personal information it collects in its counterterrorism mission.

## L.2 PRIVACY-PRESERVING DATA ANALYSIS

### L.2.1 Basic Concepts

It is intuitive that the goal of privacy-preserving data analysis is to allow the learning of particular facts or kinds of facts about individuals (units) in a data set while keeping other facts secret. The term *data set* is used loosely; it may refer to a single database or to a collection of data sources. Under various names, privacy-preserving data analysis has been addressed in various disciplines.

A statistic is a quantity computed on the basis of a sample. A major goal of official statistics is to learn broad trends about a population by studying a relatively small sample of members of the population. In many cases, such as in the case of U.S. census data and data collected by the Internal Revenue Service (IRS), privacy is legally mandated. Thus, the goal is to identify and report trends while protecting the privacy of individuals. That sort of challenge is central to medical studies: the analyst wishes to learn and report facts of life, such as "smoking causes cancer," while preserving the privacy of individual cancer patients. The analyst must be certain that the privacy of individuals is not even inadvertently compromised.

---

[9]Data stewardship is accountability for program resources being used and protected appropriately according to the defined and authorized purpose.

Providing such protection is a difficult task, and a number of seemingly obvious approaches do not work even in the best of circumstances, for example, when a trusted party holds all the confidential data in one place and can prepare a "sanitized" version of the data for release to the analyst or can monitor questions and refuse to answer when privacy might be at risk. (This point is discussed further in Section L.2.2 below.)

In the context of counterterrorism, privacy-preserving data analysis is excellent for teaching the data analyst about "normal" behavior while preserving the privacy of individuals. The task of the counterterrorism analyst is to identify "atypical" behavior, which can be defined only in contrast with what is typical. It is immediately obvious that the data on any single specific individual should have little effect on the determination of what is normal, and in fact this point precisely captures the source of the intuition that broad statistical trends do not violate individual privacy. Assuming a good knowledge of what is "normal," technology is necessary for counterterrorism that will scrutinize data in an automated or semiautomated fashion and flag any person whose data are abnormal, i.e., that satisfy a putatively "problematic" profile. In other words, the outcome of data analysis in this context must necessarily vary widely ("yes, it satisfies the profile" or "no, it does not satisfy the profile"), depending on the specific person whose data is being scrutinized. Whether the profile is genuinely "problematic" is a separate matter.

In summary, privacy-preserving data analysis may permit the analyst to learn the definition of *normal* in a privacy-preserving way, but it does not directly address the counterterrorism goal: privacy-preserving data analysis "masks" *all* individuals, whereas counterterrorism requires the exposure of selected individuals. There is no such thing as privacy-preserving examination of an individual's records or privacy-preserving examination of a database to pinpoint problematic individuals.

The question, therefore, is whether the counterterrorism goal can be satisfied while protecting the privacy of "typical" people. More precisely, suppose the existence of a perfect profile of a terrorist: the false-positive and false-negative rates are very low. (The existence of such a perfect profile is magical thinking and contrary to fact, but suppose it anyway.) Would it be possible to analyze data, probably from diverse sources and in diverse formats, in such a way that the analyst learns only information about people who satisfy the profile? As far as we know, the answer to that question is no. However, it might be possible to limit the amount of information revealed about those who do not satisfy the profile, perhaps by controlling the information and sources used or by editing them after they are acquired. That would require major efforts and attention to the quality and utility of information in integrated databases.

## L.2.2 Some Simple Ideas That Do Not Work in Practice

There are many ideas for protecting privacy, and what may seem like sensible ideas often fail. Understanding how to approach privacy protection requires rigor in two senses: spelling out what "privacy protection" means and explaining the extent to which a particular technique succeeds in providing protection.

For example, assume that all the data are held by a trustworthy curator, who answers queries about them while attempting to ensure privacy. Clearly, queries about the data on any specific person cannot be answered, for example, What is the sickle-cell status of Averill Harriman? It is therefore instructive to consider the common suggestion of insisting that queries be made only on large subsets of the complete database. A well-known differencing argument (the "set differencing" attack) demonstrates the inadequacy of the suggestion: If the database permits the user to learn exact answers, say, to the two questions, How many people in the database have the sickle-cell trait? and, How many people—not named X—in the database have the sickle-cell trait? then the user learns X's sickle-cell trait status. The example also shows that encrypting the data (another frequent suggestion) would be of no help. Encryption protects against an intruder, but in this instance the privacy compromise emerges even when the database is operated correctly, that is, in conformance with all stated security policies.

Another suggestion is to monitor query sequences to rule out attacks of the nature just described. Such a suggestion is problematic for two reasons: it may be computationally infeasible to determine whether a query sequence compromises privacy,[10] and, more surprising, the *refusal* to answer a query may itself reveal information.[11]

A different approach to preventing the set differencing attack is to add random noise to the true answer to a query; for example, the response to a query about the average income of a set of individuals is the sum of the true answer and some random noise. That approach has merit, but it must be used with care. Otherwise, the same query may be issued over and over and each time produce a different perturbation of the truth. With enough queries, the noise may cancel out and reveal the true answer. Insisting that a given query always results the same answer is problematic in that it may be impossible to decide whether two syntactically different queries

---

[10]J. Kleinberg, C. Papadimitriou, and P. Raghavan, "Auditing boolean attributes," pp. 86-91 in *Proceedings of 19th ACM Symposium on Principles of Database Systems*, Association for Computing Machinery, New York, N.Y., 2000.

[11]K. Kenthapadi, N. Mishra, and K. Nissim, "Simulatable auditing," pp. 118-127 in *Proceedings of the 24th ACM Symposium on Principles of Database Systems*, Association for Computing Machinery, New York, N.Y., 2005.

are semantically equivalent. Related lower bounds on noise (the degree of distortion) can be given as a function of the number of queries.[12]

### L.2.3 Private Computation

The cryptographic literature on private computation addresses a distinctly different goal known as secure function evaluation.[13] In this work, the term *private* has a specific technical meaning that is not intuitive and is described below. To motivate the description, recall the original description of privacy-preserving data analysis as permitting the learning of some facts in a data set while keeping other facts secret. If privacy is to be completely protected, some things simply cannot be learned. For example, suppose that the database has scholastic records of students in Middletown High School and that the Middletown school district releases the fact that no student at the school has a perfect 5.0 average. That statement compromises the privacy of every student known to be enrolled at the school—it is now known, for example, that neither Sergey nor Laticia has a 5.0 average. Arguably, that is no one else's business. (Some might try to argue that no harm comes from the release of such information, but this is defeating the example without refuting the principle that it illustrates.) Similarly, publishing the average net worth of a small set of people may reveal that at least one person has a very high net worth; a little extra information may allow that person's identity to be disclosed despite her modest lifestyle.

Private computation does not address those difficulties, and the question of which information is safe to release is not the subject of study at all.[14] Rather, it is assumed that some facts are, by fiat, going to be released, for example, a histogram of students' grade point averages or average income by block. The "privacy" requirement is that no information *that cannot be inferred from those quantities* will be leaked. The typical setting is that each person (say, each student in Middletown High School) partici-

---

[12]I. Dinur and K. Nissim, "Revealing information while preserving privacy," pp. 202-210 in *Proceedings of the 22nd ACM SIGMOD-SIGACT-SIGART Symposium on Principles of Database Systems*, Association for Computing Machinery, New York, N.Y., 2003; C. Dwork, F. McSherry, and K. Talwar, "The price of privacy and the limits of LP decoding," pp. 85-94 in *Proceedings of the 39th Annual ACM SIGACT Symposium on Theory of Computing*, Association for Computing Machinery, New York, N.Y., 2007. See also the related work on compressed sensing cited in the latter.

[13]O. Goldreich, S. Micali, and A. Wigderson, "How to solve any protocol problem," pp. 218-229 in *Proceedings of the 19th ACM SIGACT Symposium on Computing*, Association for Computing Machinery, New York, N.Y., 1987.

[14]O. Goldreich, S. Micali, and A. Wigderson, "How to solve any protocol problem," pp. 218-229 in *Proceedings of the 19th ACM SIGACT Symposium on Computing*, Association for Computing Machinery, New York, N.Y., 1987.

pates in a cryptographic protocol whose goal is the cooperative computing of the quantity of interest (the histogram of grade point averages) and that the cryptographic protocol will not cause any information to be leaked that a student cannot infer from the histogram and his or her own data (that is, from the grade point histogram and his or her own grade point average).

### L.2.4 The Need for Rigor

Privacy-preservation techniques typically involve altering raw data or the answers to queries. Those general actions are referred to as input perturbation and output perturbation,[15] depending on whether the alterations are made before the queries or in response to them.

Various methods are used for input and output perturbation. Some involve redaction of information (for example, removing "real" identifiers, the use of indirect identifiers, selective reporting, or forms of aggregation) or alteration of data elements by adding noise, swapping, recoding (for example, collapsing categories), and data simulation.[16] But no matter

---

[15]A relevant survey article is N. Adam and J. Wortmann, "Security-control methods for statistical databases: A comparative study," *ACM Computing Surveys* 21(4):515-556, 1989. Some approaches post-dating the survey are given in L. Sweeney, "Achieving k-anonymity privacy protection using generalization and suppression," *International Journal on Uncertainty, Fuzziness and Knowledge-based Systems* 10(5):557-570, 2002; A. Evfimievski, J. Gehrke, and R. Srikant, "Limiting privacy breaches in privacy preserving data mining," pp. 211-222 in *Proceedings of the Twenty-Second ACM SIGACT-SIGMOD-SIGART Symposium on Principles of Database Systems*, Association for Computing Machinery, New York, N.Y., 2003; and C. Dwork, F. McSherry, K. Nissim, and A. Smith, "Calibrating noise to sensitivity of functions in private data analysis," pp. 265-284 in *Proceedings of the Thirty-Ninth Annual ACM Symposium on Theory of Computing*, Association for Computing Machinery, New York, N.Y., 2006, and references therein.

[16]Many of these methods are described in the following papers: S.E. Fienberg, "Conflicts between the needs for access to statistical information and demands for confidentiality," *Journal of Official Statistics* 10(2):115-132, 1994; Federal Committee on Statistical Methodology, Office of Management and Budget (OMB), "Statistical Policy Working Paper 2. Report on Statistical Disclosure and Disclosure-Avoidance Techniques," OMB, Washington, D.C., 1978, available at http://www.fcsm.gov/working-papers/sw2.html; Federal Committee on Statistical Methodology, OMB, "Statistical Policy Working Paper 22 (Second version, 2005), Report on Statistical Disclosure Limitation Methodology," originally prepared by Subcommittee on Disclosure Limitation Methodology, OMB, Washington, D.C., 1994, and revised by the Confidentiality and Data Access Committee, 2005, available at http://www.fcsm.gov/working-papers/spwp22.html. Many of these techniques are characterized as belonging to the family of matrix masking methods in G.T. Duncan and R.W. Pearson, "Enhancing access to microdata while protecting confidentiality: prospects for the future (with discussion)," *Statistical Science* 6:219-239, 1991. The use of these techniques in a public-policy context is set by the following publications: National Research Council (NRC), *Private Lives and Public Policies: Confidentiality and Accessibility of Government Statistics*, G.T. Duncan, T.B. Jabine,

what the technique or approach, there are two basic questions: What does it mean to protect the data? How much alteration is required to achieve that goal?

The need for a rigorous treatment of both questions cannot be over-stated, inasmuch as "partially protecting privacy" is an oxymoron. An extremely important and often overlooked factor in ensuring privacy is the need to protect against the availability of arbitrary context information, including other databases, books, newspapers, blogs, and so on.

Consider the anonymization of a social-network graph. In a social network, nodes correspond to people or other social entities, such as organizations or Web sites, and edges correspond to social links between them, such as e-mail contact or instant-messaging. In an effort to preserve privacy, the practice of anonymization replaces names with meaningless unique identifiers. The motivation is roughly as follows: the social network labeled with actual names is sensitive and cannot be released, but there may be considerable value in enabling the study of its structure. Anonymization is intended to preserve the pure unannotated structure of the graph while suppressing the information about precisely who has contact with whom. The difficulty is that anonymous social-network data almost never exist in the absence of outside context, and an adversary can potentially combine this knowledge with the observed structure to begin compromising privacy, deanonymizing nodes and even learning the edge relations between explicitly named (deanonymized) individuals in the system.[17]

A more traditional example of the difficulties posed by context begins with the publication of redacted confidential data. The Census Bureau receives confidential information from enterprises as part of the economic census and publishes a redacted version in which identifying information on companies is suppressed. At the same time, a company may release information in its annual reports about the number of shares held by particular holders of very large numbers of shares. Although the redaction may be privacy-protective, by using very simple *linkage tools* on the redacted data and the public information, an adversary will be able to add back some of the identifying tags to the redacted confidential data. Roughly speaking, those tools allow the merging of data sets that contain, for example, different types of information about the same set of entities.

---

and V.A. de Wolf, eds., National Academy Press, Washington, D.C., 1993; NRC, *Expanding Access to Research Data: Reconciling Risks and Opportunities*, The National Academies Press, Washington, D.C., 2005.

[17]L. Backstrom, C. Dwork, and J. Kleinberg, "Wherefore art thou R3579X? Anonymized social networks, hidden patterns, and structural steganography," pp. 181-190 in *Proceedings of the 16th International Conference on World Wide Web*, 2007, available at http://www2007.org/proceedings.html.

The key point is that entities need not be directly identifiable by name to be identified. Companies can be identified by industrial code, size, region of the country, and so on. Any public company can be identified by using a small number of such variables, which may well be deduced from the company's public information and thus provide a means of matching against the confidential data.

Similarly, individuals need not be identified only by their names, addresses, or Social Security numbers. The linkage software may use any collection of data fields, or variables, to determine that records in two distinct data sets correspond to the same person. And if the "privacy-protected" or deidentified records include values for additional variables that are not yet public, simple record-linkage tools might let an intruder identify a person (that is, match files) with high probability and thus leak this additional information in the deidentified files. For example, an adversary may use publicly available data, including newspaper accounts from New Orleans on the effects of hurricane Katrina and who was rescued in what efforts, to identify people with unusual names in a confidential epidemiologic data set on rare genetic diseases gathered by the Centers for Disease Control and Prevention and thus learn all the medical and genetic information about the individuals that redaction was supposed to protect.

For a final, small-scale, example, consider records of hospital emergency-room admissions, which contain such fields as name, year of birth, ZIP code, ethinicity, and medical complaint. The combinations of fields are known to identify many people uniquely. Such a collection of attributes is called a quasi-identifier. In microaggregation, or what is known as $k$-anonymization, released data are "coarsened"; for example, ZIP codes with the same first four digits are lumped together, so for every possible value of quasi-identifier, the data set contains at least $k$ records. However, if someone sees an ambulance at his or her neighbor's house during the night and consults the published hospital emergency-room records the following day, he or she can learn a small set of complaints that contains the medical complaint of the neighbor. Additional information known to that person may allow the neighbor's precise complaint to be pinpointed.

Context also comes into play in how different privacy-preserving techniques interact when they are applied to different databases. For example, the work of Dwork et al. rigorously controlled the amount of information leaked about a single record.[18] If several databases, all containing the same record, use the same technique, and if the analyst has

---

[18]C. Dwork, F. McSherry, K. Nissim, and A. Smith, "Calibrating noise to sensitivity of functions in private data analysis," pp. 265-284 in *Proceedings of the 3rd Theory of Cryptography Conference*, Association for Computing Machinery, New York, N.Y., 2006.

access to all these databases, the cumulative erosion of privacy of the given record may be as great as the sum of the leakages suffered in the separate databases that contain it.

And that is a good case! The many methods in fields spanning computer science, operations research, economics, and statistics deal with data of different types recorded in many forms. For a targeted set of methods and specific kinds of data, although there may be results that can "guarantee" privacy in a released data file or a system responding to a series of queries, many well-known approaches fail to offer such guarantees or even weaker assurances. For example, some literature on data imputation for privacy protection never defines *privacy* at all;[19] thus, it is difficult to assess the extent to which the methods, although heuristically reasonable, actually guarantee privacy.

### L.2.5 The Effect of Data Errors on Privacy

In the real world, data records are imperfect. For example,

- Honest people make errors when providing information.
- Clerical errors yield flawed recording of correct data.
- Many data values may be measurements of quantities that regularly fluctuate or that for various other reasons are subject to measurement error.

Because of imperfections in the data, a person may be mischaracterized as problematic. That is, the profile may be perfect, but the system may be operating with bad data. That appears to be an accuracy problem, but for several reasons it also constitutes a privacy problem.

Although we have not discussed a definition of *privacy*, the recent literature studies the appropriate technical definition at length. The approach favored in the cryptography community, modified for the present context, says that for anyone whose true data do not fit the profile, there is (in a quantifiable sense) almost no difference between the behavior of a sys-

---

[19]D.B. Rubin, "Discussion: Statistical disclosure limitation," *Journal of Official Statistics* 9(2):461-468, 1993; T.E. Raghunathan, J.P. Reiter, and D.B. Rubin, "Multiple imputation for statistical disclosure limitation," Journal of Official Statistics 19(2003):1-19, 2003. However, there is also a substantial literature that does provide an operational assessment of privacy and privacy protection. For example, see G.T. Duncan and D. Lambert, "The risk of disclosure for microdata," *Journal of Business and Economic Statistics* 7:207-217, 1989; E. Fienberg, U.E. Makov, and A.P. Sanil, "A Bayesian approach to data disclosure: Optimal intruder behavior for continuous data," *Journal of Official Statistics* 13:75-89, 1997; and J.P. Reiter, "Estimating risks of identification disclosure for microdata," *Journal of the American Statistical Association* 100(2005):1103-1113, 2005.

tem that contains the person's data and the behavior of a system that does not. That is, the behavior of the system in the two cases should be *indistinguishable;* it follows that the *increase* in the risk of adverse effects of participating in a data set is small. That approach allows us to avoid subjective decisions about which type of information leakage constitutes a privacy violation. Clearly, indistinguishability can fail to hold in the case of a nonterrorist whose data are incorrectly recorded. The harm to a person of *appearing* to satisfy the perfect profile may be severe: the person may be denied credit and the freedom to travel, be prevented from being hired for some jobs, or even be prosecuted. Finally, at the very least, such a misidentification will result in further scrutiny and consequent loss of privacy. (See Gavison on protection from being brought to the attention of others.[20])

The problem of errors is magnified by linkage practices because errors tend to propagate. Consider a database, such as the one assembled by ChoicePoint by linking multiple databases. Consider, say, three separate databases created by organizations *A*, *B*, and *C*. If *A* and *B* are extremely scrupulous about preventing data errors but *C* is not, the integrated database will contain inaccuracies. The accuracy of the integrated database is only as good as the accuracy of the worst input database. Furthermore, if each database contains errors, they may well compound to create a far greater percentage of files with errors in the integrated database. Finally, there are the errors of matching themselves, which are inherent in record linkage; if these are as substantial as the literature on record linkage suggests,[21] the level of error in the merged database is magnified, and this poses greater risks of misidentification.

All the above difficulties are manifested even when a perfect profile is developed for problematic people. But imperfect profiles combined with erroneous data will lead to higher levels of false positives than either alone. Moreover, if we believe that data are of higher quality and that profiles are more accurate than they actually are, the rate of false negatives—people who are potential terrorists but go undetected—will also grow, and this endangers all of us.

Record linkage also lies at the heart of data-fusion methods and has major implications for privacy protection and harm to people. The

---

[20]R. Gavison, "Privacy and the limits of the law," pp. 332-351 in *Computers, Ethics, and Social Values*, D.G. Johnson and H. Nissenbaum, eds., Prentice Hall, Upper Saddle River, N.J., 1995.

[21]W.E. Winkler, *Overview of Record Linkage and Current Research Directions*, Statistical Research Report Series, No. RRS2006/02, U.S. Bureau of the Census, Statistical Research Division, Washington, D.C., 2006, and W.E. Winkler, "The quality of very large databases," *Proceedings of Quality in Official Statistics*, 2001, CD-ROM (also available at http://www.census.gov/srd/www/byyear.html as RR01/04).

literature on record linkage[22] makes it clear that to achieve low rates of error (high accuracy) one needs both "good" variables for linkage (such as names) and ways to organize the data by "blocks," such as city blocks in a census context or well-defined subsets of individuals characterized by variables that contain little or no measurement error. As measurement error grows, the quality of matches deteriorates rapidly in techniques based on the Fellegi-Sunter method. Similarly, as the size of blocks used for sorting data for matching purposes grows, so too do both the computational demands for comparing records in pairs and the probabilities of correct matches.

Low-quality record-linkage results will almost certainly increase the rates of both false positives and false negatives when merged databases are used to attempt to identify terrorists or potential terrorists. False negatives correspond to the failure of systems to detect terrorists when they are present and represent a systemic failure. False positives impinge on individual privacy. Government uses of such methods, either directly or indirectly, through the acquisition of commercial databases constructed with fusion technologies need to be based on adequate information on data quality especially as related to record-linkage methods.

## L.3 ENHANCING PRIVACY THROUGH INFORMATION-SYSTEM DESIGN

Some aspects of information-system design are related to the ability to protect privacy while maintaining effectiveness, and there are many designs (and tradeoffs among those designs) for potential public policies regarding data privacy for information systems. Moreover, times and technology have changed, and a new set of policies regarding privacy and information use may be needed. To be rational in debating and choosing the policies and regulations that will provide the most appropriate combination of utility (such as security) and privacy, it is helpful to consider the generic factors that influence both. This section lists the primary components of information-system design that are related to privacy and indicates the issues that are raised in considering various options.

### L.3.1 Data and Privacy

A number of factors substantially influence the effects of a deployed information system on privacy. Debates and regulations can benefit from differentiating systems and applications on the basis of the following:

---

[22]See, for example, T.N. Herzog, F.J. Scheuren, and W.E. Winkler, *Data Quality and Record Linkage Techniques*, Springer Science and Business Media, New York, N.Y., 2007.

- *Which data features are collected.* In wiretapping, recording the fact that person A telephoned person B might be less invasive than recording the conversation itself.
- *Covertness of collection.* Data may be collected *covertly* or *with the awareness* of those being monitored. For example, images of airport passengers might be collected covertly throughout the airport or with passenger awareness at the security check-in.
- *Dissemination.* Data might be collected and used only for a local application (for example, at a security checkpoint) or might be disseminated widely in a nationwide data storage facility accessible to many agencies.
- *Retention periods.* Data might be destroyed within a specified period or kept forever.
- *Use.* Data might be restricted to a particular use by policy (for example, anatomically revealing images of airport passengers might be available for the sole purpose of checking for hidden objects) or unrestricted for arbitrary future use. One policy choice of particular importance is whether the data are subject to court subpoena for arbitrary purposes or the ability to subpoena is restricted to specified purposes.
- *Audit trail.* An audit trail (showing who accessed the data and when) should be kept.
- *Control of permissions.* If data are retained, policy might specify who can grant permission for dissemination and use (for example, the collector of the data, a court, or the subject of the data).
- *Trust.* The perception of privacy violations depends heavily on the trust of the subject that the government and everyone who has access to the data will abide by the stated policy on data collection and use.
- *Analytical methods involved.* Analysis of data collected or the presentation of analytical results might be restricted by policy. For example, in searching for a weapon at a checkpoint, a scanner might generate anatomically correct images of a person's body in graphic detail. What is of interest is not those images but rather the image of a weapon, so analytical techniques that detected the presence or absence of a weapon in a particular scan could be used, and that fact (presence or absence) could be reported rather than the image itself.

### L.3.2 Information Systems and Privacy

Chapter 2 describes a framework for assessing information-based programs. But the specifics of program's implementation make a huge difference in the extent to which it protects (or can protect) privacy. The following are some of the implementation issues that arise.

- *Does the application require access to data that explicitly identify individuals?* Applications such as searching a database for all information about a particular person clearly require access to data that are associated with individual names. Other applications, such as discovering the pattern of patient symptoms that are predictive of a particular disease, need not necessarily require that individual names be present.

- *Does the application require that individually identified data be reported to its human user, and, if so, under what conditions?* Some computer applications may require personally identified data but may not need to report personal identifications to their users. For example, a program to learn which over-the-counter drug purchases predict emergency-room visits for influenza might need personally identified data of drug purchases so that it can merge them with personally identified emergency-room records, but the patterns that it learns and reports to the user need not necessarily identify individuals or associate specific data with identifiable individuals. Other systems might examine many individually identified data records but report only records that match a criterion specified by a search warrant.

- *Is the search of the data driven by a particular starting point or person, or is it an indiscriminate search of the entire data set for a more general pattern?* Searches starting with a particular lead (for example, Find all people who have communicated with person A in the preceding week) differ from searches that consider all data equally (for example, Find all groups of people who have had e-mail exchanges regarding bombs). The justification for the former hinges on the justification for suspecting person A; the latter involves a different type of justification.

- *Can the data be analyzed with privacy-enhancing methods?* Technologies in existence and under development may in some cases enable discovery of general patterns and statistics from data while providing assurances that features of individual records are not divulged.

- *Does the data analysis involve integrating multiple data sources from which additional features can be inferred, and, if so, are these features inferred and reported to the user?* In some cases, it is possible to infer data features that are not explicit in the data set, especially when multiple data sets are merged. For example, it is possible in most cases to infer the names of people associated with individual medical records that contain only birthdates and ZIP codes if that data set is merged with a census database that contains names, ZIP codes, and birthdates.

## L.4 STATISTICAL AGENCY DATA AND APPROACHES

Government statistical agencies have been concerned with confidentiality protection since early in the 20th century and work very hard to "deidentify" information gathered from establishments and individuals. They have developed methods for protecting privacy. Their goals are to remove information that could be harmful to a respondent from released data and to protect the respondents from identification. As a consequence, released statistical data, even if they may be related to individuals, are highly unlikely to be linkable with any reasonable degree of precision to other databases that are of use in prevention of terrorism. That is, the nature of redaction of individually identifiable information seems to yield redacted data that are of little value for this purpose.

### L.4.1 Confidentiality Protection and Public Data Release

Statistical agencies often promise confidentiality to their respondents regarding all data provided in connection with surveys and censuses, and, as noted above, these promises are often linked to legal statutes and provisions. But the same agencies have a mandate to report the results of their data-collection efforts to others either in summary form or in tables, reports, and public-use microdata sample (PUMS) files. PUMS files are computer-accessible files that contain records of a sample of housing units with information on the characteristics of each unit and the people in it. The data come in the form of a sample of a much larger population; as long as direct identifiers are removed and some subset of other variables "altered," there is broad agreement that sampling itself provides substantial protection. Roughly speaking, the probability of identifying an individual's record in the sample file is proportional to the probability of selection into the sample (given that it is not known whether a given individual is in the sample).[23] (In particular, if a person is not selected for the sample, the person's data are not collected and his or her privacy is protected.) It is also possible to provide privacy guarantees even in the worst case (that is, worst case over sampling).[24]

Nonetheless, many of the methods used by the agencies are ad hoc and may or may not "guarantee" privacy on their own, let alone when used with combining data from multiple databases. Nor would they satisfy the technical definitions of privacy described above. Rather, they represent an effort to balance data access with confidentiality protection—an

---

[23]See E.A.H. Elamir and C. Skinner, "Record level measures of disclosure risk for survey microdata," *Journal of Official Statistics* 22(3):525-539, 2006, and references therein.

[24]A. Evfimievski, J. Gehrke and R. Srikant, "Limiting Privacy Breaches in Privacy Preserving Data Mining," pp. 211-222 in *Proceedings of the Twenty-Second ACM SIGACT-SIGMOD-SIGART Symposium on Principles of Database Systems*, ACM, New York, N.Y., 2003; C. Dwork, F. McSherry, K. Nissim, and A. Smith, "Calibrating Noise to Sensitivity of Functions in Private Data Analysis," pp. 265-284 in *3rd Theory of Cryptography Conference*, ACM, New York, N.Y., 2006.

approach that fits with technical statistical frameworks.[25] Such trade-offs may be considered informally, but there are various formal sets of tools for their quantification.[26]

Duncan and Stokes apply such an approach to the choice of "topcoding" for income, that is, truncating the income scale at some maximum value.[27] They illustrate trade-off choices for different values of topcoding in terms of risk (of reidentification through a specific form of record linkage) and utility (in terms of the inverse mean square error of estimation for the mean or a regression coefficient).

For some other approaches to agency confidentiality and data release in the European context, see Willenborg and de Waal.[28]

### L.4.2 Record Linkage and Public Use Files

One activity that is highly developed in the context of statistical-agency data is record linkage. The original method that is still used in most approaches goes back to pioneering work by Fellegi and Sunter, who used formal probabilistic and statistical tools to decide on matches and nonmatches.[29] Inherent in the method is the need to assess accuracy of matching and error rates associated with decision rules.[30]

The same ideas are used, with refinements, by the Census Bureau

---

[25]For a discussion of the approaches to trade-offs, see the various chapters in *Confidentiality, Disclosure and Data Access: Theory and Practical Applications for Statistical Agencies*, P. Doyle, J. Lane, J. Theeuwes, and L. Zayatz, eds., North-Holland Publishing Company, Amsterdam, 2001.

[26]A framework is suggested in G.T. Duncan and D. Lambert, "Disclosure-limited data dissemination (with discussion)," *Journal of the American Statistical Association* 81:10-28, 1986. See additional discussion of the risk-utility trade-off by G.T. Duncan, S.E. Fienberg, R. Krishnan, R. Padman, and S.F. Roehrig, "Disclosure limitation methods and information loss for tabular data," pp. 135-166 in *Confidentiality, Disclosure and Data Access: Theory and Practical Applications for Statistical Agencies*, P. Doyle, J. Lane, J. Theeuwes, and L. Zayatz, eds., North-Holland Publishing Company, Amsterdam, 2001. A full decision-theoretic framework is developed in M. Trottini and S.E. Fienberg, "Modelling user uncertainty for disclosure risk and data utility," *International Journal of Uncertainty, Fuzziness, and Knowledge-Based Systems* 10(5):511-528, 2002; and M. Trottini, "A decision-theoretic approach to data disclosure problems," *Research in Official Statistics* 4(1):7-22, 2001.

[27]G.T. Duncan and S.L. Stokes, "Disclosure risk vs. data utility: The R-U confidentiality map as applied to topcoding," *Chance* 3(3):16-20, 2004.

[28]L. Willenborg and T. de Waal, *Elements of Statistical Disclosure Control*, Springer-Verlag Inc., New York, N.Y., 2001.

[29]I. Fellegi and A. Sunter, "A theory for record linkage," *Journal of the American Statistical Association* 64:1183-1210, 1969.

[30]See, for example, W. Winkler, *The State of Record Linkage and Current Research Problems*, Statistical Research Report Series, No. RR99/04, U.S. Census Bureau, Washington, D.C., 1999; W.E. Winkler, "Re-identification methods for masked microdata," pp. 216-230 in *Privacy in Statistical Databases*, J. Domingo-Ferrer, ed., Springer, New York, N.Y., 2004; M. Bilenko,

to match persons in the Current Population Survey (sample size, about 60,000 households) with IRS returns. The Census Bureau and the IRS provide the data to a group that links the records to produce a set of files that contain information from both sources. The merged files are redacted, and noise is added until neither the Census Bureau nor the IRS can rematch the linked files with their original files.[31] The data are released as a form of PUMS file. Those who prepared the PUMS file have done sufficient testing to offer specific guarantees regarding the protection of individuals whose data went into the preparation of the file. This example illustrates not only the complexity of data protection associated with record linkage but the likely lack of utility of statistical-agency data for terrorism prevention, because linked files cannot be matched to individuals.

---

R. Mooney, W.W. Cohen, P. Ravikumar, and S.E. Fienberg, "Adaptive name-matching in information integration," *IEEE Intelligent Systems* 18(5):16-23, 2003.

   [31]For more details, see J.J. Kim and W.E. Winkler, "Masking microdata files," pp. 114-119 in *Proceedings of the Survey Research Methods Section*, American Statistical Association, Alexandria, Va., 1995; J.J. Kim and W.E. Winkler, *Masking Microdata Files*, Statistical Research Report Series, No. RR97-3, U.S. Bureau of the Census, Washington, D.C., 1997.

# M

# Public Opinion Data on U.S. Attitudes Toward Government Counterterrorism Efforts

## M.1 INTRODUCTION

Since September 11, 2001 (9/11), Americans have been forced to confront conflict between the values of privacy and security more directly than at any other time in their history. On one hand, in view of the unprecedented threat of terrorism, citizens must depend on the government to provide for their own and the nation's security. On the other hand, technological advances mean that government surveillance in the interests of national security is potentially more sweeping in scope and more exhaustive in detail than at any time in the past, and thus it may represent a greater degree of intrusion on privacy and other civil liberties than the American public has ever experienced. In this appendix, we review the results of public opinion surveys that gauge the public's reaction to government surveillance measures and information-gathering activities designed to foster national security. We attempt to examine the public's view of the conflict between such surveillance measures and preservation of civil liberties.

Prior to 9/11, the American public's privacy attitudes were located in the broad context of a tradition of limited government and assertion of

*NOTE:* The material presented in this appendix was prepared by Amy Corning and Eleanor Singer of the Survey Research Center of the University of Michigan, under contract to the National Research Council, for the committee responsible for this report. Apart from some minor editorial corrections, this appendix consists entirely of the original paper provided by Corning and Singer.

the individual rights of citizens. In the past, expanded government powers have been instituted to promote security during national emergencies, but after the emergency receded, such powers have normally been rescinded.[1] Although this historical context is one crucial influence, attitudes have been further shaped by developments of the postwar period. The importance of civil rights was highlighted by the social revolutions of the 1960s and 1970s, a period also characterized by growing distrust of government; the latter decade also brought legislation designed to secure individuals' rights to privacy. During the 1980s, developments in computing and telecommunications laid the groundwork for new challenges to privacy rights. The public consistently opposed the consolidation of information on citizens in centralized files or databanks, and federal legislation attempted to preserve existing privacy protections in the context of new technological developments.[2] By the 1990s, however, technological advances—including the rise of the Internet, the widespread adoption of wireless communication, the decoding of human DNA, the development of data mining software, increasing automation of government records, the increasing speed and decreasing cost of computing and online storage power—occurred so quickly that they outpaced efforts to modify legislation to protect privacy, as well as the public's ability to fully comprehend their privacy implications, contributing to high salience of privacy considerations and concerns.[3]

The terrorist attacks of September 11, 2001, thus occurred in a charged environment, in which the public already regarded both business and government as potential threats to privacy. Almost immediately, the passage of the Patriot Act in 2001 raised questions about the appropriate nature and scope of the government's expanded powers and framed the public debate in terms of a sacrifice of civil liberties, including privacy, in the interests of national security. Citizens appeared willing to make such sacrifices at a time of national emergency, however, and in the months following 9/11, tolerance for government antiterrorism surveillance was extremely high. Nevertheless, the public did not uncritically accept government intrusions: to use Westin's term, they exhibited "rational ambivalence" by simultaneously expressing support for surveillance and

---

[1]A.F. Westin, "How the public sees the security-versus-liberty debate," pp. 19-36 in *Protecting What Matters: Technology, Security, and Liberty Since 9/11* (C. Northouse, ed.), Brookings Institution Press, Washington, D.C., 2005.

[2]A.F. Westin, "Social and political dimensions of privacy," *Journal of Social Issues* 59(2):411-429, 2003.

[3]A. Corning and E. Singer, *Survey of U.S. Privacy Attitudes*, report prepared for the Center for Democracy and Technology, Washington, D.C., 2003.

concern about protection of civil liberties as the government employed its expanded powers in investigating potential terrorist threats.[4]

Like other analysts,[5] we find that acceptance of government surveillance measures has diminished over the years since 9/11, and that people are now both less convinced of the need to cede privacy and other civil liberties in the course of terrorism investigation and personally less willing to give up their freedoms. We show that critical views are visible in the closely related domains of attitudes toward individual surveillance measures and toward recently revealed secret surveillance programs. More generally, public pessimism about protection of the right to privacy has increased.

Westin identified five influences on people's attitudes toward the balance between security and civil liberties: perceptions of terrorist threat; assessment of government effectiveness in dealing with terrorism; perceptions of how government terrorism prevention programs are affecting civil liberties; prior attitudes toward security and civil liberties; and broader political orientations, which may in turn be shaped by demographic and other social background factors.[6] This review confirms the role of these influences on public attitudes toward privacy and security in the post-9/11 era.

This examination of research on attitudes toward government surveillance since 9/11 leads us to draw the following general conclusions:

1. As time from a direct terrorist attack on U.S. soil increases, the public is growing less certain of the need to sacrifice civil liberties for terrorism prevention, less willing to make such sacrifices, and more concerned that government counterterrorism efforts will erode privacy.

2. Tolerance for most individual surveillance measures declined in the five years after 9/11. The public's attitudes toward recently revealed monitoring programs are mixed, with no clear consensus.

3. There is no strong support for health information databases that could be used to identify bioterrorist attacks or other threats to public health.

---

[4]The term is Westin's. See A.F. Westin, "How the public sees the security-versus-liberty debate," pp. 19-36 in *Protecting What Matters: Technology, Security, and Liberty Since 9/11* (C. Northouse, ed.), Brookings Institution Press, Washington, D.C., 2005.

[5]See, for example, A.F. Westin, "How the public sees the security-versus-liberty debate," pp. 19-36 in *Protecting What Matters: Technology, Security, and Liberty Since 9/11* (C. Northouse, ed.), Brookings Institution Press, Washington, D.C., 2005; S.J. Best, B.S. Krueger, and J. Ladewig, "Privacy in the Information Age," *Public Opinion Quarterly*, 70(3):375-401, 2006.

[6]A.F. Westin, "How the public sees the security-versus-liberty debate," pp. 19-36 in *Protecting What Matters: Technology, Security, and Liberty Since 9/11* (C. Northouse, ed.), Brookings Institution Press, Washington, D.C., 2005.

4. However, few citizens feel that their privacy has been affected by the government's antiterrorism efforts.

5. The public tends to defend civil liberties more vigorously in the abstract than in connection with threats for specific purposes. Despite increasingly critical attitudes toward surveillance, the public is quite willing to endorse specific measures, especially when the measures are justified as necessary to prevent terrorism.

6. However, most people are more tolerant of surveillance when it is aimed at specific racial or ethnic groups, when it concerns activities they do not engage in, or when they are not focusing on its potential personal impact. We note that people are not concerned about privacy in general, but rather with protecting the privacy of information about themselves.

7. People are concerned with control over decisions related to privacy.

8. Attitudes toward surveillance and the appropriate balance between rights and security are extremely sensitive to situational influences, particularly perceptions of threat.

9. The framing of survey questions, in terms of both wording and context, strongly influences the opinions elicited.

## M.2 DATA AND METHODOLOGY

In this appendix, we examine data from relevant questions asked by major research organizations in surveys since September 11, 2001, incorporating data from before that point when they are directly comparable to the later data or when they are pertinent. This review concentrates on trends, based on the same or closely similar questions that have been asked at multiple time points; we occasionally discuss the results from questions asked at only one point in time, when the information is illuminating or when trend data on a particular subject are not available. We restrict this review to surveys using adult national samples (or occasionally, national samples of registered voters); for the most part, these surveys are conducted by telephone using random-digit-dialed (RDD) samples,[7] although occasionally we report on surveys conducted by personal inter-

---

[7]These survey results may be biased by the fact that most or all of the surveys used did not attempt to reach cell-phone-only respondents; that is, the phone numbers called were land lines. In an era in which many individuals are using cell phones only, these surveys will not have reached many of such individuals. An article by the Pew Research Center for the People and the Press suggests that this problem is not currently biasing polls taken for the entire population, although it may very well be damaging estimates for certain subgroups (e.g., young adults) in which the use of a cell phone only is more common. (See S. Keeter, "How Serious Is Polling's Cell-Only Problem? The Landline-less Are Different and Their Numbers Are Growing Fast," Pew Research Center for the People and the Press, June 20, 2007, available at http://pewresearch.org/pubs/515/polling-cell-only-problem.)

view. We have not reviewed Web surveys. In the few instances in which samples represent groups other than the U.S. national adult population, we indicate that in the text or relevant charts or tables.

*Survey List and In-Text Citations.* The Annex at the end of this appendix lists the surveys to which we refer, identifying research organizations and sponsors as well as details on administration dates, mode, and sample design. (Response rate information is not available.) The abbreviations used in the text to identify the survey research organizations are also listed. The source citations in the text and in charts and tables are keyed to this list via the abbreviation identifying the research organization and survey date. Source citations appear as close as possible to the reported data; in other words, for data reported in figures or tables, the sources are generally indicated on the figures or tables.

*Response Rates.* We alert readers that response rates to national RDD sample surveys have declined. In a study reported in 2006, mean response rates for 20 national media surveys were estimated at 22 percent, using American Association for Public Opinion Research response rates RR3 or RR4, with a minimum of 5 percent and a maximum of 40 percent. Mean response rates for surveys done by government contractors (N = 7 for such surveys) during the same period were estimated at 46 percent, with minimums of 28 percent and maximums of 70 percent.[8]

We also note that we have no way of detecting or estimating nonresponse bias. Recent research on the relationship between nonresponse rates and nonresponse bias indicates that there is no necessary relationship between the two.[9] A 2003 Pew Research Center national study of nonresponse rates and nonresponse bias shows significant differences on only 7 of 84 items in a comparison of a survey achieving a 25 percent response rate and one achieving a 50 percent response rate through the use of more rigorous methods.[10] Two other studies also report evidence that, despite very low response rates, nonresponse bias in the surveys examined has

---

[8]A.L. Holbrook, J.A Kronsnick, and A. Pfent, "Response Rates in Surveys by the News Media and Government Survey Research Firms," paper presented at the Second Conference on Telephone Survey Methodology, Miami, Fla., January 14, 2006.

[9]R.M. Groves, "Nonresponse rates and nonresponse bias in household surveys," *Public Opinion Quarterly* 70(5):646-675, 2006.

[10]S. Keeter, C. Kennedy, M. Dimock, J. Best, and P. Craighill, "Gauging the impact of growing nonresponse on estimates from a national RDD telephone survey," *Public Opinion Quarterly* 70(5):759-779, 2006.

been negligible.[11] These findings cannot, however, be generalized to the surveys used for this current examination. Thus, the possibility of non-response bias in the findings reported cannot be ruled out, nor is there a way to estimate the direction of the bias, if it exists.

We can speculate that nonresponse bias in the surveys reviewed here might result, on one hand, in an overrepresentation of individuals especially concerned about privacy or civil liberties, if they are drawn to such survey topics; on the other hand, nonresponse might be greatest among those most worried about threats to privacy, if they refuse to participate in surveys. Of the over 100 surveys used in this review, however, most are general-purpose polls that include some questions about privacy or civil liberties among a larger number of questions on broad topics, such as current social and political affairs, health care attitudes or satisfaction with medical care, technology attitudes, terrorism, etc. Fewer than 1 in 10 of the surveys examined could be construed as focusing primarily or even substantially on privacy or civil liberties. Thus, it is unlikely that the survey topics would produce higher response among those concerned with privacy. We expect that whatever bias exists will be in the direction of excluding those most concerned about privacy and that the findings reported will tend to underestimate levels of privacy concern.

*Sources of Data and Search Strategies.* This examination draws on several different sources of survey data. First, we rely on univariate tabulations of opinion polling data that are in the public domain, available through the iPOLL Databank at the Roper Center for Public Opinion Research at the University of Connecticut (http://www.ropercenter.uconn.edu/data_access/ipoll/ipoll.html) and through the Institute for Resource and Security Studies (IRSS) repository at the University of North Carolina (http://www.irss.unc.edu/odum/jsp/content_node.jsp?nodeid=140).

We searched these repositories using combinations of the following keywords (or variants thereof): airport security, biometrics, bioterrorism, civil liberties, civil rights, data, database, data mining, health, medical, monitor, personal information, privacy, rights, safety, search, scan, screen, security, surveillance, technology, terrorism, trust, video.

Second, we searched the reports archived at the Pew Research Center for the People and the Press (http://people-press.org/reports/) and data compiled by the Polling Report (http://www.pollingreport.com/).

Searching was an iterative process, in the course of which we added

---

[11]S. Keeter, C. Miller, A. Kohut, R. Groves, and S. Presser, "Consequences of reducing nonresponse in a national telephone survey," *Public Opinion Quarterly* 64(2):125-148, 2000; R. Curtain, S. Presser, and E. Singer, "The effects of response rate changes on the index of consumer sentiment," *Public Opinion Quarterly* 64(4):413-428, 2000.

new keywords. Thus, it frequently turned out that the surveys we identified through searches of the IRSS archives, the Pew reports, and the Polling Report were also archived at the Roper Center when we searched on the new keywords. Since the Roper Center archive is more complete with respect to details on methodology, and since it allows those interested to easily obtain further data from the cited surveys, we identify it as the source of data, even when we initially identified a survey by searching other sources.

Third, when tabulations of the original survey data are not available, we draw on reports that research organizations or sponsors have prepared and posted on the Internet. These reports were identified via Internet searches using the same keywords as for the data archive searches. When referring to data drawn from such reports, the source information included in the text identifies both the survey (listed by abbreviation in subsection M.8.3 in the Annex) and the report (listed in subsection M.8.4).

Finally, we refer to several articles by researchers who have conducted their own reviews of poll results or who have conducted independent research on related topics.

## M.3  ORGANIZATION OF THIS APPENDIX

The remainder of this appendix is divided into four sections. In Section M.4, "General Privacy Attitudes," we briefly review public opinion on privacy in general, not directly related to antiterrorism efforts, in order to establish a context for understanding attitudes toward government monitoring programs. Section M.5, "Government Surveillance" begins with an overview of responses to a variety of surveillance measures, as examined in repeated surveys conducted by Harris Interactive. We then review data on attitudes toward seven specific areas of surveillance or monitoring:

- Communications monitoring
- Monitoring of financial transactions
- Video surveillance
- Travel security
- Biometric identification technologies
- Government use of databases and data mining
- Public health uses of medical information

Section M.6 is devoted to a consideration of attitudes toward the balance between defense of privacy and other civil rights that may interfere with effective terrorism investigation, on one hand, and terrorism prevention measures that may curtail liberties, on the other. Here we review survey results on public assessments of the proper balance between liberty and security, as well as trends in perceptions of the need to exchange liberty for security and personal willingness to make such sacrifices. In the concluding section, we discuss several factors that affect beliefs about the proper balance between liberty and security.

## M.4 GENERAL PRIVACY ATTITUDES

Figure M.1 displays results from a question asked by survey researchers throughout the 1990s: "How concerned are you about threats to your personal privacy in America today?" As the chart shows, respondents' concern about this issue increased steadily throughout the decade; by the last years of the 1990s, roughly 9 in 10 respondents were either "very" or "somewhat" concerned about threats to personal privacy. Once privacy issues became even more salient after September 11, 2001, the question was presumably no longer able to discriminate effectively between levels of concern about privacy, and it was not asked again by survey organizations.

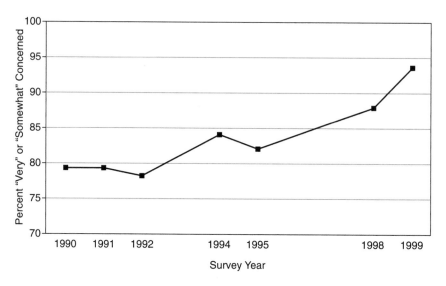

FIGURE M.1 "How concerned are you about threats to your personal privacy in America today?—very concerned, somewhat concerned, not very concerned, or not concerned at all?" (Harris Surveys, 1990-1999). SOURCE: A. Corning and E. Singer, 2003, "Surveys of U.S. Privacy Attitudes," report prepared for the Center for Democracy and Technology.

Related data for the post-9/11 period, however, suggest that general concerns about privacy have not abated. For example, public perceptions of the right to privacy are characterized by increasing pessimism. In July 2002, respondents to a survey conducted by the Public Agenda Foundation were asked "Do you believe that the right to privacy is currently under serious threat, is it basically safe, or has it already been lost?" (Table M.1). One-third of respondents thought it was basically safe, while 41 percent thought it was under serious threat and one-quarter regarded it as already lost. By September 2005, when the question was repeated in a CBS/*New York Times* poll, over half thought it was under serious threat, and 30 percent thought it had already been lost. Just 16 percent regarded it as "basically safe." Such pessimism may reflect generalized fears of privacy invasion, fueled by media reports of compromised security and ads that play to anxiety about fraud and identity theft; in addition, it may betray concerns about government intrusions on privacy in the post-9/11 era.

The perception that privacy is under threat is also due in part to concerns that the privacy of electronic information is difficult, if not impossible, to maintain. Over the past decade, survey researchers have repeated a question about online threats to privacy: "How much do you worry that computers and technology are being used to invade your privacy—is that something you worry about a lot, some, not much, or not at all?" As Figure M.2 shows, at most of the time points, half or more of respondents worried "some" or "a lot." The fluctuations from one observation to the next are probably due to house differences and to question context effects,[12] rather than to any substantive change in attitudes, and overall there appears to be a slight trend toward increasing worry about online privacy since 1994. (Considered separately, both the Princeton Survey Research Associates, PSRA, and the ABC surveys show parallel upward trends.) As Best et al. note,[13] growing concern about online privacy may be attributed to frequent reports of unauthorized access to or loss of

---

[12]The two observations of lowest levels of concern—June 1994 and January 2000—both occurred in surveys carried out by ABC. In both cases and in contrast to all the other surveys (including the March 2005 ABC/*Washington Post* survey), the question about privacy threat from computers immediately followed other questions asking about computers and privacy threat. When survey respondents are asked several questions belonging to the same domain, they tend to avoid redundancy, excluding information used in answering prior questions when answering subsequent ones (see N. Schwarz, F. Strack, and H.-P. Mai, "Assimilation and contrast effects in part-whole question sequences: A conversational logic analysis," *Public Opinion Quarterly* 55(1):3-23, 1991). Thus, the apparent lower levels of concern in the two ABC surveys may result from the fact that respondents had already expressed their concerns when answering previous questions.

[13]S.J. Best, B.S. Krueger, and J. Ladewig, "Privacy in the Information Age," *Public Opinion Quarterly* 70(3):375-401, 2006.

TABLE M.1  Right to Privacy (Public Agenda Foundation and CBS/ *New York Times* Surveys)

|  | July 2002 | September 2005 |
|---|---|---|
|  | Percent | Percent |
| "Do you believe that the right to privacy is currently under serious threat, is it basically safe, or has it already been lost?"[a] | | |
| Basically safe | 34 | 16 |
| Currently under serious threat | 41 | 52 |
| Has already been lost | 24 | 30 |
| Don't know | 2 | 2 |

[a]CBS/NYT 9/05: "Do you believe that currently the right to privacy is basically safe, under serious threat, or has already been lost?"

SOURCES: PAF/RMA 7/02; CBS/NYT 9/05.

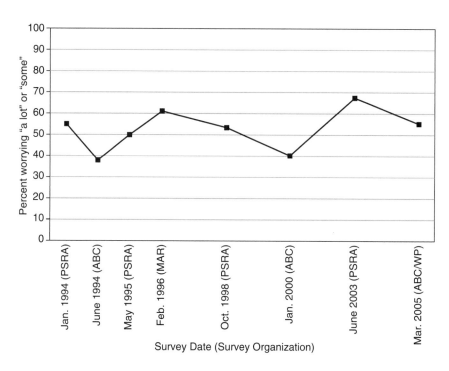

FIGURE M.2  "How much do you worry that computers and technology are being used to invade your privacy?" (surveys by PSRA, ABC News, and Marist College, 1994-2005). NOTE: Marist wording: ". . . that computers and advances in technology used to . . ." SOURCES: PSRA/TM 1/94, 5/95; ABC 6/94, 1/00; MAR 2/96; PSRA/PEW 10/98, 6/03; ABC/WP 3/05.

electronic data held by a wide variety of institutions, as well as to users' experience with spam and viruses.

These data on electronic privacy suggest that the public identifies multiple threats to privacy; surveillance by the federal government may be the most visible and controversial, but it is far from the only, or even the most important threat, in the public's view. In July 2002, respondents to a National Constitution Center survey regarded banks and credit card companies as the greatest threat to personal privacy (57 percent), while 29 percent identified the federal government as the greatest threat (PAF/RMA 7/02). When a similar question was asked in 2005 by CBS/NYT, 61 percent thought banks and credit card companies, alone or in combination with other groups, posed the greatest threat, while 28 percent named the federal government alone or in combination with other groups (CBS/NYT 9/05). (Responses cannot be compared directly, because of differences in the response options offered.)

## M.5 GOVERNMENT SURVEILLANCE

### M.5.1 Trends in Attitudes Toward Surveillance Measures

Over the years since September 11, 2001, Harris Interactive has asked a series of questions about support for specific surveillance measures that have been implemented or considered by the U.S. government as part of its terrorism prevention programs. For most of the questions, six or eight observations are available, for the period beginning just one week after the terrorist attacks in September 2001 and extending to July 2006. Table M.2 displays percentages of respondents favoring each of the measures at each time point.

Support for nearly all the measures peaked in the immediate aftermath of the 9/11 attacks, with support for stronger document and security checks and expanded undercover activities exceeding 90 percent. As the emotional response to the attacks subsided over the four years that followed, support for each of the measures declined, in many cases by more than 10 percentage points. As of the June 2005 observation, total decreases in support were fairly small for three of the more intrusive measures, which had not been as enthusiastically received in the first place: adoption of a national ID system, expanded camera surveillance in public places, and law enforcement monitoring of Internet discussions. In contrast, support for expanded monitoring of cell phone and e-mail communications—which had only barely received majority support in September 2001—had declined by 17 percentage points, to 37 percent, as of June 2005. At each time point it has been the least popular measure, by a margin of 9 or more percentage points.

TABLE M.2 Support for Government Surveillance Measures (Harris Surveys, 2001-2006)

| | September 2001[a] | March 2002 | February 2003 | February 2004 | September 2004 | June 2005 | February 2006 | July 2006 |
|---|---|---|---|---|---|---|---|---|
| | Percent | Percent | Percent | Percent | Percent | Percent | Percent | Percent |

Here are some increased powers of investigation that law enforcement agencies might use when dealing with people suspected of terrorist activity, which would also affect our civil liberties. For each, please say if you would favor or oppose it? ("Favor")

| | September 2001[a] | March 2002 | February 2003 | February 2004 | September 2004 | June 2005 | February 2006 | July 2006 |
|---|---|---|---|---|---|---|---|---|
| Stronger document and physical security checks for travelers | 93 | 89 | 84 | 84 | 83 | 81 | 84 | — |
| Stronger document and physical security checks for access to government and private office buildings | 92 | 89 | 82 | 85 | — | — | — | — |
| Expanded undercover activities to penetrate groups under suspicion | 93 | 88 | 81 | 80 | 82 | 76 | 82 | — |
| Use of facial recognition technology to scan for suspected terrorists at various locations and public events | 86 | 81 | 77 | 80 | — | — | — | — |

*continued*

TABLE M.2 Continued

| | September 2001[a] Percent | March 2002 Percent | February 2003 Percent | February 2004 Percent | September 2004 Percent | June 2005 Percent | February 2006 Percent | July 2006 Percent |
|---|---|---|---|---|---|---|---|---|
| Issuance of a secure ID technique for persons to access government and business computer systems, to avoid disruptions | 84 | 78 | 75 | 76 | — | — | — | — |
| Closer monitoring of banking and credit card transactions, to trace funding sources | 81 | 72 | 67 | 64 | 67 | 62 | 66 | 61 |
| Adoption of a national ID system for all U.S. citizens | 68 | 59 | 64 | 56 | 60 | 61 | 64 | — |
| Expanded camera surveillance on streets and in public places | 63 | 58 | 61 | 61 | 60 | 59 | 67 | 70 |
| Law enforcement monitoring of Internet discussions in chat rooms and other forums | 63 | 55 | 54 | 50 | 59 | 57 | 60 | 62 |
| Expanded government monitoring of cell phones and e-mail, to intercept communications | 54 | 44 | 44 | 36 | 39 | 37 | 44 | 52 |

[a]Fieldwork conducted September 19-24, 2001.

SOURCE: HI 9/01, 3/02, 2/03, 2/04, 9/04, 6/05, 2/06, 7/06.

Beginning with the February 2006 observation, however, most of the measures show an upturn in support, probably due to the London Underground bombings of July 2005. In particular, the growth in public approval for camera surveillance may have resulted from the role of video camera footage in establishing the identities of the London Underground bombers. A year after the London bombings, in July 2006, support for three of the measures—expanded camera surveillance, monitoring of chat rooms and other Internet forums, and expanded monitoring of cell phones and e-mail—continued to show increases.

These data suggest several generalizations. First, people appear more willing to endorse measures that they believe are unlikely to affect them. Tolerance for undercover activities targeted at suspected groups has remained at high levels. Other data support this conclusion as well: Table M.3 shows results from questions about surveillance measures asked in Pew surveys, which reveal that acceptance of racial/ethnic profiling is also comparatively high. And in surveys carried out by CBS/NYT, respondents were asked whether they "would be willing to allow government agencies to monitor the telephone calls and e-mail of ordinary Americans." Beginning in 2003, they were also asked the same question with regard to the communications "of Americans the government is suspicious of." The data, plotted in Figure M.3, indicate that support for monitoring the communications of people the government is suspicious of is much higher than support for monitoring those of ordinary Americans.

Second, people are more likely to accept measures that they do not regard as especially burdensome. Support has been highest for more rigorous security, both for travelers and for access to buildings; the added inconvenience represented by the extra checks may not seem significant to respondents. Acceptance of surveillance in public places also tends to be high. By contrast, measures intended to monitor the traditionally private domain of communications—whether Internet chat rooms or, especially, telephone and e-mail communication—have been the least accepted, at each time point.

## M.5.2 Communications Monitoring

*General Trends over Time.* Two pieces of research focusing on communications monitoring confirm the Harris data trends, mirroring both the long-term decreases in support for monitoring and sensitivity to perceptions of increased threat. First, the CBS/NYT data displayed in Figure M.3 show that the trend for "ordinary Americans" displays the same pattern visible in Table M.2, with a slow decline after 9/11 but an upturn discernible in mid-2006; here the upturn probably represents a response

TABLE M.3 Support for Government Surveillance Measures (Pew Center/PSRA Surveys, 2001-2006)

| | September 2001 | August 2002 | | January 2006 | December 2006 | |
| --- | --- | --- | --- | --- | --- | --- |
| | Personal Wording | Personal Wording | Impersonal Wording[a] | Personal Wording | Personal Wording | Impersonal Wording[a] |
| | Percent | Percent | Percent | Percent | Percent | Percent |
| "Would you favor or oppose the following measures to curb terrorism?" ("Favor") | | | | | | |
| Requiring that all citizens carry a national identity card at all times to show to a police officer on request | 70 | 59 | — | 57 | 57 | — |
| Allowing airport personnel to do extra checks on passengers who appear to be of Middle-Eastern descent | — | 59 | — | 57 | 57 | — |
| Allowing the U.S. government to monitor your personal telephone calls and e-mails | 26 | 22[b] | 33 | 24 | 22[b] | 34 |
| Allowing the U.S. government to monitor your credit card purchases | 40 | 32[b] | 43 | 29 | 26[b] | 42 |

[a] The word "your" was omitted from the question text. Asked of Form 1 half sample.
[b] Asked of Form 2 half sample.

SOURCE: PSRA/PEW 9/01, 8/02, 1/06, 12/06.

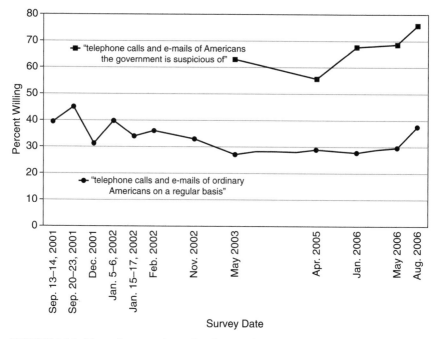

FIGURE M.3 "In order to reduce the threat of terrorism, would you be willing or not willing to allow government agencies to monitor the . . ." (surveys by CBS News, 2001-2006). SOURCES: CBS/NYT 9/01a, 9/01b, 12/01, 11/02, 1/06, 8/06; CBS 1/02a, 1/02b, 2/02, 5/03, 4/05, 5/06.

to media reports of an averted terrorist plot to bomb airplanes bound for the United States. The upturn for "Americans the government is suspicious of" shows an earlier increase as well, possibly in response to the London Underground bombings. Second, Pew Center survey questions on monitoring of communications reveal similar declines in support after the immediate post-9/11 period (Table M.3).

*Importance of Question Wording.* Taken together, the three sets of research by Harris, Pew, and CBS/NYT (shown in Tables M.2 and M.3 and Figure M.3) reveal the degree to which attitudes are dependent on specific question wording. The wording of the first Pew question, about a national ID program (Table M.3), is fairly similar in emphasis to the wording used in the Harris surveys (Table M.2). Levels of support correspond closely across the two questions, starting at 68-70 percent in September 2001 and remaining steady at about 60 percent thereafter. In the questions on monitoring of communications and credit card purchases shown in Table M.3, however, Pew highlighted the potential personal impact of the measures, asking respondents how they would react to programs that would allow their own phone calls, e-mails, and credit card purchases to be monitored. The differences are striking: even in September 2001, 54 percent supported the Harris measure allowing the government to monitor phone calls and e-mail (Harris—Table M.2), but less than half as many reacted favorably to the possibility that their own telephone and e-mail communications might be monitored (Pew—Table M.3).[14] Similarly, 81 percent approved of "closer monitoring of bank and credit card transactions" in 2001 (Harris—Table M.2), but again, just half as many were comfortable with the idea that their own purchases could be monitored (Pew—Table M.3).

At two time points, August 2002 and December 2006, Pew used a split-sample experiment that further confirms the impact of wording changes. In these experiments, half the sample was asked the questions about communications and credit card purchase monitoring in the usual form, while the other half of the sample heard the questions in a more impersonal form produced simply by omitting the word "your": "Allowing the U.S. government to monitor personal telephone calls and e-mails"; "Allowing the U.S. government to monitor credit card purchases." Results for the personal and impersonal wording are compared in Table M.3. On both measures, the impersonal wording boosted support by 11 or more percentage points at each observation.

*Telephone Records Database Program.* In May 2006, it was disclosed that the National Security Agency (NSA) was compiling a database containing the telephone call records of millions of ordinary American citizens, using information obtained from Verizon, AT&T, and BellSouth. Survey organizations responded to the ensuing controversy by asking respondents about their reactions. The public was divided: approval for the program ranged between 43 and 63 percent, with levels of support predictably varying by question wording (Table M.4). Gallup's question, which emphasized the scope of the database and the participation of the telephone companies and mentioned terrorism only at the very beginning of a long question, showed the lowest support, at 43 percent. When questions mentioned a more menacing "threat of terrorism" or "terrorist activity," as the CBS and Fox News questions did, about half of respondents favored the measure. The ABC News/WP question, which includes two mentions of terrorism and is the only question to note that

---

[14]One further difference in these questions is that the Harris wording includes the phrases "to intercept communications" and "to trace funding sources," which, by reminding respondents of the purpose of the measures, may have helped to justify them.

TABLE M.4  Four Questions on Attitudes Toward NSA's Telephone Records Database Program, May 2006

| | Approve/ Support/ Consider Acceptable | Disapprove/ Oppose/ Consider Unacceptable | Unsure |
|---|---|---|---|
| | Percent | Percent | Percent |
| "As you may know, as part of its efforts to investigate terrorism, a federal government agency obtained records from three of the largest U.S. telephone companies in order to create a database of billions of telephone numbers dialed by Americans. Based on what you have read or heard about this program to collect phone records, would you say you approve or disapprove of this government program?" (Gallup/*USA Today*) | 43 | 51 | 6 |
| "Do you approve or disapprove of the government collecting the phone call records of people in the U.S. in order to reduce the threat of terrorism?" (CBS News) | 51 | 44 | 5 |
| "As part of a larger program to detect possible terrorist activity, do you support or oppose the National Security Agency collecting data on domestic phone calls and looking at calling patterns of Americans without listening in or recording the calls?"[a] (Opinion Dynamics/Fox News) | 52 | 41 | 6 |
| "It's been reported that the National Security Agency has been collecting the phone call records of tens of millions of Americans. It then analyzes calling patterns in an effort to identify possible terrorism suspects, without listening to or recording the conversations. Would you consider this an acceptable or unacceptable way for the federal government to investigate terrorism?" (ABC/ *Washington Post*) | 63 | 35 | 2 |

[a]National sample of registered voters.

SOURCES: GAL/USA 5/06, CBS 5/06, OD/FOX 5/06, ABC/WP 5/06.

the phone calls are not listened to, found that 63 percent considered the program acceptable.

The public's ambivalence is reflected in mixed findings on concern about the program's potential personal consequences. More than half (57 percent) said they would feel their privacy had been violated if they learned that their own phone company had provided their records to the government under the program (GAL/USA 5/06). But apart from such objections to phone companies releasing information without customers' approval, respondents do not appear to be overly concerned about personal implications of the program. In answer to questions by Gallup and ABC, one-third of respondents said they would be "very" or "somewhat concerned/bothered" if they found out that the government had records of their phone calls (GAL/USA 5/06, ABC/WP 5/06). Yet despite the size of the database, most people seemed to regard this as an unlikely possibility: only one-quarter were "very" or "somewhat concerned" that the government might have their personal phone call records (CBS 5/06). Thus a minority, albeit a substantial one, expressed concern about the personal implications of the program.

When confronted with the conflicting values of investigating terrorism via the telephone records program on one hand, and the right to privacy on the other, respondents again displayed ambivalence, and different surveys showed majorities giving priority to each value. In the Gallup survey, the 43 percent who approved of the program (N = 349) were asked whether they approved because they felt the program did not "seriously violate" civil liberties or because they thought it was more important to investigate terrorism: 69 percent believed that terrorism investigation was the more important goal (GAL/USA 5/06). A survey conducted by the Winston Group (WIN 5/06; national sample of registered voters) found that 60 percent favored continuing the program because "we must do whatever we can within the law to prevent another terrorist attack," while 36 percent thought it should be discontinued because "it infringes on the right to privacy" (4 percent were unsure). A PSRA study showed that 41 percent thought the program was "a necessary tool to combat terrorism," while 53 percent found that it went "too far in invading people's privacy," and 6 percent were undecided (PSRA/NW 5/06). And when respondents to the CBS survey were asked whether phone companies should share their phone records with the government or whether that was an invasion of privacy, just 32 percent thought the phone companies should share that information, while 60 percent felt it was an invasion of privacy (CBS 5/06).

The variation in question wording, and consequently in results, makes it difficult to draw firm conclusions about public attitudes toward the program. What seems clear, however, is that, despite generally low support for surveillance of communications in the abstract, as shown by the Harris and particularly the PSRA/Pew results discussed earlier, the public exhibits greater tolerance for specific instances of surveillance, such as the telephone records database program—especially when the surveillance is justified as an antiterrorist measure.

Still, between one-third and two-thirds in each survey opposed the program. That opposition may betray public skepticism about the effectiveness and accuracy of the program. According to the CBS survey, 46 percent thought the phone call records database program would be "effective in reducing the threat of terrorism," 43 percent thought it would not be effective, and 11 percent were uncertain. And in the Gallup survey, two-thirds were concerned that the program would misidentify innocent Americans (36 percent "very concerned" and 29 percent "somewhat concerned").

Finally, the public exhibited no clear consensus even on the subject of whether the news media should disclose such secret counterterrorism efforts. In the Gallup poll, 47 percent thought the media should report on "the secret methods the government is using to fight terrorism," while 49 percent thought they should not; an ABC News/WP poll found that 56 percent thought the news media were right, and 42 percent thought they were wrong to report on the program (ABC/WP 5/06).

### M.5.3 Monitoring of Financial Transactions

Tables M.2 and M.3 show change in support for government monitoring of individuals' credit card purchases in surveys conducted by Harris and PSRA/Pew between 2001 and 2006. Although the two organizations found different overall levels of support owing to specific question wording, both trends show a total decline of 14 to 20 percentage points in the approximately five-year period since September 2001. At each time point, however, support for the monitoring of financial transactions was greater than for the monitoring of communications.

Following the revelations about the NSA's telephone call database program but prior to reports of systematic searches of international banking data carried out by the Central Intelligence Agency/Treasury Department, several other surveys asked respondents for their opinion on financial monitoring. Among the U.S. sample in a June 2006 survey sponsored by the German Marshall Fund, 39 percent supported "the government having greater authority to monitor citizens' banking transactions" as part of the effort to prevent terrorism, but 58 percent opposed such powers, and 3 percent were not sure (TNS/GMF 6/06). And in a CBS May

2006 survey, only one-quarter of respondents thought that credit card companies should "share information about the buying patterns of their customers with the government" (CBS 5/06).

After the revelations about the CIA/Treasury Department program, a *Los Angeles Times* survey found that 65 percent of respondents considered government monitoring of international bank transfers an "acceptable" way to investigate terrorism (LAT 7/06)—roughly the same percentage favoring monitoring of financial transactions in general between 2003 and 2006, as shown by the Harris surveys (Table M.2). Also in July 2006, Harris found that 61 percent of respondents favored the monitoring of financial transactions, a decline of 5 percentage points compared with the previous observation (Table M.2), while Pew also found a slight decline (Table M.3). These data are limited, but they suggest that public support for the government monitoring of financial transactions either remained stable or declined slightly in response to information about the program.

### M.5.4 Video Surveillance

Table M.2 shows that, despite initial declines in the years following September 2001, public support for video surveillance has been increasing. The percentages favoring increased video surveillance surpassed the September 2001 level (63 percent) in both February 2006 (67 percent) and July 2006 (70 percent). As noted earlier, the growing favorability may be due to the role of video cameras in identifying suspects in the London Underground bombings of 2005.

Other trend data on video surveillance attitudes also suggest that support is widespread, particularly when linked to terrorism prevention. In 1998, a CBS News poll asked respondents whether installing video cameras on city streets was "a good idea because they may help to reduce crime," or "a bad idea because [they] may infringe on people's privacy rights." Although more than half thought such cameras were a good idea, 34 percent regarded the cameras as an infringement on privacy (CBS 3/98). The same question, repeated in 2002, generated similar results (CBS 4/02). In July 2005, after the London Underground bombings, the question was rephrased to mention reducing "the threat of terrorism" instead of reducing crime. This time, 71 percent of respondents considered video surveillance a good idea, and just 23 percent thought it a bad idea (CBS 7/05). Other data on attitudes to video surveillance at national monuments—not explicitly linked to crime or to terrorism—showed that 81 percent support such surveillance, with only 17 percent finding it an invasion of privacy (CBS 4/02).

## M.5.5 Travel Security

Tighter airport security has been a source of frustration for travelers, and media reports perennially question the effectiveness of the measures. Nevertheless, the Harris data show higher levels of support for passenger screening and searches than for any other measure, both immediately after 9/11 and continuing through 2006 (see Table M.2).

While respondents are by no means fully convinced of airport security's effectiveness, public confidence does not appear to be waning, perhaps because of the absence, since 9/11, of terrorism involving airliners. In 2002, Fox News asked a sample of registered voters whether they thought the "random frisks and bag searches at airport security checkpoints are mostly for show" or whether they were "effective ways to prevent future terrorist attacks." The poll found that 41 percent thought the searches were for show and 45 percent thought they were effective, with 14 percent unsure (OD/FOX 4/02). In January and August 2006, CBS respondents were asked to evaluate the effectiveness of the government's "screening and searches of passengers who travel on airplanes in the U.S." While only 24 and 21 percent thought they were "very effective" in January and August, respectively, 53 percent in January and 61 percent in August found them "somewhat effective" (CBS 1/06 and 8/06). The Fox and CBS questions are of course not directly comparable, but there is no evidence of a decline in public confidence.

However, travel security now extends well beyond such airport searches to encompass such issues as what information airlines may collect and share with the government. When asked in 2006 whether airport security officials should have access to "passengers' personal data like their previous travel, credit card information, email addresses, telephone numbers and hotel or car reservations linked to their flight," just over half of respondents agreed that officials should have such access, while 43 percent said they should not (SRBI/TIME 8/06). When the public is asked whether "the government should have the right to collect personal information about travelers," support is somewhat lower (IR/QNS 6/06). One-quarter thought the government should have the right under any circumstances, and another 17 percent only with the traveler's consent. Still, a further 39 percent favored collecting such information if the traveler was suspected of some wrongdoing. A 2003 survey for the Council for Excellence in Government proposed a "smart card" that would store personal information digitally and could facilitate check-in, but it might also lead to the abuse of information; only 27 percent felt that the benefits of such a card outweighed the concerns, while 54 percent thought the concerns outweighed the benefits (H&T 2/03).

In addition, there is the question of what the airlines or the government may do with information they have collected. A Council for Excellence in Government survey in February 2004 (H&T 2/04) showed that 59 percent of respondents supported airline companies' sharing of information with the government "if there is any chance that it will help prevent terrorism," but 36 percent thought the government should not have access to the information "because that information is private and there are other things the government can do to prevent terrorism." In the Ipsos-Reid survey (IR/QNS 6/06), 73 percent would allow the government to share traveler information with foreign governments—but only 21 percent thought the government should be allowed to share information about any traveler, while 52 percent would restrict such sharing to information about travelers suspected of wrongdoing.

### M.5.6 Biometric Identification Technologies

A small handful of studies have attempted to gauge public attitudes toward biometric technologies that may be used for the identification of terrorists. As indicated in Table M.2, public support for the use of facial recognition technology declined somewhat after 9/11 but remained at high levels. In February 2004, the most recent observation available, 80 percent favored the use of such technology.

Other surveys have examined attitudes toward biometrics in the context of enhancing airport security. In a survey conducted in late September 2001 (HI/ID 9/01a), respondents were read the following description of an electronic fingerprint scanning process that could facilitate check-in and security procedures:

> I would like to read you a description of a new airport security solution and get your opinion. This new solution uses an electronic image of a fingerprint for a "real-time" background check to ensure that passengers, airline personnel, and airport employees are not linked with criminal or terrorist activities. The fingerprint images of people with no criminal or terrorist associations are immediately destroyed to protect the individual's privacy. The fingerprint image is used to link passengers to their boarding pass, baggage and passport control for better security.

Based on this description, and in the tense atmosphere immediately following 9/11, public support was substantial: 76 percent said such a new system would be "extremely" or "very valuable," and a further 16 percent thought it would be "somewhat valuable." Respondents were also overwhelmingly willing to have their own fingerprints scanned for airport security: 82 percent would be "very willing," and an additional 13 percent "would do it reluctantly." Only 4 percent "would not do it under any circumstances."

Several days later, in a subsequent survey by the same sponsor and organization, respondents were asked to choose between the fingerprint scan and an electronic facial scan:

> I want to explain two technologies that are being offered up as important solutions for airport security. The first is an electronic finger scan. Fingerprints are recognized as a highly accurate means of identification—even among identical twins. This new solution uses an electronic image of a fingerprint for a "real-time" background check to ensure that passengers, airline personnel, and airport employees are not linked with criminal or terrorist activities. The fingerprint images of people with no criminal or terrorist associations are immediately destroyed to protect the individual's privacy.
>
> The second is an electronic facial scan. With this solution, a camera captures images of all people in the airport within range of the camera to provide a "real-time" background check against known criminals or terrorists. The solution is automatic and does not require a person's permission or knowledge that it is occurring. This solution is convenient. However, there are more likely to be errors in distinguishing between people with very similar appearance, especially identical twins. Additionally, changes in facial hair or cosmetic surgery may make it difficult to provide an accurate match.

After hearing these descriptions, respondents rated the value of each method. Attitudes toward the fingerprint scan were again very favorable, closely matching the previous results. Clearly, the question portrays the facial scan as the less palatable option—not only is it more susceptible to errors, but it also can be used without consent or even knowledge. Not surprisingly, the facial scan ratings were substantially lower, with just 28 percent considering it "extremely" or "very valuable," and 44 percent finding it "somewhat valuable" (HI/ID 9/01b).

More recent data on attitudes toward the use of biometric technology for security purposes are not available, but data from 2006 do indicate that this is an area about which the public is still not well informed. The Ipsos-Reid study found that, in the United States, just 5 percent of respondents considered themselves "very knowledgeable" about "biometrics for facial and other bodily recognition," and only 24 percent said they were "somewhat knowledgeable" about the technology (IR/QNS 6/06).

### M.5.7 Government Use of Databases and Data Mining

Reports about the telephone records database program, discussed separately above, offer a view of public reaction to a specific instance of government compilation and searching of data. But in general, does the public feel it is appropriate for the government to use such methods? The few surveys that have examined this issue suggest that there is support for database searches by the government, particularly when presented as instrumental to counterterrorism efforts. In December 2002, respondents to a poll by the *Los Angeles Times* were told that

> The Department of Defense is developing a program which could compile information from sources such as phone calls, e-mails, web searches, financial records, purchases, school records, medical records and travel histories to provide a database of information about individuals in the United States. Supporters of the system say that it will provide a powerful tool for hunting terrorists. Opponents say it is an invasion of individual privacy by the government. (LAT 12/02)

Roughly equal proportions expressed support for the program (31 percent) and opposition to it (36 percent). However, respondents' lack of knowledge about data mining was reflected in the large percentage saying they hadn't heard enough to judge (28 percent). (Indeed, the Ipsos-Reid survey [IR/QNS 6/06] indicates that respondents were somewhat more knowledgeable about "data mining of personal information" than about biometrics, but still not well informed. In all, 11 percent said they were "very knowledgeable" about it, and 30 percent "somewhat knowledgeable," leaving more than half "not very" or "not at all knowledgeable.")

In early 2003, another study asked respondents to make a similar choice between the competing priorities of terrorism investigation and privacy with respect to government searches of "existing databases, such as those for Social Security" (H&T 2/03). Again, respondents were divided, with 49 percent finding it "appropriate" for government to carry out such searches, and 42 percent finding it "not appropriate." (Both percentages are higher than in the *Los Angeles Times* survey because no "don't know enough" option was explicitly offered.)

When respondents are not forced to choose between terrorism prevention and privacy, they express substantial concern about such efforts. In May 2006, in the context of questions about the telephone call records database program, Gallup asked respondents, "How concerned are you that the government is gathering other information on the general public, such as their bank records or Internet usage?" (GAL/USA 5/06). This question mentions only two of the possible personal information sources listed in the *Los Angeles Times* description of the Defense Department program. Nonetheless, 45 percent were "very concerned" and 22 percent "somewhat concerned" about such information-gathering.

## M.5.8 Public Health Uses of Medical Information

*Privacy of Medical Information.* Previous studies indicate high levels of concern about the privacy of health care information.[15] In 1999, Harris found that 54 percent of respondents were "very concerned" and 29 percent "somewhat concerned" about protecting the privacy of their health and medical information (HARRIS 4/99). A Gallup survey in 2000 found that over three-quarters of respondents thought it was "very important" that their medical records be kept confidential (Corning and Singer 2003). (The questions are worded differently, so no conclusions about trends can be drawn from these data).

The Health Insurance Portability and Accountability Act (HIPAA), with provisions designed to protect the privacy of individuals' health information, took effect in 2003. The 2005 National Consumer Health Privacy Survey was partly devoted to an evaluation of the impact of HIPAA on public attitudes, but the results were not encouraging. The study found that, although 67 percent of respondents claimed to be aware of federal laws protecting the privacy and confidentiality of medical records and 59 percent could recall receiving a privacy notice, only 27 percent thought they now had more rights than before. The study recorded high levels of concern about medical privacy: 67 percent of respondents overall and 73 percent of those belonging to an ethnic minority were "very" or "somewhat concerned" about the privacy of their "personal medical records."[16] And 52 percent of respondents were worried that insurance claims information might be used against them by their employers—an increase of 16 percentage points over the 1999 figure (FOR/CHCF Summer/05; California Health Care Foundation 2005).

Concern about medical privacy may in part reflect a lack of trust in the confidentiality of shared information. The Health Confidence Survey, conducted in 1999 and 2001-2003, found that just under half of respondents had high confidence that their medical records were kept confidential (GRN/EBRI 5/99, 4/01, 4/02, 4/03; there is no evidence of systematic change over the four observations available). In the National Consumer Health Privacy Survey, one-quarter of respondents were aware of incidents in which the privacy of personal information had been compromised, and those who were aware of such privacy breaches said that such

---

[15]E. Singer, R.Y. Shapiro, and L.R. Jacobs, "Privacy of health care data: What does the public know? How much do they care?," pp. 393-418 in *Health Care and Information Ethics: Protecting Fundamental Human Rights* (A.R. Chapman, ed.), Sheed and Ward, Kansas City, Mo., 1997; A. Corning and E. Singer, *Survey of U.S. Privacy Attitudes*, report prepared for the Center for Democracy and Technology, Washington, D.C., 2003.

[16]The same question was not asked in the 1999 survey, so no over-time comparison is possible for these data on medical privacy concern.

incidents had contributed to their concern about the privacy of their own health records (FOR/CHCF Summer/05; CHCF 2005).

Several surveys have compared concern about privacy in different domains, finding that levels of concern with regard to medical information are high. Even in 1978, Harris reported that 65 percent of respondents thought that it was important for Congress to pass additional privacy legislation in the area of medicine and health, as well as in the area of insurance—a larger proportion than favored such legislation for employment, mailing lists, credit cards, telephone call records, or public opinion polling (HARRIS 11/78). More recently, financial privacy concerns have exceeded concerns about medical records. As mentioned above, 54 percent of respondents in the 1999 Harris survey were concerned about protection of health and medical privacy, compared with 64 percent who were concerned about protecting privacy of information about their financial assets (HARRIS 4/99). And in 1995, PSRA found that more than half of respondents were "very" or "somewhat concerned" about "threats to privacy from growing computer use" in the areas of bank accounts (65 percent), credit cards (69 percent), and job and health records (59 percent; PSRA/NW 2/95).

*Attitudes Toward Electronic Medical Records.* Indeed, Corning and Singer (2003) note that the public's concerns about the privacy of health and medical information are due in part to the computerization of health and medical records and to perceptions of the vulnerability of computerized records to hacking or other unauthorized use. A 1999 survey found that 59 percent of respondents were worried "that some unauthorized person might gain access to your financial records or personal information such as health records on the Internet" (ICR/NPR 11/99). Of those, 36 percent were "very worried" about such unauthorized access. The Pew Research Center in 2000 found that 60 percent thought it would be "a bad thing" if "your health care provider put your medical records on a secure Internet Web site that only you could access with a personal password," because "you would worry about other people seeing your health records" (PSRA/PEW 7/00). The 2005 National Consumer Health Privacy Survey found that 58 percent of respondents thought medical records were "very" or "somewhat secure" in electronic format, compared with 66 percent in paper format (FOR/CHCF Summer/05; CHCF 2005). And in the 2005 Health Confidence Survey, just 10 percent said they were "extremely" or "very confident" that their medical records would remain confidential if they were "stored electronically and shared through the Internet," 20 percent were "somewhat confident," and 69 percent were "not too" or "not at all confident" (GRN/EBRI 6/05).

Incidents in which the privacy of personal information stored electronically has been compromised have tended to increase concern about online medical record-keeping (FOR/CHCF Summer/05; CHCF 2005). A Markle Foundation study in 2006[17] found that 65 percent of respondents were interested in storing and accessing their medical records in electronic format, but 80 percent were worried about identity theft or fraud, and 77 percent were worried about the information being used for marketing purposes (LRP/AV 11/06; Markle Foundation 2006).

*Opposition to National Medical Databases.* Such concerns are likely to have contributed to public opposition to the establishment of national databases that would store medical information. Opposition to such databases and to proposed systems of medical identification numbers ranges from moderate to nearly unanimous, depending on the question asked. For example, in 1992, 56 percent of respondents had "a great deal" of concern about "a health insurance company putting medical information about you into a computer information bank that others have access to" (RA/ACLUF 11/92, survey conducted via personal interview). Similarly, a 1998 PSRA survey examined attitudes toward a system of medical identification numbers. After answering a series of questions about potential risks and benefits of the proposed system, respondents answered a summary question, which showed that 52 percent would oppose such a system (PSRA/CHCF 11/98). In 2000, Gallup asked respondents, "Would you support a plan that requires every American, including you, to be assigned a medical identification number, similar to a social security number, to track your medical records and place them in a national computer database without your permission?" In response to that question, 91 percent of respondents opposed the plan (GAL/IHF 8/00).

*Support for Public Health Uses of Medical Records.* There have been few attempts to gauge attitudes toward the sharing of medical information for public health purposes, such as the conduct of research on health care, detection of disease outbreaks, or identification of bioterrorist attacks. The limited data available suggest that public support for such uses of medical information varies substantially depending on the safeguards specified, but it is far from universal. In the National Consumer Health Privacy Survey, only 30 percent of respondents were willing to share their medical information with doctors not involved in their care, and only 20 percent with government agencies (FOR/CHCF Summer/05; CHCF 2005). More-

---

[17]Markle Foundation, *Survey Finds Americans Want Electronic Personal Health Information to Improve Own Health Care*, 2006. Available at http://www.markle.org/downloadable_assets/research_doc_120706.pdf. [Accessed 3/10/07]

over, just half of respondents in that survey believed they had a "duty" to share medical information in order to improve health care.

Guarantees of anonymity may boost support: in 2003, *Parade Magazine* asked respondents whether, "assuming that there is no way that anyone will have access to your identity," they would be willing to release health information for various purposes. A total of 69 percent said they would share health information "so that doctors and hospitals can try to improve their services"; 67 percent, in order for "researchers to learn about the quality of health care, disease treatment, and prevention, and other related issues"; and 56 percent, so that "public health officials can scan for bio-terrorist attacks" (CRC/PAR 12/03). The Markle Foundation's most recent survey questions also provided for the protection of patient identity and found somewhat greater enthusiasm for the sharing of medical data: 73 percent would be willing to release their information to detect outbreaks of disease, 72 percent for research on improving the quality of care, and 58 percent to detect bioterrorist attacks (LRP/AV 11/06; Markle Foundation 2006).[18]

*Control and Consent.* The desire for control over personal medical information is a recurrent theme in the research on attitudes toward online medical record-keeping. In general, those who are willing to accept the online storage of medical records appear to be motivated by perceived personal benefits (FOR/CHCF Summer/05; CHCF 2005). Some of these benefits take the form of increased control over the content of medical records: in the Markle Foundation survey, 91 percent of respondents wanted to have access to electronic health records in order to "see what their doctors write down," and 84 percent in order to check for errors (LRP/AV 11/06; Markle Foundation 2006). Other types of perceived personal benefits, such as better coordination of medical treatment (FOR/CHCF Summer/05; CHCF 2005) or reductions in unnecessary procedures (LRP/AV 11/06; Markle Foundation 2006) also tend to incline respondents more positively toward electronic medical record-keeping.

In addition to seeking greater control over what is in the medical record, survey respondents also express a desire for control over decisions about the release of medical information. In data from the 1990s, Singer, Shapiro, and Jacobs (1997)[19] found broad support for individual consent

---

[18]The actual question wording is not available, but it is possible that the greater support found in the Markle Foundation survey may be due in part to provisions for consent prior to release of the information.

[19]E. Singer, R.Y. Shapiro, and L.R. Jacobs, "Privacy of health care data: What does the public know? How much do they care?," pp. 393-418 in *Health Care and Information Ethics: Protecting Fundamental Human Rights* (A.R. Chapman, ed.), Sheed and Ward, Kansas City, Mo., 1997.

prior to the release of medical data to those not involved in treatment. Respondents also preferred to require that the patient's permission be obtained for the use of medical records in research, even when the patient was not personally identified; 56 percent thought that general advance consent was not satisfactory and that permission should be required each time the record was accessed. Corning and Singer (2003) note that a strong majority of respondents to the 1998 CHCF survey thought that requiring individual consent before using data would be an effective way to protect privacy (PSRA/CHCF 11/98; CHCF 1999). Finally, the Markle Foundation's (2006) report noted that respondents "want to have some control over the use of their information" for research or public health purposes.

The implications of these findings for public support of databases designed to monitor public health threats are threefold. First, concerns about privacy make respondents hesitant about any online health database system. Second, respondents expect to exert no small degree of control over how their medical information is used and to whom it is released. Third, when respondents perceive personal benefits, they are more willing to consider online storage and sharing of information, but they do not appear to be motivated to share information by broader concerns about social well-being or by any sense of civic duty. Thus, to the extent that members of the public regard disease outbreaks or bioterrorist attacks as remote possibilities that will probably not affect them directly, they are unlikely to wish to share medical information to help track such occurrences.

## M.6  THE BALANCE BETWEEN CIVIL LIBERTIES AND TERRORISM INVESTIGATION

In recent years, survey organizations have used several broad questions asking respondents to weigh the competing priorities of terrorism investigation, on one hand, and protection of privacy or civil liberties, on the other. Although such questions are artificial in that they present the conflict between protection of individual rights and security in extreme, all-or-nothing terms, they do reflect the reality that support for civil rights is not an absolute value, but is dependent on judgments about the importance of other strongly held values.[20] In this section we review data from such forced-choice questions to examine public willingness to exchange privacy for security. We also examine public perceptions of the

---

[20]D.W. Davis and B.D. Silver, "Civil liberties vs. security: Public opinion in the context of the terrorist attacks on America," *American Journal of Political Science* 48(1):28-46, 2004.

need to sacrifice civil liberties, as well as personal willingness to make such sacrifices.

### M.6.1 Civil Liberties Versus Terrorism Prevention

Between 2002 and 2006, the following question was included in nine different surveys, mostly conducted by Gallup in conjunction with CNN and *USA Today*, but on two occasions conducted by Quinnipiac University:

> Which comes closer to your view? The government should take all steps necessary to prevent additional acts of terrorism in the U.S., even if it means your basic civil liberties would be violated. OR: The government should take steps to prevent additional acts of terrorism, but not if those steps would violate your basic civil liberties.

And, focusing more specifically on privacy, ABC News/WP asked:

> What do you think is more important right now—for the FBI [federal government] to investigate possible terrorist threats, even if that intrudes on personal privacy, or for the FBI [federal government] not to intrude on personal privacy, even if that limits its ability to investigate possible terrorist threats?"[21]

For each question, the trends for percentages choosing the civil liberties–oriented options are plotted in Figure M.4, which shows graphically the increasing affirmation of civil liberties since 9/11. Shortly after 9/11, in January 2002, 47 percent thought that "the government should take all steps necessary" for terrorism prevention (data not shown), but roughly half of respondents defended the preservation of civil liberties. By December 2005, just after the government had confirmed the existence of its warrantless monitoring program, 65 percent favored protection of civil liberties in the course of terrorism prevention. When "personal privacy" is singled out, as in the ABC/WP question, the overall percentages defending privacy against investigative measures are lower, but the trend is similar.[22]

The trend for another forced-choice question is plotted in Figure M.5. Respondents were asked, "What concerns you more right now? That

---

[21]Before January 2006, the question asked about the "FBI." As of January 2006, the wording was changed, replacing "FBI" with "federal government." Thus, the magnitude of the change between September 2003 and January 2006 may result in part from the change in wording. Nevertheless, the overall trend corresponds to that identified in other data.

[22]It is possible that the phrase "personal privacy" tends to minimize the scope and nature of the violation. When the phrase "privacy rights" is used in another forced-choice question, results correspond closely to those from a similar question about "civil liberties" (see below).

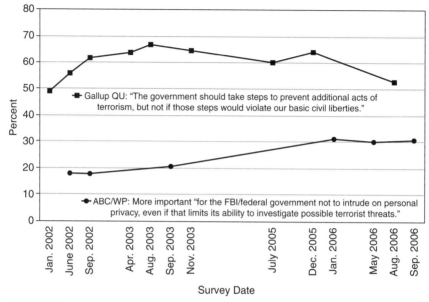

FIGURE M.4 Support for preserving privacy/civil liberties in the course of terrorism prevention (surveys by Gallup, Quinnipiac University, ABC, 2002-2006). SOURCES: Gallup/QU: GAL/CNN/USA 1/02, 6/02, 9/02, 4/03, 8/03, 12/05; GAL 11/03; QU 7/05, 8/06. ABC/*Washington Post*: ABC/WP 6/02, 1/06, 5/06; ABC 9/02, 9/03, 9/06.

the government will fail to enact strong anti-terrorism laws, or that the government will enact new anti-terrorism laws which excessively restrict the average person's civil liberties?" (Responses to the second option are plotted.) Figure M.5 shows that since September 2001, concern for preserving civil liberties has increased, and has remained at high levels or even grown slightly since 2002. A similar question, examining concerns specifically about privacy, was included in NBC/WSJ polls: "Which worries you more—that the United States will not go far enough in monitoring the activities and communications of potential terrorists living in the United States, or that the United States will go too far and violate the privacy rights of average citizens?" In December 2002, 31 percent were more worried that the United States would go too far; by July 2006, that figure had increased to 45 percent (H&T/NBC/WSJ 12/02, H&I/NBC/WSJ 7/06). Other observations are not available, but the trend conforms to that for the question asking more broadly about civil rights.

The public's concern does not appear to be based on any personal experience of privacy intrusions resulting from government efforts at

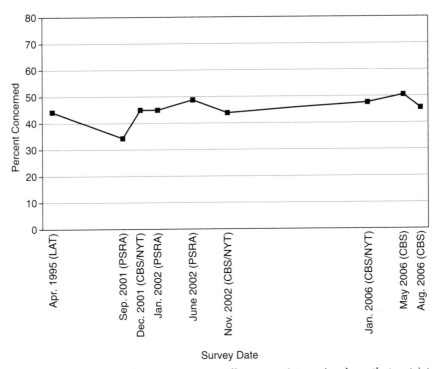

FIGURE M.5 Concern that government will enact anti-terrorism laws that restrict civil liberties (surveys by *Los Angeles Times*, PSRA, and CBS News, 1995-2006). SOURCES: LAT 4/95, PSRA/PEW 9/01, 1/02, 6/02; CBS/NYT 12/01, 11/02, 1/06; CBS 5/06, 8/06.

terrorism prevention. When Harris asked a question phrased in more personal terms—"How much do you feel government anti-terrorist programs have taken your own personal privacy away since September 11, 2001?"—perceptions showed stability over the same period (Table M.5). At each time point, a majority felt that their privacy had not been affected at all or had been affected "only a little."

*Sensitivity to Perceptions of Threat.* Attitudes toward the proper balance between terrorism investigation and protection of civil liberties are clearly responsive to changes in threat perception. Such volatility is especially visible in responses to the Gallup/QU question (Figure M.4), which show declines in support for civil liberties at the July 2005 and August 2006 observations, which occurred just after the London Underground bombings and the reports of planned terrorist attacks on transatlantic

TABLE M.5  Impact on Personal Privacy of Government Antiterrorist Programs (Harris Surveys, 2004-2006)

|  | February 2004 | September 2004 | June 2005 | February 2006 |
|---|---|---|---|---|
|  | Percent | Percent | Percent | Percent |
| "How much do you feel government anti-terrorist programs have taken your own personal privacy away since September 11, 2001 (the date of the terrorist attacks on the World Trade Center and the Pentagon)?" | | | | |
| A great deal | 8 | 8 | 10 | 7 |
| Quite a lot | 6 | 9 | 7 | 7 |
| A moderate amount | 22 | 21 | 24 | 23 |
| Only a little | 29 | 26 | 25 | 28 |
| None at all | 35 | 35 | 32 | 35 |
| Not sure/NA | 1 | 1 | 1 | — |

SOURCE: HI 2/04, 9/04, 6/05, 2/06.

flights, respectively.[23] Responses to the ABC/WP question, which asked specifically about privacy, appear less sensitive, perhaps partly as a result of the timing of the observations. It may also be that the public regards such rights as due process and personal freedom as greater obstacles to terrorism investigation than privacy as such, but of course those rights have important privacy dimensions as well. Concern for preserving civil liberties (Figure M.5) has also been more stable, though we note that the peak in May 2006 coincided with reports on the NSA telephone records database.[24] Thus, it is not only attitudes toward specific surveillance measures that are responsive to perceptions of increased threat (see Table M.2

---

[23]It should also be noted that these two observations showing lower support are both from studies carried out by Quinnipiac University, in contrast to all other observations, which are from Gallup surveys. However, the decreases make substantive sense and correspond to trends identified elsewhere.

[24]Change in the percentages of respondents who are worried that strong laws will not be enacted usually correspond to changes in concern about civil liberties, but this is not always the case. In the mid-1990s, 44 percent expressed greater concern that the government would restrict civil liberties, while 40 percent (data not shown) were more concerned that strong laws would not be enacted. In September 2001, concern about civil liberties dropped to 34 percent, but there was no corresponding increase in concern that strong laws would not be enacted; rather, the percentage who were concerned about both possibilities increased, as did the percentage who couldn't say. Concern that strong laws would fail to be enacted has remained at 35-40 percent (data not shown) since June 2002. Percentages saying "don't know" have also been stable, so that, since then, increases in concern about civil liberties have been matched by decreases in concern about enactment of laws (and vice versa).

and Figure M.3), but also broader prioritizations of individual rights and terrorism investigation.

### M.6.2 Privacy Costs of Terrorism Investigation

In 1996, well before the events of 9/11 and even prior to several terrorist attacks on U.S. interests, 69 percent of respondents to an NBC News/*Wall Street Journal* survey said they would support "new laws to strengthen security measures against terrorism, even if that meant reducing privacy protections such as limits on government searches and wiretapping" (H&T/NBC/WSJ 8/96). Thus it should come as no surprise that, in recent years, majorities of respondents recognize that terrorism investigation comes at a cost to privacy. In surveys conducted in September 2003, January 2006, and September 2006, between 58 and 64 percent agreed that, "in investigating terrorism . . . federal agencies like the FBI are intruding on some Americans' privacy rights" (ABC 9/03, 9/06; ABC/WP 1/06). Yet between 49 and 63 percent of those who regarded the investigations as infringing on privacy rights thought the loss of privacy was justified (ABC 9/03, 9/06; ABC/WP 1/06).

Further evidence of the public's belief that sacrifices of civil liberties or personal freedoms will be needed in order to combat terrorism comes from two questions asked between March 1996 and August 2006. During that period, the Pew Research Center/PSRA asked, "In order to curb terrorism in this country, do you think it will be necessary for the average person to give up some civil liberties, or not?"[25] And beginning in September 2001, CBS News asked, "Do you think Americans will have to give up some of their personal freedoms in order to make the country safe from terrorist attacks, or not?" The two trends are shown in Figure M.6. The overall difference between proportions agreeing with the two different propositions can probably be attributed to the difference between CBS News' higher bar of making "the country safe" from threatening-sounding "terrorist attacks" (versus the more measured "curb terrorism" in the Pew/PSRA studies).

Both trends show the same pattern, however: percentages believing that sacrifices would be necessary were highest immediately after September 11, 2001. With increasing distance from those events, the public became less convinced of the need for sacrifices. Yet even when the percentages agreeing that sacrifices of civil liberties would be called for were at their lowest post-9/11 level, in July 2004, they had still not returned to the levels of the mid-1990s.

---

[25]In July 2004 and July 2005, the question read, ". . . do you think it is necessary . . . ."

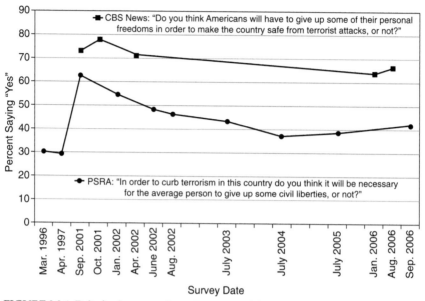

FIGURE M.6 Beliefs about need to give up civil liberties in order to curb terrorism (surveys by PSRA and CBS News, 1996-2006). SOURCES: PSRA: PSRA/PEW 3/96, 4/97, 1/02, 6/02, 7/03, 7/04, 7/05, 9/06; PSRA/NW 9/01, 8/02. CBS: CBS/NYT 9/01a, 8/06; CBS 10/01, 4/02, 1/06.

### M.6.3 Personal Willingness to Sacrifice Freedoms

Public beliefs about the need for sacrifice at the national level appear to translate into personal willingness to make sacrifices as well. When respondents are asked whether they themselves would "give up some of [their] personal freedom in order to reduce the threat of terrorism," substantial proportions say they are willing to do so. Figure M.7 shows that the trend on this question, too, conforms to the pattern discerned earlier. Again, there is an early observation, in May 2001, that serves as a pre-9/11 baseline: at that point, 33 percent said they would be willing to sacrifice some freedom, a figure that leaped to 71 percent after 9/11. The curve shows a decline over the next 12 months to 61 percent, where it remains until dropping again in January and May 2006. At the time of those surveys, respondents may have felt less inclined to consider further sacrifices, perhaps having become aware—after hearing reports about the government's warrantless monitoring and telephone call records database programs—that they were already giving up more freedoms than they had realized. Still, the low point in May 2006 of 54 percent does not approach the pre-9/11 figure of 33 percent, suggesting that 9/11 may have

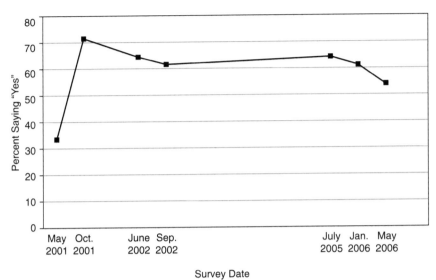

Survey Date

FIGURE M.7 "Would you be willing to give up some of your personal freedom in order to reduce the threat of terrorism?" (national samples of registered voters; Opinion Dynamics Surveys, 2001-2006). SOURCES: OD/FOX 5/01, 10/01, 6/02, 9/02, 7/05, 1/06, 5/06.

brought about real change in the extent to which Americans are willing to assert their right to customary freedoms.

### M.6.4 Concerns About Uses of Expanded Powers

Have the public's concerns about how expanded powers would be used changed over the period since 2001? Table M.6 shows the percentages expressing "high" or "moderate concern" about possible problems in the implementation of surveillance measures, both at the level of oversight and at the level of application of those measures. In February 2006, roughly three-quarters of respondents were concerned about the potential for abuses of civil liberties by the courts and Congress, with slightly fewer concerned about inappropriate use of powers by law enforcement. Concerns about abuses by law enforcement changed little over the five-year period, while concerns about lapses by the courts and Congress declined slightly. In fact, "high" concern about abuses by the courts decreased substantially over the period, from 44 to 34 percent (data not shown).

In contrast to these trends, a new question added in 2004 about the adequacy of White House oversight shows an increase in concern,

TABLE M.6 Concern About Uses of Expanded Powers (Harris Surveys, 2001-2006)

| | September 2001 | February 2004 | September 2004 | June 2005 | February 2006 |
|---|---|---|---|---|---|
| | Percent | Percent | Percent | Percent | Percent |

"Now, here are some concerns that people might have about the way these increased powers might be used by law enforcement. Would you say you have high concern, moderate concern, not much concern, or no concerns at all about each of the following possibilities?" (percent "high" or "moderate" concern).

| | September 2001 | February 2004 | September 2004 | June 2005 | February 2006 |
|---|---|---|---|---|---|
| Judges who authorize investigations would not look closely enough at the justifications for that surveillance | 79 | 78 | 77 | 75 | 76 |
| Congress would not include adequate safeguards for civil liberties when authorizing these increased powers | 78 | 75 | 74 | 75 | 75 |
| Law enforcement would investigate legitimate political and social groups | 68 | 67 | 68 | 68 | 68 |
| The White House would not issue the proper rules for legal due process for government surveillance programs | — | — | 69 | 72 | 75 |
| The mail, telephone, e-mails, or cell-phone calls of innocent people would be checked | 72 | 76 | — | — | — |
| Non-violent critics of government policies would have their mail, telephone, e-mails, or cell-phone calls checked | 71 | 76 | — | — | — |
| New surveillance powers would be used to investigate crimes other than terrorism | 67 | 71 | | | |
| There would be broad profiling of people and searching them based on their nationality, race, or religion | 77 | 73 | — | — | — |

SOURCE: HI 9/01, 2/04, 9/04, 6/05, 2/06.

from 69 to 75 percent over the three observations available. Most of this increase was the result of expanding "high" concern, from 35 to 41 percent (data not shown). With respect to the groups responsible for regulating increased powers, then, public confidence in Congress, the courts, and law enforcement has remained stable or increased slightly since 2001, but citizens are growing less sanguine about how those powers are used by the executive branch of government.

Respondents were also asked how concerned they were about some specific inappropriate uses of the increased powers (Table M.6). Between September 2001 and February 2004, concerns that the communications of innocent people would be monitored, that nonviolent groups would be investigated, and that new surveillance powers would be used for purposes other than terrorism investigation each showed small increases, while concerns about racial and religious profiling decreased slightly; more recent data for these questions are not available.

The trends discussed above point to growing support for defending privacy and other civil liberties, even at some cost to terrorism investigation; to sustained high levels of concern, after the immediate post-9/11 period, that antiterrorism laws will restrict civil liberties; to declining public conviction that sacrifice of civil liberties is truly necessary for terrorism prevention; to decreasing personal willingness to sacrifice freedom for the sake of terrorism investigation; and to stability or slight increases in concern about abuses of expanded powers. The influence of specific events, such as terrorist incidents, news about terrorist activity, and reports of surveillance programs, can be discerned as high and low points in the overall trends. But the long-term trends themselves can probably best be attributed to distance in time from the most recent instance of terrorism within the United States and to dissatisfaction with the balance the government has achieved between protecting civil liberties and combating terrorism.

## M.7 CONCLUSIONS

In the introduction to this appendix, we noted the main conclusions to emerge from this review of trends in attitudes toward government surveillance and the associated loss of privacy. Here, we discuss some of the relevant findings from more in-depth research on support for civil liberties in the post-9/11 period.

This literature offers insights into demographic and other attitudinal correlates of opinions about civil liberties versus security, which we have not been able to explore in this review because bivariate tabulations of data by demographic group are not readily available and a reanalysis of data sets is beyond the scope of this research. A recent study shows that demographic influences on attitudes toward the civil liberties versus security balance are important, however. Davis and Silver (2004) carried out a national survey shortly after the attacks of 9/11, in which they studied people's willingness to exchange civil liberties for security.[26] In a multivariate analysis that controls for the effects of other variables, the authors found that African Americans showed a stronger preference for preserving civil liberties, even at the expense of security, compared with whites (OSR/MSU 11/01). Similarly, affirmation of civil liberties is stronger among young people (ages 18-24) and urban residents. Political ideology is also significant, with liberals more likely than conservatives to favor protection of civil rights over terrorism investigation; dogmatic people, who are characterized by intolerance, inflexibility, and insecurity, are more likely to favor security over the defense of civil liberties. These findings support and expand on the demographic relationships noted by Westin.[27]

The literature also provides corroboration of the trends described above—in particular, the finding that abstract support for civil rights tends to dissolve in specific situations.[28] This phenomenon manifests itself in two ways in the data discussed in this review. First, however strongly respondents avow their support for civil liberties, they are willing to make concessions when called on to choose between preserving rights and supporting specific security measures that may violate those rights. Such willingness is most clearly visible in the contrast between the high levels of support for restricting terrorism prevention steps to those that preserve civil liberties, shown in Figure M.4, and the even higher levels of support for individual surveillance measures, shown in Table M.2. The widespread belief that sacrifices of civil liberties will be required in order to combat terrorism, as well as respondents' willingness to countenance such sacrifices (see Figures M.6 and M.7) also reveal public willingness to compromise on privacy and other rights.

Second, the phenomenon can be observed in the vulnerability of support for privacy and other personal freedoms to the external influence of

---

[26]The survey was conducted between November 14, 2001, and January 15, 2002, by means of telephone interviews with an adult national RDD sample. The response rate (calculated as RR4 in the "standard definitions" of the American Association for Public Opinion Research [AAPOR]) was 52.3 percent, and the refusal rate was 19.0 percent. The survey was carried out by the Office for Survey Research of the Institute for Public Policy and Social Research at Michigan State University. See D.W. Davis and B.D. Silver, op. cit.

[27]A.F. Westin, "How the public sees the security-versus-liberty debate," pp. 19-36 in *Protecting What Matters: Technology, Security, and Liberty Since 9/11* (C. Northhouse, ed.), Brookings Institution Press, Washington, D.C., 2005.

[28]See, for example, L. Huddy, N. Khatib, and T. Capelos, "Trends: Reactions to the Terrorist Attacks of September 11, 2001," *Public Opinion Quarterly* 66(3):418-450, 2002; and Davis and Silver, 2004, op. cit.

terrorist threat. Nearly all of the trends that contain sufficient data points display a response, in the form of reduced support for civil liberties and/ or greater support for surveillance measures, to two periods of increased threat perception: July 2005, when the London Underground bombings took place, and August 2006, when it was reported that a major terrorist attack on transatlantic airliners had been averted.[29] This inverse relationship between perceptions of threat and support for civil rights was examined more systematically in the research by Davis and Silver (2004). The authors conclude that "when they feel threatened, people who previously protected civil liberties and personal freedom may compromise on these values for greater security."[30]

The influence of a sense of threat extends beyond a straightforward negative association with support for civil liberties, however, to important interaction effects. Davis and Silver[31] found that the perception of threat conditions the relationship between other variables, such as trust in government and liberal-conservative ideology and civil liberties support. Trust in government is negatively associated with affirmation of civil rights: those with greater trust in government are more willing to sacrifice freedoms, compared with those with less trust. And, other things being equal, liberals are more likely to defend civil liberties than conservatives. However, the effect of both variables on attitudes toward civil liberties is moderated by the sense of threat. For example, there is a substantial reduction in support for civil liberties among those who have strong trust in government *and* who perceive high threat. And liberals with a strong sense of threat may be less supportive of civil liberties than conservatives who perceive no threat.

This examination of trend data, of course, does not allow us to definitively identify associations between variables, much less interactions. However, we have noted above that reductions in support for civil liberties appear to correspond to periods when threat perception is high. Following Davis and Silver,[32] we would expect declines in trust in government to be reflected in increased concern for preservation of civil liberties,

---

[29]Whether numerous other instances of terrorism—the Madrid train bombings of 2004, the attacks on a residential compound for foreigners in Saudi Arabia in 2003, and the Bali bombing of 2002, among others—may likewise have increased threat perception among the American public, we cannot say, because pollsters did not conduct surveys (or did not ask questions about surveillance and/or civil liberties versus security) in the immediate aftermath of those incidents. This is a reminder that knowledge of public attitudes and the ability to discern trends based on public opinion data are strongly dependent on judgments by the media and pollsters about what events offer worthwhile material for the study of public reaction.

[30]Davis and Silver, 2004, op. cit, p. 38.

[31]Davis and Silver, 2004, op. cit.

[32]Ibid.

particularly if part of the reason for the reduction in trust is a perception of indifference to individual rights on the part of the government. Yet concern for civil liberties can be easily, and dramatically, suppressed by heightened threat.

## M.8 ANNEX

### M.8.1 Details of Cited Surveys

Each survey cited in this appendix is identified in the text by an abbreviation referring to the research organization and, in most cases, the sponsor and the date; these abbreviations and dates are included at the beginning of each entry.

Surveys are grouped alphabetically by abbreviation identifying the research organization/sponsor; within the groups for each research organization/sponsor, surveys are listed in chronological order.

Methodological details include sample design, sample size, survey method, and fieldwork dates. Not all information is available for all surveys, however, and response rates are not available. All surveys were conducted in the United States, and all were conducted either by telephone (random digit dialed) or via personal interview. Internet surveys are not included in this review.

Unless otherwise indicated in the entries, the data cited can be found at the iPOLL Databank at the Roper Center for Public Opinion Research, University of Connecticut: http://www.ropercenter.uconn.edu/ipoll. html. The survey title given here is the title listed in the iPOLL archive. When data were obtained from sources other than the iPOLL Databank, the survey entry identifies the report or Web site from which the data were obtained.

### M.8.2 Research Organization/Sponsor Name Abbreviations

| | |
|---|---|
| ABC | ABC News |
| ABC/WP | ABC News/*Washington Post* |
| CBS | CBS News |
| CBS/NYT | CBS News/*New York Times* |
| CRC/PAR | Charleton Research Company for *Parade Magazine* |
| FOR/CHCF | Forrester Research for the California HealthCare Foundation |
| GAL | Gallup Organization |
| GAL/CNN/USA | Gallup Organization for CNN/*USA Today* |

| GAL/USA | Gallup Organization for *USA Today* |
| GRN/EBRI | Matthew Greenwald and Associates for Employee Benefit Research Institute, Consumer Heath Education Council |
| HARRIS | Louis Harris and Associates |
| HI | Harris Interactive |
| HI/ID | Harris Interactive for Identix |
| H&M/NBC/WSJ | Hart and McInturff Research Companies for NBC News/*Wall Street Journal* |
| H&T | Hart and Teeter Research Companies |
| H&T/NBC/WSJ | Hart and Teeter Research Companies for NBC News/*Wall Street Journal* |
| ICR/NPR | International Communications Research for National Public Radio, the Henry J. Kaiser Family Foundation, and Harvard University's Kennedy School of Government |
| LAT | *Los Angeles Times* |
| LRP/AV | Lake Research Partners and American Viewpoint for the Markle Foundation |
| MAR | Marist College Institute for Public Opinion |
| OD/FOX | Opinion Dynamics for Fox News |
| OSR/MSU | Office for Survey Research of the Institute for Public Policy and Social Research at Michigan State University |
| PAF/RMA | Public Agenda Foundation and Robinson and Muenster Associates for the National Constitution Center |
| PSRA | Princeton Survey Research Associates |
| PSRA/CHCF | Princeton Survey Research Associates for the California Health Care Foundation |
| PSRA/NW | Princeton Survey Research Associates for *Newsweek* |
| PSRA/PEW | Princeton Survey Research Associates for the Pew Research Center |
| QNS/IR | Ipsos-Reid for Queens University, Canada |
| QU | Quinnipiac University Polling Institute |
| RA/ACLUF | Response Analysis for the American Civil Liberties Union Foundation |
| SRBI/TIME | Schulman, Ronca and Bucuvalas for *Time* |
| TNS/GMF | TNS Opinion and Social Institutes for the German Marshall Fund of the U.S. and the Compagnia di San Paolo, Italy |
| WIN | Winston Group for New Models |

## M.8.3  List of Surveys

ABC 6/94. **ABC News Poll.** Telephone survey conducted by ABC News, with a national adult sample of 813. Fieldwork carried out June 7-8, 1994.

ABC 1/00. **ABC News Poll.** Telephone survey conducted by ABC News, with a national adult sample of 1,006. Fieldwork carried out January 21-26, 2000.

ABC 9/02. **ABC News Poll.** Telephone survey conducted by TNS Intersearch for ABC News, with a national adult sample of 1,011. Fieldwork carried out September 5-8, 2002.

ABC 9/03. **ABC News Poll.** Telephone survey conducted by TNS Intersearch for ABC News, with a national adult sample of 1,004. Fieldwork carried out September 4-7, 2003.

ABC 9/06. **ABC News Poll.** Telephone survey conducted by TNS Intersearch for ABC News, with a national adult sample of 1,003. Fieldwork carried out September 5-7, 2006.

ABC/WP 6/02. **ABC News/*Washington Post* Poll.** Telephone survey conducted by TNS Intersearch for ABC News/*Washington Post*, with a national adult sample of 1,004. Fieldwork carried out June 7-9, 2002.

ABC/WP 3/05. **ABC News/*Washington Post* Poll.** Telephone survey conducted by TNS Intersearch for ABC News/*Washington Post*, with a national adult sample of 1,001. Fieldwork carried out March 10-13, 2005.

ABC/WP 1/06. **ABC News/*Washington Post* Poll.** Telephone survey conducted by TNS Intersearch for ABC News/*Washington Post* with a national adult sample of 1,001. Fieldwork carried out January 5-8, 2006.

ABC/WP 5/06. **ABC News/*Washington Post* Poll.** Telephone survey conducted by TNS Intersearch for ABC News/*Washington Post* with a national adult sample of 502. Fieldwork carried out May 11, 2006.

CBS 3/98. **CBS News Poll.** Telephone survey conducted by CBS News with a national adult sample of 994. Fieldwork carried out March 30-April 1, 1998.

CBS 10/01. **CBS News Poll.** Telephone survey conducted by CBS News with a national adult sample of 436. Fieldwork carried out on October 8, 2001.

CBS 1/02a. **CBS News Poll.** Telephone survey conducted by CBS News, with a national adult sample of 1,000. Fieldwork carried out January 5-6, 2002.

CBS 1/02b. **CBS News Poll.** Telephone survey conducted by CBS News, with a national adult sample of 1,030. Fieldwork carried out January 15-17, 2002.

CBS 2/02. **CBS News Poll.** Telephone survey conducted by TNS Research for CBS News, with a national adult sample of 861. Fieldwork carried out February 24-26, 2002.

CBS 4/02. **CBS News Poll.** Telephone survey conducted by CBS News, with a national adult sample of 1,119. Fieldwork carried out April 15-18, 2002.

CBS 5/03. **CBS News Poll.** Telephone survey conducted by CBS News, with a national adult sample of 758. Fieldwork carried out May 27-28, 2003.

CBS 4/05. **CBS News Poll.** Telephone survey conducted by CBS News, with a national adult sample of 1,149. Fieldwork carried out April 13-16, 2005.

CBS 7/05. **CBS News Poll.** Telephone survey conducted by CBS News, with a national adult sample of 632. Fieldwork carried out July 13-14, 2005.

CBS 1/06. **CBS News Poll.** Telephone survey conducted by CBS News, with a national adult sample of 1,151. Fieldwork carried out January 5-8, 2006.

CBS 5/06. **CBS News Poll.** Telephone survey conducted by CBS News, with a national adult sample of 636. Fieldwork carried out May 16-17, 2006.

CBS 8/06. **CBS News Poll.** Telephone survey conducted by CBS News, with a national adult sample of 974. Fieldwork carried out August 11-13, 2006.

CBS/NYT 9/01a. **CBS News/*New York Times* Poll.** Telephone survey conducted by CBS News/*New York Times*, with a national adult sample of 959. Fieldwork carried out September 13-14, 2001.

CBS/NYT 9/01b. **CBS News/*New York Times* Poll.** Telephone survey conducted by CBS News/*New York Times*, with a national adult sample of 1,216. Fieldwork carried out September 20-23, 2001.

CBS/NYT 12/01. **CBS News/*New York Times* Poll.** Telephone survey conducted by CBS News/*New York Times*, with a national adult sample of 1,052. Fieldwork carried out December 7-10, 2001.

CBS/NYT 11/02. **CBS News/*New York Times* Poll.** Telephone survey conducted by CBS News/*New York Times*, with a national adult sample of 996. Fieldwork carried out November 20-24, 2002.

CBS/NYT 9/05. **CBS News/*New York Times* Poll.** Telephone survey conducted by CBS News/*New York Times*, with a national adult sample of 1,167. (An oversample of African Americans was employed, but results are weighted to be representative of the national adult population.) Fieldwork carried out September 9-13, 2005.

CBS/NYT 1/06. **CBS News/*New York Times* Poll.** Telephone survey conducted by CBS News/*New York Times*, with a national adult sample of 1,229. Fieldwork carried out January 20-25, 2006.

CBS/NYT 8/06. **CBS News/*New York Times* Poll.** Telephone survey conducted by CBS News/*New York Times*, with a national adult sample of 1,206. Fieldwork carried out August 17-21, 2006.

CRC/PAR 12/03. ***Parade*/ResearchAmerica Health Poll.** Telephone survey conducted by Charleton Research Company for *Parade Magazine*, with a national adult sample of 800. Fieldwork carried out during December 2003.

FOR/CHCF Summer/05. **National Consumer Health Privacy Survey 2005.** Telephone survey conducted by Forrester Research for the California HealthCare Foundation, with a sample of 1.000 adults. The total sample size of 2,100 includes an oversample (N = 1,000) of California residents and an oversample of respondents with HIV or substance abuse (N = 100). Results cited here are for the national sample only. Specific fieldwork dates are not provided; materials indicate that the survey was conducted in "Summer 2005." Data reported in "Executive Summary," retrieved March 29, 2006, from http://www.chcf.org/topics/view.cfm?itemID=115694.

GAL 11/03. **Gallup Poll.** Telephone survey conducted by Gallup Organization, with a national adult sample of 1,004. Fieldwork carried out November 10-12, 2003.

GAL/CNN/USA 1/02. **Gallup/CNN/*USA Today* Poll.** Telephone survey conducted by Gallup Organization for CNN and *USA Today*, with a national adult sample of 1,011. Fieldwork carried out January 25-27, 2002.

GAL/CNN/USA 6/02. **Gallup/CNN/*USA Today* Poll.** Telephone survey conducted by Gallup Organization for CNN and *USA Today*, with a national adult sample of 1,020. Split sample employed so that only half of the sample responded to some questions reported here. Fieldwork carried out June 21-23, 2002.

GAL/CNN/USA 9/02. **Gallup/CNN/*USA Today* Poll.** Telephone survey conducted by Gallup Organization for CNN and *USA Today*, with a national adult sample of 1,003. Split sample employed so that some questions reported here were asked of only half the sample. Fieldwork carried out September 2-4, 2002.

GAL/CNN/USA 4/03. **Gallup/CNN/*USA Today* Poll.** Telephone survey conducted by Gallup Organization for CNN and *USA Today*, with a national adult sample of 1,001. Fieldwork carried out April 22-23, 2003.

GAL/CNN/USA 8/03. **Gallup/CNN/*USA Today* Poll.** Telephone survey conducted by Gallup Organization for CNN and *USA Today*, with a national adult sample of 1,009. Split sample employed so that some questions reported here were asked of only half the sample. Fieldwork carried out August 25-26, 2003.

GAL/CNN/USA 12/05. **Gallup/CNN/USA Today Poll.** Telephone survey conducted by Gallup Organization for CNN and *USA Today*, with a national adult sample of 1,003. Fieldwork carried out December 16-18, 2005.

GAL/IHF 8/00. **Public Attitudes toward Medical Privacy.** Telephone survey conducted by Gallup Organization for the Institute for Health Freedom, with a national adult sample of 1,000. Fieldwork carried out August 11-26, 2000. Data reported in Corning and Singer, 2003.

GAL/USA 5/06. **Gallup/USA Today Poll.** Telephone survey conducted by Gallup Organization for *USA Today*, with a national adult sample of 809. Fieldwork carried out May 12-13, 2006.

GRN/EBRI 5/99. **Health Confidence Survey 1999.** Telephone survey conducted by Matthew Greenwald and Associates for Employee Benefit Research Institute, Consumer Education Council, with a national adult sample of 1,001. Fieldwork carried out May 13-June 14, 1999.

GRN/EBRI 4/01. **Health Confidence Survey 2001.** Telephone survey conducted by Matthew Greenwald and Associates for Employee Benefit Research Institute, Consumer Education Council, with a national adult sample of 1,001. Fieldwork carried out April 17-May 27, 2001.

GRN/EBRI 4/02. **Health Confidence Survey 2002.** Telephone survey conducted by Matthew Greenwald and Associates for Employee Benefit Research Institute, Consumer Education Council, with a national adult sample of 1,000. Fieldwork carried out April 18-May 19, 2002.

GRN/EBRI 4/03. **Health Confidence Survey 2003.** Telephone survey conducted by Matthew Greenwald and Associates for Employee Benefit Research Institute, Consumer Education Council, with a national adult sample of 1,002. Fieldwork carried out April 24-May 24, 2003.

GRN/EBRI 6/05. **Health Confidence Survey 2005.** Telephone survey conducted by Matthew Greenwald and Associates for Employee Benefit Research Institute, Consumer Education Council, with a national adult sample of 1,003. Fieldwork carried out June 30-August 6, 2005.

HARRIS 11/78. **Dimensions of Privacy.** Survey conducted by Louis Harris and Associates for Sentry Insurance, with a national adult sample of 1,513. The survey was conducted by personal interview November 30-December 10, 1978.

HARRIS 4/99. **Consumers and the 21st Century Survey.** Telephone survey conducted by Louis Harris and Associates for the National Consumers League, with a national adult sample of 1,006. Fieldwork carried out April 22-May 3, 1999.

HI 9/01. **Harris Poll.** Telephone survey conducted by Harris Interactive with a national adult sample of 1,012. Fieldwork carried out September 19-24, 2001.

HI 3/02. **Harris Poll.** Telephone survey conducted by Harris Interactive with a national adult sample of 1,017. Fieldwork carried out March 13-19, 2002.

HI 2/03. **Harris Poll.** Telephone survey conducted by Harris Interactive with a national adult sample of 1,010. Fieldwork carried out February 12-16, 2003.

HI 2/04. **Harris Poll.** Telephone survey conducted by Harris Interactive with a national adult sample of 1,020. Fieldwork carried out February 9-16, 2004.

HI 9/04. **Harris Poll.** Telephone survey conducted by Harris Interactive with a national adult sample of 1,018. Fieldwork carried out September 9-13, 2004.

HI 6/05. **Harris Poll.** Telephone survey conducted by Harris Interactive with a national adult sample of 1,015. Fieldwork carried out June 7-12, 2005.

HI 2/06. **Harris Poll.** Telephone survey conducted by Harris Interactive with a national adult sample of 1,016. Fieldwork carried out February 7-14, 2006.

HI 7/06. **Harris Poll.** Telephone survey conducted by Harris Interactive with a national adult sample of 1,000. Fieldwork carried out July 21-24, 2006.

HI/ID 9/01a. **Airport Security Survey.** Telephone survey conducted by Harris Interactive for Identix, with a national adult sample of 1,015. Fieldwork carried out September 21-24, 2001.

HI/ID 9/01b. **Airport Security Survey.** Telephone survey conducted by Harris Interactive for Identix, with a national adult sample of 1,009. Fieldwork carried out September 26-29, 2001.

H&M/NBC/WSJ 7/06. **NBC News/*Wall Street Journal* Poll.** Telephone survey conducted by Hart and McInturff Research Companies with a national adult sample of 1,010. Fieldwork carried out July 21-24, 2006.

H&T 2/03. **E-Government Survey.** Telephone survey conducted by Hart and Teeter Research Companies for the Council for Excellence in Government, with a national adult sample of 1,023. Fieldwork carried out February 19-25, 2003.

H&T 2/04. **America Speaks Out About Homeland Security Survey.** Telephone survey conducted by Hart and Teeter Research Companies for the Council for Excellence in Government, with a national adult sample of 1,633. Fieldwork carried out February 5-8, 2004.

H&T/NBC/WSJ 8/96. **NBC News/***Wall Street Journal* **Poll.** Telephone survey conducted by Hart and Teeter Research Companies with a national adult sample of 1,203. Fieldwork carried out August 2-6, 1996.

H&T/NBC/WSJ 12/02. **NBC News/***Wall Street Journal* **Poll.** Telephone survey conducted by Hart and Teeter Research Companies with a national adult sample of 1,005. Fieldwork carried out December 7-9, 2002.

ICR/NPR 11/99. **NPR/Kaiser/Harvard Technology Survey.** Telephone survey conducted by International Communications Research for National Public Radio, the Henry J. Kaiser Family Foundation, and Harvard University's Kennedy School of Government, with a national adult sample of 1,506. The sample included an oversample of black respondents, but results are weighted to represent the national adult population. Fieldwork carried out November 15-December 19, 1999.

IR/QNS 6/06. **Global Privacy of Data International Survey.** Telephone survey conducted by Ipsos-Reid for Queens University, Canada/The Surveillance Project, with a U.S. national adult sample of 1,000. Fieldwork carried out June 27-July 28, 2006. Report retrieved March 29, 2006, from http://www.queensu.ca/sociology/Surveillance/files/Ipsos_Report_Nov_2006.pdf.

LAT 4/95. *Los Angeles Times* **Poll.** Telephone survey conducted by the *Los Angeles Times* with a national adult sample of 1,032. Fieldwork carried out April 26-27, 1995.

LAT 12/02. *Los Angeles Times* **Poll.** Telephone survey conducted by the *Los Angeles Times* with a national adult sample of 1,305. Fieldwork carried out December 12-15, 2002.

LAT 7/06. *Los Angeles Times*/**Bloomberg Poll.** Telephone survey conducted by the *Los Angeles Times*/Bloomberg with a national adult sample of 1,478. Fieldwork carried out July 28-August 1, 2006.

LRP/AV 11/06. **Connecting Americans to Their Health Care.** Telephone survey conducted by Lake Research Partners and American Viewpoint for the Markle Foundation, with a national adult sample of 1,003. Fieldwork carried out November 11-15, 2006. Data reported in Markle 2006, retrieved March 10, 2007, from http://www.markle.org/downloadable_assets/research_doc_120706.pdf.

MAR 2/96. **Marist College Institute for Public Opinion Poll.** Telephone survey conducted by Marist College Institute for Public Opinion, with a national adult sample of approximately 900. Fieldwork carried out during February 1996.

OD/FOX 5/01. **Fox News/Opinion Dynamics Poll.** Telephone survey conducted by Opinion Dynamics for Fox News, with a national registered voters sample of 900. Fieldwork carried out May 9-10, 2001.

OD/FOX 10/01. **Fox News/Opinion Dynamics Poll.** Telephone survey conducted by Opinion Dynamics for Fox News, with a national registered voters sample of 900. Fieldwork carried out October 17-18, 2001.

OD/FOX 4/02. **Fox News/Opinion Dynamics Poll.** Telephone survey conducted by Opinion Dynamics for Fox News, with a national registered voters sample of 900. Fieldwork carried out April 16-17, 2002.

OD/FOX 6/02. **Fox News/Opinion Dynamics Poll.** Telephone survey conducted by Opinion Dynamics for Fox News, with a national registered voters sample of 900. Fieldwork carried out June 4-5, 2002.

OD/FOX 9/02. **Fox News/Opinion Dynamics Poll.** Telephone survey conducted by Opinion Dynamics for Fox News, with a national registered voters sample of 900. Fieldwork carried out September 8-9, 2002.

OD/FOX 7/05. **Fox News/Opinion Dynamics Poll.** Telephone survey conducted by Opinion Dynamics for Fox News, with a national registered voters sample of 900. Fieldwork carried out July 26-27, 2005.

OD/FOX 1/06. **Fox News/Opinion Dynamics Poll.** Telephone survey conducted by Opinion Dynamics for Fox News, with a national registered voters sample of 900. Fieldwork carried out January 10-11, 2006.

OD/FOX 5/06. **Fox News/Opinion Dynamics Poll.** Telephone survey conducted by Opinion Dynamics for Fox News, with a national registered voters sample of 900. Fieldwork carried out May 16-18, 2006.

OSR/MSU 11/01. **Civil Liberties Survey.** Telephone survey conducted by the Office for Survey Research of the Institute for Public Policy and Social Research at Michigan State University, with a national adult sample of 1,448. An oversample of African American and Hispanic respondents was included, but results reported here are weighted to be representative of the national adult population. The response rate (calculated as RR4 in the "standard definitions" of the American Association for Public Opinion Research) was 52.3 percent, and the refusal rate was 19.0 percent. Fieldwork was carried out between November 14, 2001 and January 15, 2002. Study results reported in Darren W. Davis and Brian D. Silver, "Civil Liberties vs. Security: Public Opinion in the Context of the Terrorist Attacks on America," *American Journal of Political Science*, 48(1):28-46, 2004.

PAF/RMA 7/02. **Knowing It By Heart: The Constitution and Its Meaning Survey.** Telephone survey conducted by Public Agenda Foundation/Robinson and Muenster Associates, Inc. for the National Constitution Center, with a national adult sample of 1,520. Fieldwork carried out July 10-24, 2002.

PSRA/CHCF 11/98. **Medical Privacy and Confidentiality Survey.** Telephone survey conducted by Princeton Survey Research Associates for the California Health Care Foundation, using a national adult sample of 1,000. A separate sample of California residents was also included, but results reported here are for the national sample only. Fieldwork carried out November 12-December 22, 1998. Data reported in "Topline Report," retrieved April 7, 2007, from http://www.chcf. org/topics/view.cfm?itemID=12500

PSRA/NW 2/95. **Princeton Survey Research Associates/***Newsweek* **Poll.** Telephone survey conducted by PSRA for *Newsweek*, with a national adult sample of 752. Fieldwork carried out February 16-17, 1995.

PSRA/NW 9/01. **Princeton Survey Research Associates/***Newsweek* **Poll.** Telephone survey conducted by Princeton Survey Research Associates for *Newsweek*, with a national adult sample of 1,005. Fieldwork carried out September 20-21, 2001.

PSRA/NW 8/02. **Princeton Survey Research Associates/***Newsweek* **Poll.** Telephone survey conducted by Princeton Survey Research Associates for *Newsweek*, with a national adult sample of 1,005. Fieldwork was carried out August 28-29, 2002.

PSRA/NW 5/06. **Princeton Survey Research Associates International/** *Newsweek* **Poll.** Telephone survey conducted by Princeton Survey Research Associates for *Newsweek* with a national adult sample of 1,007. Fieldwork carried out May 11-12, 2006.

PSRA/PEW 3/96. **Pew News Interest Index Poll.** Telephone survey conducted by Princeton Survey Research Associates for the Pew Research Center, with a national adult sample of 1,500. A split sample was used for some of the questions reported here, so that they were asked only of half the sample. Fieldwork carried out March 28-31, 1996.

PSRA/PEW 4/97. **Pew News Interest Index Poll.** Telephone survey conducted by Princeton Survey Research Associates for the Pew Research Center, with a national adult sample of 1,206. Fieldwork carried out April 3-6, 1997.

PSRA/PEW 10/98. **People and the Press 1998 Technology Survey.** Telephone survey conducted by Princeton Survey Research Associates for the Pew Research Center, using a national adult sample of 3,184. (An oversample of 1,184 Internet users was included, but results are weighted to be representative of the national adult population.) Fieldwork carried out October 26-December 1, 1998.

PSRA/PEW 7/00. **Tracking Online Life Survey.** Telephone survey conducted by Princeton Survey Research Associates for the Pew Research

Center, with a national adult sample of 2,109. Fieldwork carried out July 24-August 20, 2000.

PSRA/PEW 9/01. **People and the Press Post-Terrorist Attack Poll.** Telephone survey conducted by Princeton Survey Research Associates for the Pew Research Center, with a national adult sample of 1,200. Fieldwork was carried out September 13-17, 2001, but data cited here are from questions asked September 14-17 only.

PSRA/PEW 1/02. **Pew News Interest Index Poll.** Telephone survey conducted by Princeton Survey Research Associates for the Pew Research Center, with a national adult sample of 1,201. Fieldwork carried out January 9-13, 2002.

PSRA/PEW 6/02. **Pew News Interest Index Poll.** Telephone survey conducted by Princeton Survey Research Associates for the Pew Research Center, with a national adult sample of 1,212. Fieldwork carried out June 19-23, 2002.

PSRA/PEW 8/02. **People and the Press 2002 Year-After-9/11 Poll.** Telephone survey conducted by Princeton Survey Research Associates for the Pew Research Center, with a national adult sample of 1,001. Fieldwork was carried out August 14-25, 2002.

PSRA/PEW 6/03. **2003 Methodology Study Poll 1.** Telephone survey conducted by Princeton Survey Research Associates for the Pew Research Center, with a national adult sample of 1,000. The study included in this review was a standard survey; another study (the 2003 Methodology Study Poll 2) incorporated procedures designed to maximize response rates, but those results are not reported here. Fieldwork was carried out June 4-8, 2003.

PSRA/PEW 7/03. **2003 Values Update Survey.** Telephone survey conducted by Princeton Survey Research Associates for the Pew Research Center, with a national adult sample of 2,528. The sample included an oversample of blacks, but results are weighted to be representative of the national adult population. Fieldwork was carried out July 14-August 5, 2003.

PSRA/PEW 7/04. **Foreign Policy and Party Images Poll.** Telephone survey conducted by Princeton Survey Research Associates for the Pew Research Center and the Chicago Council on Foreign Relations, with a national adult sample of 2,009. Fieldwork was carried out July 8-18, 2004.

PSRA/PEW 7/05. **Pew News Interest Index Poll.** Telephone survey conducted by Princeton Survey Research Associates for the Pew Research Center, with a national adult sample of 1,502. Fieldwork was carried out July 13-17, 2005.

PSRA/PEW 1/06. **Pew News Interest Index Poll.** Telephone survey conducted by Princeton Survey Research Associates for the Pew Research

Center, with a national adult sample of 1,503. Fieldwork carried out January 4-8, 2006.

PSRA/PEW 9/06. **Pew News Interest Index Poll.** Telephone survey conducted by Princeton Survey Research Associates for the Pew Research Center, with a national adult sample of 1,507. Fieldwork carried out September 6-10, 2006.

PSRA/PEW 12/06. **Pew News Interest Index Poll.** Telephone survey conducted by Princeton Survey Research Associates for the Pew Research Center, with a national adults sample of 1,502. Fieldwork carried out December 6-10, 2006.

PSRA/TM 1/94. **Technology in the American Household.** Telephone survey conducted by Princeton Survey Research Associates for Times Mirror, with a national adult sample of 3,667, including an oversample of 207 modem users. Fieldwork carried out January 4-February 17, 1994.

PSRA/TM 5/95. **Technology and Online Use Survey.** Telephone survey conducted by Princeton Survey Research Associates for Times Mirror, using a national adult sample of 3,603. (An oversample of 402 online users was employed, but results are weighted to be representative of the national adult population). Fieldwork carried out May 25-June 22, 1995.

QU 7/05. **Quinnipiac University Poll.** Telephone survey conducted by Quinnipiac University Polling Institute, with a national registered voters sample of 920. Fieldwork conducted July 21-25, 2005.

QU 8/06. **Quinnipiac University Poll.** Telephone survey conducted by Quinnipiac University Polling Institute, with a national registered voters sample of 1,080. Fieldwork conducted August 17-23, 2006.

RA/ACLUF 11/92. **American Public Opinion about Privacy at Home and at Work.** Personal interview survey conducted by Response Analysis for the American Civil Liberties Union Foundation, using a national adult sample of 993. Fieldwork carried out November 13-December 13, 1992.

SRBI/TIME 8/06. *Time*/**SRBI Poll**. Telephone survey conducted by Schulman, Ronca and Bucuvalas for *Time*, with a national adult sample of 1,002. Fieldwork carried out August 22-24, 2006.

TNS/GMF 6/06. **Transatlantic Trends 2006 Survey.** Telephone survey conducted by TNS Opinion and Social Institutes for the German Marshall Fund of the U.S. and the Compagnia di San Paolo, Italy, with a U.S. national adult sample of 1,000. Surveys were conducted in thirteen nations; data are reported for the U.S. sample only. Fieldwork carried out June 6-24, 2006.

WIN 5/06. **New Models National Brand Poll.** Telephone survey conducted by Winston Group for New Models with a national sample of 1,000 registered voters. Fieldwork carried out May 16-17, 2006.

## M.8.4 References

California Health Care Foundation (CHCF). 2005. "National Consumer Health Privacy Survey 2005: Executive Summary." Retrieved March 10, 2007, from http://www.chcf.org/documents/ihealth/Consumer Privacy2005ExecSum.pdf

California Health Care Foundation (CHCF). 1999. "Medical Privacy and Confidentiality Survey: Topline Report." Retrieved April 7, 2007, from http://www.chcf.org/topics/view.cfm?itemID=12500

Ipsos-Reid, 2006. "Global Privacy of Data International Survey Summary Report." November 2006. Queens University. Retrieved March 8, 2007, from http://www.queensu.ca/sociology/Surveillance/files/Ipsos_Report_Nov_2006.pdf.

# N

# Committee and Staff Biographical Information

## COMMITTEE MEMBERS

WILLIAM J. PERRY, *Co-chair*, is the Michael and Barbara Berberian Professor at Stanford University, with a joint appointment at the Stanford Institute for International Studies (SIIS) and the School of Engineering. He is also a senior fellow at SIIS and the Hoover Institution and serves as co-director of the Preventive Defense Project, a research collaboration of Stanford and Harvard universities. He was the co-director of the Center for International Security and Arms Control from 1988 to 1993, during which time he was also a half-time professor at Stanford. Dr. Perry was the 19th secretary of defense for the United States, serving from February 1994 to January 1997. He previously served as deputy secretary of defense (1993-1994) and as under secretary of defense for research and engineering (1977-1981). Dr. Perry is on the board of directors of several emerging high-tech companies and is chair of Global Technology Partners. His previous business experience includes serving as a laboratory director for General Telephone and Electronics (1954-1964); founder and president of ESL Inc. (1964-1977); executive vice-president of Hambrecht & Quist, Inc. (1981-1985); and founder and chairman of Technology Strategies and Alliances (1985-1993). He is an expert in U.S. foreign policy, national security, and arms control. He is a member of the National Academy of Engineering (NAE) and a fellow of the American Academy of Arts and Sciences (AAAS). He received a B.S. and M.S. from Stanford University and a Ph.D. from Pennsylvania State University, all in mathematics.

CHARLES M. VEST, *Co-chair*, is president of the National Academy of Engineering and president emeritus of the Massachusetts Institute of Technology (MIT). He became president of MIT in 1990 and served in that position until December 2004. Dr. Vest was a director of DuPont for 14 years and of IBM for 13 years, was vice chair of the U.S. Council on Competitiveness for eight years, and served on various federal committees and commissions, including the President's Committee of Advisors on Science and Technology during the Clinton and Bush administrations, the Commission on the Intelligence Capabilities of the United States Regarding Weapons of Mass Destruction, the Secretary of Education's Commission on the Future of Higher Education, the Secretary of State's Advisory Committee on Transformational Diplomacy, and the Rice-Chertoff Secure Borders and Open Doors Advisory Committee. He serves on the boards of several non-profit organizations and foundations devoted to education, science, and technology. In July 2007 he was elected to serve as president of NAE for six years. He has authored a book on holographic interferometry and two books on higher education. He has received honorary doctoral degrees from ten universities and was awarded the 2006 National Medal of Technology by President George W. Bush.

W. EARL BOEBERT is an expert on information security, with experience in national security and intelligence as well as commercial applications. Currently retired, he was a senior scientist at Sandia National Laboratories. He has 30 years experience in communications and computer security, is the holder or co-holder of 12 patents, and has participated in National Research Council (NRC) studies on security matters. Prior to joining Sandia, he was the technical founder and chief scientist of Secure Computing Corporation, where he developed the Sidewinder security server, a system that currently protects several thousand sites. Prior to that, he worked for 22 years at Honeywell, rising to the position of senior research fellow. At Honeywell he worked on secure systems, cryptographic devices, flight software, and a variety of real-time simulation and control systems, and he won Honeywell's highest award for technical achievement for his part in developing a very-large-scale radar landmass simulator. He developed and presented a course on systems engineering and project management that was eventually given to over 3,000 students in 13 countries. He served on the NRC committees that produced *Computers at Risk: Computing in the Information Age; For the Record: Protecting Electronic Health Information;* and *Information Technology for Counterterrorism: Immediate Actions and Future Possibilities.* He participated in the NRC workshops on "Cyber-Attack" and "Insider Threat."

MICHAEL L. BRODIE is the chief scientist of Verizon Services Operations at Verizon Communications and is an adjunct professor at the National University of Ireland, Galway. Dr. Brodie works on large-scale strategic information technology (IT) challenges for Verizon Communications Corporation's senior executives. His primary interest is delivering business value from advanced and emerging technologies and practices to enable business objectives while optimizing and transforming IT. He also investigates the relationships between economics, business, and technology and computing-communications convergence. His long-term industrial research focus is on advanced computational models and architectures and the large-scale information systems that they support. He is concerned with the "big picture," business and technical contexts, core technologies, and "integration" within a large-scale, operational telecommunications environment. Dr. Brodie has authored over 150 books, chapters, journal articles, and conference papers. He has presented keynote talks, invited lectures, and short courses on many topics in over 30 countries. He is a member of the boards of several research foundations including the Semantic Technology Institutes International (2007-present); the European Research Consortium for Informatics and Mathematics (2007-present); the advisory board of the School of Computer and Communication Sciences, École Polytechnique Fédérale de Lausanne, Switzerland (2001-present); the advisory board of the Digital Enterprise Research Institute, National University of Ireland (2003-present); Forrester Research, Inc. (2006-present); expert advisor to the Information Society Technologies priority of the European Commission's Sixth and Seventh Framework Programmes (2003-present); the VLDB (Very Large Databases) Endowment (1992-2004); and he is on the editorial board of several research journals. He received his Ph.D. in computer science from the University of Toronto in 1978.

DUNCAN A. BROWN is a member of the principal staff and director of the Strategic Assessments Office (SAO) at the Johns Hopkins University (JHU) Applied Physics Laboratory (APL). The SAO conducts broad-ranging analyses and assessments of national security strategy, policy, and technology trends that may affect APL. Recent efforts have included conducting an alternative futures exercise to examine potential geopolitical strategic futures and their impact on the military and related research and development, conducting an effort for the Office of the Secretary of Defense and the Office of the Secretary of the Navy to examine the principles of war, providing technical analysis and advice to the Defense Advanced Research Projects Agency, and serving on a study panel sponsored by the National Reconnaissance Office and the Navy to assess the

338 PROTECTING INDIVIDUAL PRIVACY IN THE STRUGGLE AGAINST TERRORISTS

future use of space. Prior efforts have included Submarine Force wartime readiness assessments, creation of the U.S. Navy's Unmanned Combat Aerial Vehicle Program, serving on a Naval Research and Advisory Committee Panel to examine issues associated with transitioning technology, and serving on the NRC Naval Studies Board to examine the role of experimentation in building future naval forces. Mr. Brown has also served on the Navy staff in the Pentagon as the science advisor to the Deputy Chief of Naval Operations, in the Pacific as the science advisor to the Commander in Chief Pacific Fleet, and in the Pentagon as the director for Submarine Technology. Mr. Brown also headed the Hydrodynamics Branch at the Naval Undersea Warfare Center in Newport, R.I. Mr. Brown has received three Navy Superior Civilian Service awards. Mr. Brown's formal education includes graduate work in national security studies at Georgetown and MIT. He was a fellow in MIT's Seminar XXI Foreign Politics and International Relations in the National Interest Program. Mr. Brown also holds an M.S. degree from Johns Hopkins University in engineering management, an M.S. degree in ocean engineering from the University of Rhode Island, and a B.S. degree in engineering science from Hofstra University.

FRED H. CATE is a distinguished professor, the C. Ben Dutton Professor of Law, adjunct professor of informatics, and director of the Center for Applied Cybersecurity Research at Indiana University. He is a senior policy advisor to the Center for Information Policy Leadership at Hunton & Williams LLP, a member of Microsoft's Trustworthy Computing Academic Advisory Board, and a member of the board of editors of *Privacy & Information Law Report*. He also serves as reporter for the American Law Institute's project on Principles of the Law on Government Access to and Use of Personal Digital Information. Previously, he served as counsel to the Department of Defense Technology and Privacy Advisory Committee, was a reporter for the third report of the Markle Task Force on National Security in the Information Age, and a member of the Federal Trade Commission's Advisory Committee on Online Access and Security. He directed the Electronic Information Privacy and Commerce Study for the Brookings Institution, chaired the International Telecommunication Union's High-Level Experts on Electronic Signatures and Certification Authorities, and served as a member of the United Nations Working Group on Emergency Telecommunications. Professor Cate is the author of many articles and books, including *Privacy in the Information Age*, *The Internet and the First Amendment*, and *Privacy in Perspective*. He researches and teaches in the areas of privacy, security, and other information policy and law issues. An elected member of the American Law Institute, he attended Oxford University and received his J.D. and his A.B. with honors and distinction from Stanford University.

RUTH A. DAVID is the president and chief executive officer of ANSER, a not-for-profit, public-service research institution that provides research and analytic support on issues relating to international and domestic terrorist threats. Dr. David is a member of the Department of Homeland Security Advisory Council (HSAC), NAE, and the Corporation for the Charles Stark Draper Laboratory, Inc. She is vice chair of the HSAC Senior Advisory Committee of Academia and Policy Research and serves on the National Security Agency Advisory Board, the NRC Naval Studies Board, the NAE Committee on Engineering Education, the AAAS Committee on Scientific Freedom and Responsibility, the Jet Propulsion Laboratory's Technical Division Advisory Board, and the External Advisory Committee for Purdue University's Homeland Security Institute. From September 1995 to September 1998, Dr. David was deputy director for science and technology at the Central Intelligence Agency (CIA). As technical advisor to the director of central intelligence, she was responsible for research, development, and deployment of technologies in support of all phases of the intelligence process. She represented the CIA on numerous national committees and advisory bodies, including the National Science and Technology Council and the Committee on National Security. Prior to moving to the CIA, she was director of advance information technologies at Sandia National Laboratories where she began her professional career. She is the recipient of many awards including the CIA's Distinguished Intelligence Medal, the CIA Director's Award, and the Director of NSA Distinguished Service Medal. She is a former adjunct professor at the University of New Mexico. Her research interests include digital and microprocessor-based system design, digital signal analysis, adaptive signal analysis, and system integration. Dr. David received her Ph.D. in electrical engineering from Stanford University.

RUTH M. DAVIS is president and chief executive officer of the Pymatuning Group, Inc., which specializes in industrial modernization strategies and technology development. She has served on the boards of 12 corporations and private organizations and was a member of the board of regents of the National Library of Medicine from 1989 to 1992. She has chaired the board of trustees of the Aerospace Corporation. Dr. Davis served as assistant secretary of energy for resource applications and deputy undersecretary of defense for research and advanced technology. She has taught at Harvard University and at the University of Pennsylvania, and she has served on the University of Pennsylvania's board of overseers of the

School of Engineering and Applied Science. She has served on a number of advisory committees to the federal government and was on the NAE Council. She was elected to NAE in 1976. She has served on more than two dozen NRC panels and committees. She has a Ph.D. in mathematics from the University of Maryland. Her research interests include expediting the development process for law enforcement technologies, and she has worked extensively on means of identifying meaningful requirements for law enforcement technologies and ensuring adequacy of life cycle functions. She has studied and written on the technical and managerial features of the technology-based threat to information assets.

WILLIAM H. DuMOUCHEL is chief statistical scientist at the Lincoln Safety Group of Phase Forward, Inc. His current research focuses on statistical computing and Bayesian hierarchical models, including applications to meta-analysis and data mining. Dr. DuMouchel is the inventor of the empirical Bayesian data mining algorithm known as GPS and its successor MGPS, which have been applied to the detection of safety signals in databases of spontaneous adverse drug event reports. These methods are now used within the Food and Drug Administration and industry. From 1996 through 2004, he was a senior member of the data mining research group at AT&T Laboratories. Before that, he was chief statistical scientist at BBN Software Products, where he was lead statistical designer of a software advisory system for data analysis and experimental design called RS/Discover and RS/Explore. Dr. DuMouchel has been on the faculties of the University of California at Berkeley, the University of Michigan, MIT, and most recently was professor of biostatistics and medical informatics at Columbia University from 1994-1996. He has also been an associate editor of the *Journal of the American Statistical Association, Statistics in Medicine, Statistics and Computing*, and the *Journal of Computational and Graphical Statistics*. Dr. DuMouchel is an elected fellow of the American Statistical Association and of the Institute of Mathematical Statistics, and he has served previously on the NRC Committee on Applied and Theoretical Statistics. Recently he served on the Institute of Medicine's Committee on Postmarket Surveillance of Pediatric Medical Devices and is currently a member of the NRC Committee on National Statistics. He received his Ph.D. in statistics from Yale University.

CYNTHIA DWORK is a principal researcher at Microsoft Research's Silicon Valley Laboratory, which she joined at its inception in 2001. Prior to that, she was a staff fellow at Compaq in 2000 to 2001, and from 1985 to 2000, she was a research staff member at the IBM Almaden Research Center. She has made seminal contributions in three areas of theoretical computer science: distributed computing, cryptography, and, most recently, privacy-preserving analysis of data. A founding member of the *Journal of Privacy and Confidentiality*, she is on the editorial boards of *Information and Computation*, the *Journal of Cryptology, Internet Mathematics*, and the *Journal of Theoretical Computer Science*. Dr. Dwork has been a member of the advisory board for the Center for Discrete Mathematics and Theoretical Computer Science at Rutgers University for more than a decade and is still serving on the advisory board of the Bertinoro International Center for Informatics in Italy and on the Fellows Selection Committee of the International Association of Cryptologic Researchers. She received the Edsger W. Dijkstra Award in 2007 and was elected to both NAE and AAAS in 2008.

STEPHEN E. FIENBERG is the Maurice Falk University Professor of Statistics and Social Science at Carnegie Mellon University in the Department of Statistics, the Machine Learning Department, and Cylab. He has served as dean of the College of Humanities and Social Sciences at Carnegie Mellon University and as academic vice president of York University in Toronto, Canada. His current research interests include approaches to data confidentiality, record linkage, and disclosure limitation; modeling of network data; causation; machine learning and Bayesian mixed-membership models; foundations of statistical inference; sample surveys and randomized experiments; statistics and the law. He has participated in a wide array of NRC committees and workshops, including as chair of the Committee to Review the Scientific Evidence on the Polygraph, chair of the Committee on National Statistics, and as a member of the National Academies' Division of Behavioral and Social Sciences and Education. Dr. Fienberg has also served as president of the Institute of Mathematical Statistics and of the International Society for Bayesian Analysis. He is the author, co-author, or editor of numerous books including *Discrete Multivariate Analysis: Theory and Practice; The Analysis of Cross-classified Categorical Data*, and *Who Counts? The Politics of Census-Taking in Contemporary America*. He is an editor of the *Annals of Applied Statistics* and a founder of the *Journal of Privacy and Confidentiality*. He is a former editor of the *Journal of the American Statistical Association*, founding co-editor of *Chance*, and served as co-editor of the "Section for Statistics" of the *International Encyclopedia of the Social and Behavioral Sciences*. He holds a Ph.D. in statistics from Harvard University and is a member of the National Academy of Sciences and a fellow of AAAS and the Royal Society of Canada.

ROBERT J. HERMANN is senior partner of Global Technology Partners, LLC, which specializes in providing strategic advice on national security and technology issues. In 1998, Dr. Hermann retired from United Technologies Corporation where he held the position of senior vice president, science and technology. In this role, he was responsible for assuring the development of the company's technical resources and the full exploitation of science and technology by the corporation. He was also responsible for the United Technologies Research Center (UTRC). Dr. Hermann joined the company in 1982 as vice president, systems technology, in the electronics sector and later served in a series of assignments in the defense and space systems groups prior to being named vice president, science and technology. Prior to joining UTRC, he served for 20 years with NSA with assignments in research and development, operations, and the North Atlantic Treaty Organization. In 1977, he was appointed principal deputy assistant secretary of defense for communications, command, control, and intelligence. In 1979, he was named assistant secretary of the U.S. Air Force for research, development, and logistics and, in parallel, was director of the National Reconnaissance Office. He received B.S., M.S., and Ph.D. degrees in electrical engineering from Iowa State University.

R. GIL KERLIKOWSKE is a 32-year law enforcement veteran and was appointed as the chief of police for the Seattle Police Department in August 2000. He is a former deputy director for the U.S. Department of Justice's Office of Community Oriented Policing Services, which provides federal grants to local police agencies in support of community policing services. He served as the police commissioner for Buffalo, New York, where his selection by the mayor became the first outside appointment in 30 years. Mr. Kerlikowske also served as the chief of police for two Florida cities—Fort Pierce and Port St. Lucie—both of which received the Attorney General's Crime Prevention Award. In 1985 he was a visiting fellow at the National Institute of Justice where he designed an evaluation of police procedures throughout the country. He began his law enforcement career in 1972 as a police officer for the St. Petersburg Police in Florida. Mr. Kerlikowske also served in the U.S. Army Military Police. He holds B.A. and M.A. degrees in criminal justice from the University of South Florida and is a graduate of the National Executive Institute at the Federal Bureau of Investigations Academy.

ORIN S. KERR is a professor of law at the George Washington University of School of Law. He is a prolific scholar in the area of criminal procedure and computer crime law. Professor Kerr's articles have appeared in *Harvard Law Review, Columbia Law Review, Stanford Law Review, Michigan Law Review, New York University Law Review, Georgetown Law Journal, Northwestern University Law Review,* and many other journals. From 1998 to 2001, he was an honors program trial attorney in the Computer Crime and Intellectual Property Section of the Criminal Division at the U.S. Department of Justice. He is a former law clerk for Judge Leonard I. Garth of the U.S. Court of Appeals for the Third Circuit and for Justice Anthony M. Kennedy of the U.S. Supreme Court. Professor Kerr received a B.S.E. in mechanical and aerospace engineering from Princeton University, an M.S. in mechanical engineering from Stanford University, and a J.D. from Harvard Law School.

ROBERT W. LEVENSON is a professor in the Department of Psychology at the University of California, Berkeley, and is the director of the Institute of Personality and Social Research and the Berkeley Psychophysiology Laboratory. His research interests include the physiological, facial expressive, and subjective aspects of emotion, and the emotional changes that occur in neurodegenerative disorders and normal aging. He has published numerous papers on the autonomic nervous system. He has served as president of the Society for Psychophysiological Research, as president of the Association for Psychological Science, and as co-chair of the Behavioral Sciences Workgroup at the National Institute of Mental Health. Dr. Levenson received his B.A. in psychology from Georgetown University and his Ph.D. from Vanderbilt University in clinical psychology.

TOM M. MITCHELL is the E. Fredkin Professor and founding head of the Machine Learning Department at Carnegie Mellon University. His research interests are generally in machine learning, artificial intelligence, and cognitive neuroscience. His recent research has focused both on machine learning approaches to extracting structured information from unstructured text and on studying the neural representation of language in the human brain using functional magnetic resonance imaging. Dr. Mitchell is a past president of the American Association of Artificial Intelligence, past chair of the AAAS Section on Information, Computing, and Communication, and author of the textbook *Machine Learning*. From 1999 to 2000, he served as chief scientist and vice president for WhizBang Labs, a company that employed machine learning to extract information from the Web. Dr. Mitchell has served on the NRC's Computer Science and Telecommunication Board and on the committee that produced the report *Information Technology for Counterterrorism: Immediate Actions and Future Possibilities*. He testified at the U.S. House Committee on Veterans' Affairs hearing on the potential uses of artificial intelligence to improve benefits claims processing at the Veterans' Administration. Dr. Mitchell received his Ph.D. in electrical engineering with a computer science minor from Stanford University.

TARA O'TOOLE is chief executive officer and director of the Center for Biosecurity at the University of Pittsburgh Medical Center (UPMC), and professor of medicine and of public health at the University of Pittsburgh.

Dr. O'Toole was one of the original members of the Johns Hopkins Center for Civilian Biodefense Strategies and served as its director from 2001 to 2003. She was one of the principal authors and producers of "Dark Winter," an influential exercise conducted in June 2001 to alert national leaders to the dangers of bioterrorist attacks. From 1993 to 1997, she served as the assistant secretary of energy for environment safety and health. As assistant secretary, Dr. O'Toole was the principal advisor to the secretary of energy on matters pertaining to protecting the environment and worker and public health from the U.S. nuclear weapons complex and Department of Energy laboratories. From 1989 to 1993, Dr. O'Toole was a senior analyst at the Office of Technology Assessment (OTA), where she directed and participated in studies of health impacts on workers and the public due to environmental pollution resulting from nuclear weapons production. Dr. O'Toole is a board-certified internist and occupational medicine physician. She received her bachelor's degree from Vassar College, her M.D. from George Washington University, and an M.P.H. from Johns Hopkins University. She completed a residency in internal medicine at the Yale School of Medicine and a fellowship in occupational and environmental medicine at Johns Hopkins University.

DARYL PREGIBON is a research scientist at Google, Inc. He is a recognized leader in data mining, the interdisciplinary field that combines statistics, artificial intelligence, and database research. From 1981 to 2004, he worked at Bell Labs and AT&T Labs and served as head of statistics research for fifteen years. He is a past member of the NRC Committee on National Statistics, the Committee on the Feasibility of a National Ballistics Database, and the Committee on Applied and Theoretical Statistics (past chair). He is a member of the National Advisory Committee for the Statistical and Applied Mathematical Sciences Institute and a former director of the Association for Computer Machinery's (ACM's) Special Interest Group on Knowledge Development and Data Mining. In 1985 he co-founded (with Bill Gale) the Society for Artificial Intelligence and Statistics. He has authored more than 60 publications and holds four patents. Dr. Pregibon received his Ph.D. in statistics from the University of Toronto and his M.Math. degree in statistics from the University of Waterloo.

LOUISE RICHARDSON is executive dean of the Radcliffe Institute for Advanced Study at Harvard University. She received her bachelor's and master's degrees in history from Trinity College, Dublin, and an M.A. and a Ph.D. in government from Harvard University. From 1989 to 2001, Dr. Richardson was assistant and associate professor of government at Harvard. She teaches courses on terrorism at Harvard College, Graduate School, and Law School. A political scientist by training, Dr. Richardson has specialized in international security with an emphasis on terrorist movements. Her recent publications include *What Terrorists Want: Understanding the Enemy, Containing the Threat* (2006); *The Roots of Terrorism* (2006); *Democracy and Counterterrorism: Lessons from the Past* (2007); and *When Allies Differ* (1996), along with numerous articles on international terrorism, British foreign and defense policy, security institutions, and international relations. She is co-editor of the SUNY Press series on terrorism. Dr. Richardson's current research projects involve a study of patterns of terrorist violence and a study on counter-terrorism lessons to be derived from earlier experiences with terrorism.

BEN A. SHNEIDERMAN is a professor in the Department of Computer Science and the Institute for Advanced Computer Studies at the University of Maryland, College Park. He is a founding director of the Human-Computer Interaction Laboratory (1983-2000). He has taught previously at the State University of New York (SUNY) and at Indiana University. He is a fellow of the ACM and the AAAS and received the ACM CHI (Computer Human Interaction) Lifetime Achievement Award in 2001. He has co-authored two textbooks, edited three technical books, and published more than 300 technical papers and book chapters. He co-authored *Readings in Information Visualization: Using Vision to Think* with Stu Card and Jock Mackinlay and *The Craft of Information Visualization: Readings and Reflections* with Ben Bederson. Dr. Shneiderman's vision of the future is presented in his book *Leonardo's Laptop: Human Needs and the New Computing Technologies*, which won the IEEE 2003 award for Distinguished Literary Contribution. He has consulted and lectured for many organizations including Apple, AT&T, Citicorp, General Electric, Honeywell, IBM Corporation, Intel Corporation, Microsoft, NCR, the Library of Congress, the National Aeronautics and Space Administration, and university research groups. He received his Ph.D. from SUNY at Stony Brook.

DANIEL J. WEITZNER is co-director of MIT's CSAIL Decentralized Information Group (DIG) and teaches Internet public policy in the MIT Electrical Engineering and Computer Science Department. He is also policy director of the World Wide Web Consortium's (W3C's) technology and society activities. At DIG he leads research on the development of new technology and public policy models for addressing legal challenges raised by the Web, including privacy, intellectual property, identity management, and new regulatory models for the Web. At W3C he is responsible for Web standards needed to address public policy requirements, including the Platform for Privacy Preference (P3P) and XML Security technologies. He was the first to advocate user control technologies such as content filtering to protect children and to avoid government censorship, and he played a critical role in the landmark Internet freedom of expression case in the U.S. Supreme Court, *Reno v. ACLU* (1997). In 1994, his advocacy work won legal protections for e-mail and Web logs in the *U.S. Electronic Communications Privacy Act*. Mr. Weitzner was co-founder and deputy director of the Center for Democracy and Technology and deputy policy director of the Electronic Frontier Foundation. He serves on the board of directors of the Center for Democracy and Technology, the Software Freedom Law Center, and the Internet Education Foundation. He has a law degree from Buffalo Law School and a B.A. in philosophy from Swarthmore College. His writings have appeared in *Science, Yale Law Review, Communications of the ACM, Computerworld, IEEE Internet Computing, Wired Magazine, Social Research,* and *The Whole Earth Review*.

## STAFF MEMBERS

BETTY M. CHEMERS is a senior project officer at the National Research Council, which she joined in May 2005 after spending 30 years in the public and not-for-profit sectors working on criminal justice and juvenile justice issues. She currently directs two studies: one on terrorism prevention funded by the Department of Homeland Security and the National Science Foundation and a second study on an assessment of the research program of the National Institute of Justice (NIJ) funded by NIJ. Prior to this, she held numerous positions at the U.S. Department of Justice including director of the evaluation division of the NIJ (2002-2005) and deputy administrator for discretionary programs at the Office of Juvenile Justice and Delinquency Prevention (1995-2001), where she oversaw its $100 million budget of research, demonstration, and training and technical assistance activities. Her non-federal service includes directing the planning and policy analysis division for the Maryland Department of Public Safety and Correctional Services and consulting on strategic planning, finance, and management issues with nonprofits. She holds an M.A. in history from Boston University and a B.A. in education/sociology from the University of Maryland.

MICHAEL L. COHEN is a senior program officer for the NRC Committee on National Statistics. Previously, he was a mathematical statistician at the Energy Information Administration, an assistant professor in the School of Public Affairs at the University of Maryland, a research associate at the Committee on National Statistics, and a visiting lecturer at the Department of Statistics, Princeton University. His general area of research is the use of statistics in public policy, with particular interest in census undercount and model validation. He is also interested in robust estimation.

He has a B.S. degree in mathematics from the University of Michigan and M.S. and Ph.D. degrees in statistics from Stanford University.

HERBERT S. LIN is chief scientist of the Computer Science and Telecommunications Board, National Research Council of the National Academies, where he has been the study director of major projects on public policy and information technology. These studies include a 1996 study on national cryptography policy (*Cryptography's Role in Securing the Information Society*); a 1992 study on the future of computer science (*Computing the Future: A Broader Agenda for Computer Science and Engineering*); a 1999 study of the U.S. Department of Defense systems for command, control, communications, computing, and intelligence (*Realizing the Potential of C4I: Fundamental Challenges*); a 2001 study on workforce issues in high technology (*Building a Workforce for the Information Economy*); and a 2002 study on protecting children from Internet pornography and sexual exploitation (*Youth, Pornography, and the Internet*). Prior to his NRC service, he was a professional staff member and staff scientist for the House Armed Services Committee (1986-1990), where his portfolio included defense policy and arms control issues. He also has significant expertise in mathematics and science education. He received his doctorate in physics from the Massachusetts Institute of Technology.

CAROL PETRIE is director of the NRC Committee on Law and Justice, a standing committee within the Division of Behavioral and Social Sciences and Education. In this capacity since 1997, she has developed and supervised a wide range of projects resulting in NRC reports in such areas as juvenile crime, pathological gambling, transnational organized crime, prosecution, crime victimization, improving drug research, school violence, firearms, policing, and forensic science. Prior to 1997, she served as the director of planning and management at the National Institute of Justice, where she was responsible for policy development, budget, and administration. In 1994, she served as the acting director of the NIJ. Throughout her career she has worked in the area of criminal justice research, statistics, and public policy at the NIJ and at the Bureau of Justice Statistics. She has conducted research on violence and public policy and managed numerous research projects on the development of criminal behavior, domestic violence, child abuse and neglect, and improving the operations of the criminal justice system.

JULIE ANNE SCHUCK has been a research associate at the NRC for over six years in the Division of Behavioral and Social Sciences and Education. She has worked on a number of different projects and workshops, including those on improving undergraduate instruction in science, technology, engineering, and mathematics; understanding the technical and privacy dimensions of information for terrorism prevention; and assessing the research program of the NIJ. Prior to coming to the NRC, she was a research support specialist at Cornell University, where she conducted a study examining the under-representation of women in physics-based engineering majors. She holds an M.S. in education from Cornell University and a B.S. in engineering physics from the University of California, San Diego.

# O

# Meeting Participants and Other Contributors

## MEETING PARTICIPANTS

The Committee on Technical and Privacy Dimensions of Information for Terrorism Prevention and Other National Goals held five meetings starting in 2006. These meetings included information-gathering sessions open to the public, as well as closed segments for committee deliberation. The committee heard from numerous presenters at these meetings. They include the following by meeting date and session.

### April 27-28, 2006

**Session 1: Deception Detection and Reducing Errors**

Paul Ekman, University of California, San Francisco
Henry Greely, Stanford University School of Law
Barry Steinhardt, Technology and Liberty Program, American Civil Liberties Union
John Woodward, Intelligence Policy Center, Rand Corporation
Tom Zeffiro, Center for Functional and Molecular Imaging, Georgetown University

**Session 2: Communications**

Clint C. Brooks, National Security Agency (retired)
Whitfield Diffie, Sun Microsystems
John Pike, Director, GlobalSecurity.Org
Jody Westby, Global Cyber Risk, University of California

## Session 3: Data Mining

Randy Ferryman, U.S. National Counter Terrorism Center
John Hollywood, Rand Corporation
David Jensen, Knowledge Discovery Laboratory, University of
 Massachusetts, Amherst
Jeff Jonas, Entity Analytic Systems, IBM Corporation
David Scott, Rice University
Kim Taipale, Center for Advanced Studies in Science and Technology
 Policy

## July 27-28, 2006

### Session 1: Privacy Laws and Concepts; Law and Policy Revision Efforts

Lee Tien, Electronic Frontier Foundation

### Session 2: Law Enforcement, Counter-Terrorism, and Privacy

Philip R. Reitinger, Trustworthy Computing, Microsoft Corporation

### Session 3: Data Mining in the Commercial World

Scott Loftesness, Glenbrook Partners
Dan Schutzer, Financial Services Technology Consortium

## October 26-27, 2006

### Session 1: Providing a National Perspective

Adm. Scott Redd, National Counter Terrorism Center

### Session 2: Law Enforcement Intelligence

Michael Fedarcyk, Bearingpoint and Federal Bureau of Investigation
 (retired)
Roy I. Apseloff, National Media Exploitation Center
Joe Connell, Counter-Terrorist Command, New Scotland Yard

### Session 3: Status of Research on Deception Detection Technologies

Mark Frank, University at Buffalo
Rafi Ron, Ben Gurion Airport, Israel (retired) and Boston Logan Airport

## Session 4: Bio-Surveillance Technology and Privacy Issues

James V. Lawler, Homeland Security Council, White House
Lynn Steele, Emergency Preparedness and Response, Centers for Disease
 Control and Prevention (CDC)
Barry Rhodes, Emergency Preparedness and Response, CDC
Farzad Mostashari, New York City Public Health Department
Patricia Quinlisk, State of Iowa

## Session 5: Data Linkages

William E. Winkler, U.S. Census Bureau

## Session 6: Presentation on DHS Data System Activities

Lisa J. Walby, Transportation Security Administration, Department of
 Homeland Security (DHS)
Sandy Landsberg, Science and Technology Directorate, DHS

## January 18-19, 2007

*Closed Meeting*

## March 29-30, 2007

*Closed Meeting*

## OTHER CONTRIBUTIONS

From January 1 to March 1, 2007, the committee solicited well-reasoned white papers that identified and discussed issues relevant to the use of data mining, information fusion, and deception detection technologies as they relate to the twin goals of protecting privacy and pursuing terrorism prevention, law enforcement, and public health. The following papers were submitted for the committee's review:

Michael D. Larsen. 2007. "Record Linkage, Nondisclosure,
 Counterterrorism, and Statistics." Department of Statistics and
 Center for Survey Statistics and Methodology, Iowa State University.
Peter Swire. 2006. "Privacy and information sharing in the war on
 terrorism." *Villanova Law Review* 51, available at http://ssrn.
 com/abstract=899626.

In response to the call for papers, the DHS Data Privacy and Integrity Advisory Committee[1] transmitted the following five reports:

Data Privacy and Integrity Advisory Committee. 2006. *The Use of RFID for Human Identity Verification.* Report No. 2006-02. Adopted December 6, 2006. DHS, Washington, D.C.

Data Privacy and Integrity Advisory Committee. 2006. *The Use of Commercial Data.* Report No. 2006-03. Adopted December 6, 2006. DHS, Washington, D.C.

Data Privacy and Integrity Advisory Committee. 2006. *Framework for Privacy Analysis of Programs, Technologies, and Applications.* Report No. 2006-01. Adopted March 7, 2006. DHS, Washington, D.C.

Data Privacy and Integrity Advisory Committee. 2006. *Recommendations on the Secure Flight Program.* Report No. 2005-02. Adopted December 6, 2005. DHS, Washington, D.C.

Data Privacy and Integrity Advisory Committee. 2005. *The Use of Commercial Data to Reduce False Positives in Screening Programs.* Report No. 2005-01. Adopted September 28, 2005. DHS, Washington, D.C.

---

[1]See http://www.dhs.gov/xinfoshare/committees/editorial_0512.shtm for more information on the DHS Data Privacy and Integrity Advisory Committee.